# SOCIAL-ECOLOGICAL RESILIENCE AND LAW

# Social-Ecological Resilience and Law

AHJOND S. GARMESTANI AND CRAIG R. ALLEN,
EDITORS

Columbia University Press
*New York*

Columbia University Press
*Publishers Since 1893*
New York Chichester, West Sussex
cup.columbia.edu


The Nebraska Cooperative Fish and Wildlife Research Unit is jointly supported by a cooperative agreement between the U.S. Geological Survey, the Nebraska Game and Parks Commission, the University of Nebraska–Lincoln, the United States Fish and Wildlife Service and the Wildlife Management Institute. The views expressed herein are those of the authors and do not necessarily represent those of the U.S. Environmental Protection Agency.

Library of Congress Cataloging-in-Publication Data
Social-ecological resilience and law / Ahjond S. Garmestani and Craig R. Allen, editors.
p. cm.
Includes bibliographical references and index.
ISBN 978-0-231-16058-2 (cloth : alk. paper) — ISBN 978-0-231-16059-9 (pbk. : alk. paper) —
ISBN 978-0-231-53635-6 (ebook)
1. Environmental law. 2. Law reform. 3. Environmental policy. 4. Ecosystem management.
5. Ecosystem health. 6. Resilience (Ecology) I. Garmestani, Ahjond S., editor of compilation.
II. Allen, Craig R., editor of compilation.
K3585.S863 2013
344.04'6—dc23
2013024271

Columbia University Press books are printed on permanent and durable acid-free paper.
This book is printed on paper with recycled content.
Printed in the United States of America

c 10 9 8 7 6 5 4 3 2 1
p 10 9 8 7 6 5 4 3 2 1

COVER PHOTO: Johann Schumacher © Getty Images
COVER DESIGN: Milenda Nan Lee

# Contents

# Acknowledgments

We would like to thank Marty Anderies, Brad Autrey, Melinda Harm Benson, Alex Camacho, Barb Cosens, Robin Craig, Rob Glicksman, Olivia Odom Green, J. B. Ruhl, and Sandi Zellmer for reviewing chapters in this book. Their reviews were invaluable and have made the book a much stronger contribution to the resilience science literature.

This work is not a product of the U.S. Environmental Protection Agency, and the views expressed are those of the authors and do not necessarily represent those of the U.S. Environmental Protection Agency. The Nebraska Cooperative Fish and Wildlife Research Unit is jointly supported by a cooperative agreement between the U.S. Geological Survey, the Nebraska Game and Parks Commission, the University of Nebraska–Lincoln, the United States Fish and Wildlife Service, and the Wildlife Management Institute.

# SOCIAL-ECOLOGICAL RESILIENCE AND LAW

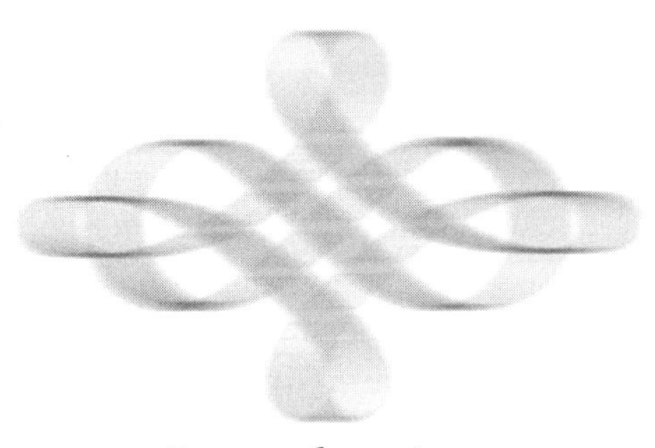

# Introduction

## *Social-Ecological Resilience and Law*

AHJOND S. GARMESTANI, CRAIG R. ALLEN,
CRAIG ANTHONY (TONY) ARNOLD, AND LANCE H. GUNDERSON

Environmental law is intimately connected to ecological concepts and understanding. The legal instruments, institutions, and administration of law in the United States are predicated on assumptions that nature is globally stable and that the inherent variability in ecological systems is bounded. This current legal framework is based upon an understanding of ecological systems operating near an equilibrium, or if disturbed, moving back toward an equilibrium. Such assumptions make much current environmental law ill-suited for many pressing environmental issues (Ruhl 1999; Garmestani et al. 2009; Craig 2010; Verchick 2010; Benson and Garmestani 2011). Emerging environmental challenges, such as cross-boundary water governance or climate change, are not easily addressed within the current legal framework, because although the problems may be easily identified, the solutions require frequent recalibration of the policy used to manage the environmental issue.

The legal system functions to create and sustain certainty and security in the distribution of resources among humans in society: power and authority, land and natural resources, financial capital and income, the fruits of transactions and innovation, physical safety, risk, etc. (Weber 1923; Unger 1983; Nedelsky 1990; Delgado 1991; Eagle 2006; Dick 2011). A dramatic paradigm shift in American law occurred in the 1970s, when Congress targeted hazardous waste, water pollution, and protection of endangered species with sweeping new legislation (Lazarus 2004). However, many of the federal and state laws directed toward the environment enacted in the late 1960s to early 1970s have been applied for more than a generation, and there is a growing question of relevancy and efficacy of these laws (Arnold 2011). One example is the legal approach to control and constrain ozone-depleting compounds, formulated in the 1970s. While successful at mitigating ozone degradation, this approach failed miserably when applied to current issues of climate change and greenhouse gas emissions (Nordhaus 2006). In addition to numerous challenges to laws that regularly test their meaning and effectiveness, laws are also predicated on social values and scientific understanding, both of which are subject to change. The uncertainty associated with the dynamics of social-ecological systems presents a set of problems outside the scope of classic environmental law and has led to a fundamental conclusion regarding the interaction of environmental law and social-ecological systems: rigid legal standards that protect property rights and advance economic considerations are largely incompatible with our current understanding of the dynamics of social-ecological systems.

The legal system, particularly in the fields of environmental, natural resource, and property law, assumes a globally stable nature (Ruhl 2008; Craig 2010). That is, the environment, ecosystems, and natural resources are presumed to exist in a particular condition or state. These conditions can be described generally for a range of ecosystems as clean water, productive fisheries, healthy forests, or viable populations. Many of the impacts of human activities, such as harvesting fish or timber or changing land cover types and patterns, are thought to move the environmental system away from a particular state. Therefore, nature and human relationships with nature are to be regulated and managed based on historic conditions and linear patterns of change that are designed to return these systems to a particular (generally predisturbance) state. For example, many environmental laws are set up to control pollution by regulating discharge of pollutants into the ground, air, or water (e.g., Comprehensive Environmental Response, Compensation, and Liability Act; Clean Air Act; Clean Water Act). The implicit model in this regulatory approach is that by limiting these inputs, the ecosystem will return to a prior state—one of contaminant-free soil, clean air, or clear water.

However, during the past forty years, ecologists have developed a substantial amount of data and examples that indicate ecosystems can exist in a variety of stable configurations (Gunderson et al. 2009). Examples include clear lakes or estuaries that suddenly became turbid due to algal blooms (Scheffer 2009), grasslands that switched into shrublands (Folke et al. 2004), forests with periodic eruptions of pests (Holling 1986), coral reefs that have become algal reefs, and healthy populations that became endangered and vulnerable (Hughes 1994). In all of these cases, the ecosystem is characterized by alternate configurations, states, or regimes. Ecological resilience is the emergent property of ecosystems that mediates the transition among alternative states (Gunderson et al. 2009). Resilience theory has been developed to explain environmental systems that are complex, dynamic, and subject to abrupt and unpredictable change (Gunderson and Holling 2002; Gunderson and Pritchard 2002; Walker and Salt 2006; Gunderson et al. 2009).

The U.S. legal system is resistant to change (Hathaway 2001). It is governed by principles of stare decisis (the common law's following of prior decisions or precedents in deciding new cases), checks and balances on government authority, judicial self-restraint, protection of individual rights and freedoms, and similar concepts. A variety of legal rules and procedures make it very difficult for the U.S. Constitution to be amended and, to a lesser degree, for legislation to be enacted or amended (Ruhl 2012). The U.S. legal system favors the finality of decisions through principles of res judicata and collateral estoppel. It creates procedural and jurisdictional obstacles to appeals of or challenges to decisions. Rapid and often nonlinear transformations in ecosystems and social systems, though, require social institutions—including legal institutions—that are flexible and adaptive to change (Gunderson and Holling 2002; Gunderson et al. 2006; Olsson et al. 2006; Ruhl 2010). Although the change-slowing effect of law helps society to absorb shocks and disturbances up to a point, law can be brittle and maladaptive if it cannot keep up with the pace, scale, and direction of social-ecological change, such as drought and flooding patterns and their effects. Likewise, law is brittle and maladaptive if it assumes a static state that does not match ecological and/or social change.

The legal system is more flexible with respect to societal conditions than to ecological conditions. From time to time, the legal system balances its functions as a relatively conservative (i.e., change-resistant) institution with its principles and tools of equity, human rights, and social justice. When it does so, though, the legal system is focused either on nonenvironmental human interests or on human interests in their environment and natural resources, not on the well-being of nonhuman species or of nature's ecological features and processes. In the U.S. legal system, nonhuman species or ecosystems do not have rights or legal standing (Stone 1972). Few, if any, truly ecocentric principles are imbedded in current laws (Tarlock 2012). To the extent that the law protects or advances ecological health or welfare, it is a by-product of political choice through the enactment of legislation and its implementation through regulations. In essence, the law deems ecological health and welfare a subsidiary of human health and welfare. Science, though, demonstrates that the functioning of social systems are intricately interconnected with the functioning of ecosystems at multiple scales (Gunderson and Holling 2002; Walker and Salt 2006). The feedbacks between the two systems render elevation of either human welfare or nature's welfare, to the subordination of the other, an exercise in futility. Aldo Leopold made this point more than sixty years ago (Leopold 1949).

Throughout much of its history, the U.S. legal system has been an arena of contest between two competing ideas of instrumentalist choice: relatively autonomous, self-contained decision making based on a certain kind of deductive human reasoning from principles and theories of law, which we might loosely and perhaps a little imprecisely call "legal analysis" or "legal doctrine" (Wells 2010), versus contextual, fact-specific, pragmatic decision making based on mediation among different interests and ideas about human welfare, which we might call "legal pragmatism" (Grey 1996). However, natural systems and forces do not obey the laws of humans. Legal decision making disconnected from the underlying ecological and social context will likely produce unintended adverse consequences: solutions that are mismatched to problems; ineffective results due to uncooperative forces of nature and society; or even harm to people, ecosystems, or social institutions that could have been avoided by less abstraction and more concrete contextualism. Nonetheless, even contextual decision making based on pragmatic considerations puts too much faith in human capacity—including the capacity of legal institutions—to control nature and society. Human cognition and action, whether individually or collectively through organizations and institutions, are bounded (Lindblom 1959; Holling 1978). Legal institutions cannot unilaterally stop hurricanes, prevent droughts, dictate sea levels, or alter changes in climate.

Observers from nearly every discipline and ideological perspective have recognized the glaring, urgent need for U.S. law to improve its adaptive capacity and role in supporting the resilience of both social and ecological systems (Thompson 2000; Driesen 2003; Garmestani et al. 2009; Zellmer and Gunderson 2009; Craig 2010; Miller 2010; Benson and Garmestani 2011; Ruhl 2011; van Rijswick and Salet 2012). The maladaptive nature of law can allow, facilitate, or even mandate pathological choices and behaviors with respect to ecosystems. It can contribute to incidents of ecological collapse, which in turn lead to incidents of social collapse: humans and societies depend on resilient ecosystems if they are to survive and thrive (Gunderson et al. 2006). At regional scales, policies and actions that focus on ecosystem stabilization in order to optimize particular social goals (such as controlling floods through dams and levees on river systems), have led to ecosystems that are much less resilient and more vulnerable to various shocks—ecological or economic (Gunderson and Holling 2002). Take, for example, the levees that were constructed along the Mississippi River in New Orleans over the past 300 years; these structures have reduced the impact of minor flooding, yet increased the vulnerability to and cost of extreme events, such as Hurricane Katrina (Kates et al. 2006).

The term "resilience" has developed different meanings since Holling (1973) defined it as the capacity of an ecosystem to withstand a disturbance and maintain the same basic processes and structures. Since that time, ecologists have observed the degradation of resilience in many ecosystems (Folke et al. 2004). In most of these cases, resilience was eroded (mostly by human activities) to the point where these systems crossed a threshold that represents the limits of a particular system state, into an alternative state. That is, the processes and structures that characterize a system can rapidly change and self-organize around an alternate state characterized by a different set of processes and structures. It is the focus on alternative states that prompted Holling (1996) to distinguish between ecological resilience and engineering resilience. Ecological resilience is the amount of disturbance required to flip the system into an alternative state, whereas engineering resilience is the capacity of a system to absorb a disturbance and return to a stable equilibrium state. With respect to social-ecological systems, resilience is the amount of disturbance a linked social-ecological system can absorb before reorganizing into a new state characterized by a different set of processes and structures. A resilient system may or may not be highly variable or may or may not be in a state that is desirable to humans, but it is defined by the ability to withstand disturbance. A disturbance could primarily affect the social domain or the ecological domain, but because social-ecological systems are linked, a disturbance in one domain will affect the other domain. Contributors to this book all follow the definition of social-ecological resilience with caveats and specifics particular to their needs. For example, Benson and Hopton use a definition of resilience from Carpenter et al. (2001) that states that resilience is: (1) the amount of change the system can undergo and still retain the same controls on function and structure; (2) the degree to which the system is capable of self-organization; and (3) the ability to build and increase the capacity for learning and adaptation. Cosens and Stow follow the definition offered by Walker and Salt (2006), which states that resilience is the ability of a system to provide or shift to a regime that can provide necessary ecosystem functions and services. Ebbesson and Folke define resilience as the capacity of a social-ecological system to absorb disturbances and reorganize, while retaining essential function, structure, identity, and feedbacks. Similarly, Eason and colleagues define resilience as the capacity of a system to remain within a dynamic regime and maintain system function (self-organization). According to this definition, human activities can erode the persistence of a favorable regime that provides essential ecosystem services. Regardless of these differences, the contributors address social-ecological resilience rather than the simpler, return time of a system following disturbance associated with engineering resilience.

This volume seeks to explore and perhaps offer solutions to what we see as the fundamental issue with respect to the law and social-ecological resilience: Can a legal framework be reformed or designed to accommodate regime shifts associated with social-ecological resilience while maintaining enforceability? Regime shifts are a type of change that has been proposed to describe system-level change. Other theorists

(Schumpeter 1942; Holling 1986; Gunderson and Holling 2002) use the phrase *creative destruction* to indicate a period of rapid, system transition in which some forms of capital are destroyed and new types of capital are created. Creative destruction is endemic to many systems (e.g., economic, ecological, and social) of particular concern to humankind and thus must be considered in the calculus of managing for social-ecological resilience. By attempting to stem the inevitable process of creative destruction, humans have created the conditions for change that has the capacity to far exceed the scale (spatial and temporal) of the naturally occurring pulses of change (Odum 2007). This presents great challenges for environmental governance as we attempt to manage environmental problems within the context of an outdated legal framework, because ecosystems are not readily managed with cookie-cutter, front-end management proscriptions (Benson and Garmestani 2011). This obviously presents problems from the legal perspective, and one of the ways in which this conflict can be resolved is through application of adaptive management to ecosystem management. Likewise, principles of adaptive governance could be applied to the broader field of environmental governance and policy.

If environmental law is to be improved, there are at least three nonexclusive general approaches available. First, laws could be developed that foster resilience in systems of people and nature. Second, current law could be made more flexible and adaptive to changing ecological or social conditions. And third, law could be made to facilitate adaptive management, both as a framework for the generation and testing of laws themselves and as an approach to the management of our natural resources and heritage. Adaptive management is a key aspect of managing for social-ecological resilience. However, under the current administrative law paradigm, which is focused on the "front end" of policy, adaptive management is very limited. Part of the problem with the application of resilience theory is that it is difficult to translate fluid concepts into law, as there are aspects of social-ecological resilience that are not directly observable (Carpenter et al. 2005).

Shapiro and Glicksman (2004) suggest that in order to better accommodate adaptive management, "back-end" adjustments to regulation could serve as a more effective mechanism. They argue that administrative law would need to be amended via deadline extensions, exceptions, waivers, or variances for back-end adjustments to be incorporated into the regulatory framework. A notice and comment process is the recommended mechanism for back-end adjustments with the caveat, noted by Shapiro and Glicksman (2004), that in addition to a notice and comment process, Congress would also need to require agencies to establish electronic dockets and create annual reports on the back-end adjustment process. The back-end adjustment process is viewed as an improvement to the "considerable guess work that is involved in rationalizing regulations at the front-end of the process" (Shapiro and Glicksman 2004). Glicksman and Shapiro (2004) recognize that a regulatory system that has some flexibility, in appropriate cases, has the potential to generate the necessary conditions to manage for social-ecological resilience. This idea has great promise, as resilience-based management requires an iterative process that improves management as the process unfolds.

An adaptive system of law will need to focus on maintaining the resilience and adaptive capacity of both social and ecological systems, including such subsystems as institutions and communities. The failure of legal institutions to value and facilitate the resilience of ecosystems will likely reduce the sustainability of social systems that depend on ecosystems. At times, the legal system seems to operate as though its primary function is to promote the robustness and resilience of the legal system itself (e.g., Arnold 2010). Alternative conceptions of law focus too narrowly on the resilience of ecosystems, without adequate attention to the vitality and adaptability of the social systems and institutions that often seem to be at odds with the natural environment. For example, legal changes that give primacy to ecosystems or biodiversity, particularly if they require substantial transformations to social systems and institutions, may produce a variety of unintended consequences, including political backlash, nonimplementation or underimplementation of reforms, political and social conflict, and fiscal or economic hardships (e.g., Doremus and Tarlock 2008). Also, disturbances to social systems and institutions often adversely affect ecosystems and biological communities, while ecocentric legal reforms may fail to address the most significant pathologies of the interconnections among nature, society, and law. In contrast, an adaptive legal system aims for structures, methods, and processes that build the resilience and adaptive capacity of both nature and society—a range of ecosystems, social systems, and institutions. In order to manifest a transition to a sustainability paradigm, we must develop a better understanding of resilience, as sustainability is dependent upon social-ecological resilience. By this, we mean that sustainability is not a static state or an end point, but rather a dynamic process. Such processes have the capacity for nonlinear change and cross-scale interactions, and presents a tremendous challenge to laws based upon the conception of systems existing in a "balance of nature." In essence, the law will need to be reformed or new law will need to be crafted in order to allow us to manage for social-ecological resilience.

This book is primarily concerned with trying to explore our capacity to manage for social-ecological resilience within the context of existing laws and organizations. Sandi Zellmer and Marty Anderies investigate the historical role of and provide an assessment of the current state of wilderness preserves in the United States, characterizing wilderness laws and policies within the context of resilience theory. Zellmer and Anderies argue that adaptation will sometimes require active intervention in "untouched" wilderness preserves. Melinda Harm Benson and Matt Hopton analyze the legal frameworks and institutions that can accommodate a resilience-based approach to biodiversity protection; they accomplish this by conducting an overview of the history and the current state of wildlife management and biodiversity protection in the United States. Benson and Hopton conclude that the current legal framework in the United States is in need of reform, and they provide recommendations for legal and institutional reform allowing for resilience-based management of biodiversity. Rob Glicksman and Graeme Cumming provide an analysis of the management of parks, refuges, and preserves for resilience, and a review of U.S. laws that have perpetuated a preservation paradigm for nature. They argue that there is now broad consensus that resource management laws need to shift to fostering and managing for resilience. Glicksman and Cumming conclude that while some of the flexibility in current law is being utilized, the existing legal framework, developed under the assumption of a "balance of nature" not in concert with ecological reality, is still too rigid in its approach. Robin Craig and Terry Hughes explore the governance of the oceans as it relates to marine protected areas, spatial planning, and resilience. Craig and Hughes review the threats to marine ecosystems and identify the fragmentation of regulatory authority as the critical governance challenge for these systems. They conclude that coastal nations could foster resilience in marine ecosystems by incorporating place-based marine management (e.g., marine protected areas) into their governance regimes. Barb Cosens and Craig Stow conduct an analysis of water governance and how it affects water allocation and water quality. They assert that there are two areas of the law that be must reformed if the goals of the Clean Water Act are to be met: fragmentation of policy and addressing uncertainty in data that policy is based upon. They conclude that integration of water governance and creating the capacity for adaptation requires new approaches to environmental management. Olivia Odom Green and Charles Perrings describe transboundary water agreements and the role they play in social-ecological resilience. Their analysis of social-ecological resilience concludes that treaties must create the institutional capacity to manage conflicts and integrate iterative governance mechanisms. J. B. Ruhl and Terry Chapin integrate resilience theory and ecosystem services theory. They highlight that the resilience of ecosystems and the resilience of policy are two different animals and that the resilience of one does not, nor should it necessarily, guarantee the resilience of the other aspect of sustainability. Ruhl and Chapin conclude by offering suggestions on where ecosystem services theory is useful for supporting the resilience of ecosystems and the resilience of ecosystem policy. Alex Camacho and Doug Beard offer a chapter that considers the interaction between resilience and climate change. They conclude that existing government institutions lack the adaptive capacity to effectively manage a problem as complex as climate change. Regulators and managers lack information on effects of and management strategies for climate change, as well as the institutional infrastructure to obtain this critical information. Jonas Ebbesson and Carl Folke make the observation that legal scales and jurisdictions do not align well with the complex nature of social-ecological systems, and this mismatch erodes resilience. Ebbesson and Folke argue that the distinction between national and international needs to be demoted in order to manifest transboundary cooperation in environmental matters. The chapter by Tarsha Eason, Alyson Flournoy, Heriberto Cabezas, and Michael Gonzalez incorporates resilience and innovation into American law and policy via two proposed new laws: the National Environmental Legacy Act (NELA) and the Environmental Competition Statute (ECS).

The authors provide an analysis of both model laws based upon recent sustainability research in the United States. Eason and coauthors view the problem of resilience and law as one that will require new law in order to manifest a transition to sustainability. Tony Arnold and Lance Gunderson undertook the herculean task of creating a system of adaptive law, which is necessary, because legal decision making detached from the dynamics of social-ecological systems is likely to result in adverse consequences. They conclude that an adaptive system of law: has multiple goals, is multimodal and integrated, has the capacity to adapt to context, and has iterative legal processes with accountability. Garmestani and colleagues conclude with a chapter that integrates lessons learned from the book and synthesizes our current understanding of the interaction between social-ecological resilience and law.

The rigidity of our current legal framework is not well-suited to the complexity of social-ecological systems (Garmestani et al. 2009). In order to account for resilience in social-ecological systems, environmental law likely must evolve in response to changing environmental conditions (Garmestani et al. 2009). Law is incremental by design, and broad-scale change is therefore unlikely (Lazarus 2004). Thus, the law will likely have to evolve in an incremental manner, interspersed with dramatic change, in response to environmental problems. The chapters in this book emphasize different aspects of law and governance and highlight the features of law that allow for and hinder the capacity to manage for social-ecological resilience.

The authors in this book have analyzed the law specific to their chapter topic and suggested legal reform (see Zellmer and Anderies; Benson and Hopton; Glicksman and Cumming; Camacho and Beard), reform in governance (Craig and Hughes; Cosens and Stow; Ebbesson and Folke), reform to treaties (Green and Perrings), integration of theories (Ruhl and Chapin), and new law (Eason et al.; Arnold and Gunderson) as the means to deal with the tension between social-ecological resilience and law. It is our hope that by teaming top legal scholars with leaders in resilience science, we have contributed to a step forward in creating improved understanding of the barriers and bridges to resilience-based governance, illuminating the interaction between social-ecological resilience and law.

## References


Arnold, C. A. "Adaptive Watershed Planning and Climate Change." *Environmental & Energy Law & Policy Journal* 5 (2010): 417–487.

Arnold, C. A. "Fourth-Generation Environmental Law: Integrationist And Multimodal." *William and Mary Environmental Law & Policy Review* 35 (2011): 771–886.

Benson, M. H., and A. S. Garmestani. "Embracing Panarchy, Building Resilience and Integrating Adaptive Management Through a Rebirth of the National Environmental and Policy Act." *Journal of Environmental Management* 92 (2011): 1420–1427.

Carpenter, S., B. Walker, J. M. Anderies, and N. Abel. "From Metaphor to Measurement: Resilience of What to What?" *Ecosystems* 4 (2001): 765–781.

Carpenter, S. R., F. Westley, and M. G. Turner. Surrogates for Resilience of Social-Ecological Systems." *Ecosystems* 8 (2005): 941–944.

Clean Air Act, 42 U.S.C. § 7401 et seq.

Comprehensive Environmental Response, Compensation and Liability Act (CERCLA), 42 U.S.C. § 1906 et seq.

Craig, R. K. "'Stationarity Is Dead'—Long Live Transformation: Five Principles for Climate Change Adaptation Law." *Harvard Environmental Law Review* 34 (2010): 9–73.

Delgado, R. "Our Better Natures: A Revisionist View Of Joseph Sax's Public Trust Theory of Environmental Protection, and Some Dark Thoughts on the Possibility of Law Reform." *Vanderbilt Law Review* 44 (1991): 1209–1227.

Dick, D. L. 2011. "Confronting the Certainty Imperative in Corporate Finance Jurisprudence." *Utah Law Review* 4 (2011): 1461–1527.

Doremus, H., and A. D. Tarlock. *Water War in the Klamath Basin: Macho Law, Combat Biology, and Dirty Politics.* Washington, D.C.: Island Press, 2008.

Driesen, D. M. *The Economic Dynamics of Environmental Law*. Cambridge, Mass.: MIT Press, 2003.

Eagle, S. J. "Private Property, Development, and Freedom: On Taking Our Own Advice." *SMU Law Review* 59 (2006): 345–383.

Federal Water Pollution Control Act [commonly known as the Clean Water Act of 1972] 33 U.S.C. §1251 1387.

Folke, C., S. Carpenter, B. Walker, M. Scheffer, T. Elmqvist, L. Gunderson, and C. S. Holling. "Regime Shifts, Resilience, and Biodiversity in Ecosystem Management." *Annual Review of Ecology Evolution and Systematics* 35 (2004): 557–581.

Garmestani, A. S., C. R. Allen, and H. Cabezas. "Panarchy, Adaptive Management and Governance: Policy Options for Building Resilience." *Nebraska Law Review* 87 (2009): 1036–1054.

Glicksman, R. L., and S. A. Shapiro. "Improving Regulation Through Incremental Adjustment." *University of Kansas Law Review* 52 (2004): 1179–1248.

Grey, T. C. "Freestanding Legal Pragmatism." *Cardozo Law Review* 18 (1996): 21–42.

Gunderson, L. H., and C. S. Holling, eds. *Panarchy: Understanding Transformations in Human and Natural Systems.* Washington, D.C.: Island Press, 2002.

Gunderson, L. H., and L. Pritchard, Jr., eds. *Resilience and the Behavior of Large-Scale Systems.* Washington, D.C.: Island Press, 2002.

Gunderson, L. H., S. R. Carpenter, C. Folke, P. Olsson, and G. Peterson. "Water RATs (Resilience, Adaptability, and Transformability) in Lake and Wetland Social-Ecological Systems." *Ecology and Society* 11, no. 1 (2006): 16 [online]. http://www.ecologyandsociety.org/vol11/iss1/art16/.

Gunderson, L. H., C. R. Allen, and C. S. Holling, eds. *Foundations of Ecological Resilience.* Washington, D.C.: Island Press, 2009.

Hathaway, O. A. "Path Dependence in the Law: The Course and Pattern of Legal Change in a Common Law System." *Iowa Law Review* 86 (2001): 101–165.

Holling, C. S. "Resilience and Stability of Ecological Systems." *Annual Review of Ecology and Systematics* 4 (1973): 1–23.

Holling, C. S. *Adaptive Environmental Assessment and Management.* London: John Wiley, 1978.

Holling, C. S. "The Resilience of Terrestrial Ecosystems: Local Surprise and Global Change." In *Sustainable Development of the Biosphere.* eds. W. C. Clark and R. E. Munn, 292–317. Cambridge: Cambridge University Press, 1986.

Holling C. S. "Engineering Resilience vs. Ecological Resilience." In *Engineering within Ecological Constraints.* ed. P. C. Schulze, 31–43. Washington, D.C.: National Academies Press, 1996.

Hughes, T. P. "Catastrophes, Phase Shifts, and Large-Scale Degradation of a Caribbean Coral Reef." *Science* 265 (1994): 1547–1551.

Kates, R. W., C. E. Colten, S. Laska, and S. P. Leatherman. "Reconstruction of New Orleans After Hurricane Katrina: A Research Perspective." *Proceedings of the National Academy of Sciences USA* 103 (2006): 14653–14660.

Lazarus, R. J. *The Making of Environmental Law.* Chicago: University of Chicago Press, 2004.

Leopold, A. *A Sand County Almanac and Sketches Here and There.* New York: Oxford University Press, 1949.

Lindblom, C. E. "The Science of 'Muddling Through.'" *Public Administration Review* 19 (1959): 79–88.

Miller, K. A. "Grappling with Uncertainty: Water Planning and Policy in a Changing Climate." *Environmental & Energy Law & Policy Journal* 5 (2010): 395–416.

Nedelsky, J. *Private Property and the Limits of American Constitutionalism: The Madisonian Framework and Its Legacy.* Chicago: University of Chicago Press, 1990.

Nordhaus, W. D. "After Kyoto: Alternative Mechanisms to Control Global Warming." *American Economic Review* 96 (2006): 31–34.

Odum, H. T. *Environment, Power and Society for the Twenty-first Century: The Hierarchy of Energy.* New York: Columbia University Press, 2007.

Olsson, P., L. H. Gunderson, S. R. Carpenter, P. Ryan, L. Lebel, C. Folke, and C. S. Holling. "Shooting the Rapids: Navigating Transitions to Adaptive Governance of Social-Ecological Systems." *Ecology and Society* 11, no. 1 (2006): 8 [online]. http://www.ecologyandsociety.org/vol11/iss1/art18/.

Ruhl, J. B. "The Co-Evolution of Sustainable Development and Environmental Justice: Cooperation, then Competition, then Conflict." *Duke Environmental Law & Policy Forum* 9 (1999): 161–185.

Ruhl, J. B. "Climate Change and the Endangered Species Act: Building Bridges to the No-Analog Future." *Boston University Law Review* 88 (2008): 1–62.

Ruhl, J. B. "Climate Change Adaptation and the Structural Transformation of Environmental Law." *Environmental Law* 40 (2010): 363–431.

Ruhl, J. B. "General Design Principles for Resilience and Adaptive Capacity in Legal Systems—with Applications to Climate Change Adaptation." *North Carolina Law Review* 89 (2011): 1373–1401.

Ruhl, J. B. "Panarchy and the Law." *Ecology and Society* 17, no. 3 (2012): 31 [online]. http://dx.doi.org/10.5751/ES-05109-170331.

Scheffer, M. *Critical Transitions in Nature and Society*. Princeton, N.J.: Princeton University Press, 2009.

Schumpeter, J. A. *Capitalism, Socialism And Democracy*. New York: Harper Perennial, 1942.

Shapiro, S. A., and R. L. Glicksman. "The APA and the Back-End of Regulation: Procedures for Informal Adjudication." *Administrative Law Review* 56 (2004): 1159–1178.

Stone, C. D. "Should Trees Have Standing—Toward Legal Rights for Natural Objects." *Southern California Law Review* 45 (1972): 450–501.

Tarlock, D. "Is a Substantive, Non-Positivist U.S. Environmental Law Possible?" *Michigan Journal of Environmental & Administrative Law* 1 (2012): 159–207.

Thompson, B. H., Jr., "Markets for Nature." *William & Mary Environmental Law and Policy Review* 25 (2000): 261–316.

Unger, R. M. "The Critical Legal Studies Movement." *Harvard Law Review* 96 (1983): 561–675.

van Rijswick, M., and W. Salet. "Enabling the Contextualization of Legal Rules in Responsive Strategies to Climate Change." *Ecology and Society* 17, no. 2 (2012): 18 [online]. http://dx.doi.org/10.5751/ES-04895-170218.

Verchick, R. R. M. *Facing Catastrophe: Environmental Action for a Post-Katrina World*. Cambridge, Mass.: Harvard University Press, 2010.

Walker, B., and D. Salt. *Resilience Thinking: Sustaining Ecosystems and People in a Changing World*. Washington, D.C.: Island Press, 2006.

Weber, M. *General Economic History* [Wirtschaftsgeschichte]. New York: Cosimo, 1923.

Wells, C. P. "Langdell and the Invention of Legal Doctrine." *Buffalo Law Review* 58 (2010): 551–618.

Zellmer, S., and L. Gunderson. "Why Resilience May Not Always Be a Good Thing: Lessons in Ecosystem Restoration from Glen Canyon and the Everglades." *Nebraska Law Review* 87 (2009): 893–949.

ONE

# Wilderness Preserves

## *Still Relevant and Resilient After All These Years*

SANDRA B. ZELLMER AND JOHN M. ANDERIES

Since the late nineteenth century, policy makers and conservation groups in the United States have devoted a great deal of attention to preserving natural places (Hays 1959). Wilderness preserves, in particular, represent both the legacy of America's past—remnant patches of the vast lands occupied for millennia by Native Americans and wild creatures—and our options and hopes for a biologically and culturally resilient future (Scott 2004). Wilderness areas provide many ecological and anthropocentric benefits, including habitat for a diverse array of species, watershed protection, carbon sequestration, recreational opportunities, beauty, and quiet sanctuary. In describing one "lovely and terrible," "harshly and beautifully colored" wild area in Utah, author Wallace Stegner explained, "We simply need that wild country . . . for it can be a means of reassuring ourselves of our sanity as creatures, a part of the geography of hope" (Stegner 2004, 11).

Stegner wrote those lines a half-century ago. Although the strategy of setting aside certain wild or natural areas has served the nation well in the past, it is not clear that it will prove to be a viable conservation strategy in the future. Scientists have sounded the alarm: rapid and dramatic changes in climate are threatening the ability of ecological communities and processes to persist (Intergovernmental Panel on Climate

Change 2007). Adaptation strategies that promote resilient local and regional ecosystem responses to climate change will be imperative. In some areas, such strategies may include active intervention to foster transitions to more resilient ecological communities (Galatowitsch et al. 2009). For wilderness preserves, the desire for adaptation strategies raises a compelling question: Does it still make sense to protect wilderness areas from human intrusion?

This chapter explores the continuing relevance of preserving wilderness by preventing active human intervention. It concludes that the symbolic and ecological benefits of wilderness are as significant today as they were fifty years ago. Indeed, the importance of preserving wilderness areas will only increase as the climate changes. Land managers face complex challenges, however, when managing wilderness resources already degraded due to climate change or other human impacts that may require intervention to prevent further degradation. Deciding whether and how to intervene with active management tools, while maintaining the overarching "wild" values of wilderness, is difficult but not impossible. It is a fair bet, though, that historic characteristics and variability can no longer be the primary reference points for decision making, and that strategic approaches to monitoring and managing existing, expanded, and new preserves will be necessary (Craig 2010).

Based on a combination of lessons learned from several case studies of wilderness areas in which interventions were undertaken to address threats, and ideas from resilience theory and adaptive management, we propose three threshold inquiries that should be answered in the affirmative before a wilderness restoration project is undertaken. First, is there sufficient understanding about reference conditions and processes and the long-term effects of restoration actions? Second, is restoration even possible in a particular wilderness area, given the complexity of ecosystem dynamics and the pervasiveness of ecological change? Finally, can humans extricate themselves from the system within some discrete period of time and allow the ecological processes indicative of pre-degraded characteristics to resume functioning? If the answer to all of these questions is yes, then it may be acceptable to prioritize the need of the natural system for active restoration-oriented interventions over society's need to keep wilderness areas wild and untrammeled.

## Naturalness and Wilderness

The Wilderness Act of 1964 is widely known as one of the nation's preeminent preservation statutes (Rodgers 1994). Today, federally designated wilderness areas are found within each major category of federal lands—national forests, national parks, wildlife refuges, and public lands managed by the Bureau of Land Management. There are nearly 700 federally designated wilderness areas in forty-four states, covering 109 million acres of land or around 5 percent of the U.S. land base (Gorte 2008). About 75 percent of the wilderness in the lower forty-eight states is located within only five ecoregions—one desert ecoregion, the Mojave Desert of California, and four high-elevation ecoregions, the southern and middle Rocky Mountains, California's Sierra Nevada Mountains, and the Cascade Mountains of the Pacific Northwest (Wilderness Society 2011).

Over the years, the Wilderness Act has been remarkably stable and robust, with few legislative revisions to its substantive requirements. The Act is so well loved that, as Bill Rodgers notes, it is "virtually repeal-proof" (Rodgers 1994). During almost every congressional session since 1964, new wilderness areas have been added to the system or existing areas have been expanded. Once a wilderness area is established, Congress rarely de-designates it, although land exchanges that release land from wilderness study are occasionally authorized (Scott 2004).

The Wilderness Act directs that a wilderness area must be "protected and managed so as to preserve its *natural conditions*" (Wilderness Act § 1131(c)). Neither "natural" nor "wild" is specifically defined in the Act. "Natural" is commonly understood as "produced or existing in nature" (*Merriam-Webster's Collegiate Dictionary* 2003), as opposed to artificial or human-made. In ordinary parlance, "wild" means free, untamed, autonomous, and "in a state of nature" (*Merriam-Webster's Collegiate Dictionary* 2003). The principal author of the Wilderness Act, Howard Zahniser, defined the term "wild" as "untrammeled": "not subject to human controls and manipulations that hamper the free play of natural forces" (Zahniser 1959). The Act specifies that only those lands retaining a "primeval character and influence," which are "affected primarily by the forces of nature, with the imprint of man's work substantially unnoticeable," qualify as wilderness (Wilderness Act § 1131(c)).

Because federal wilderness areas are to remain both natural and free of human manipulation, wilderness designations impose the most restrictive management directives in federal law, far more so than the directives that apply to national parks, national forests, wildlife refuges, and other federal land categories (*Wilderness Society v. U.S. Fish and Wildlife Service* 2003). In fact, when surveyed about their ability to implement climate adaptation policies, federal land managers indicated that the constraints imposed by the Wilderness Act could act as a potential barrier (Jantarasami et al. 2010).

While the congressional mission for the National Park System provides the closest analogy to the Wilderness Act's mission, even the national parks (i.e., nonwilderness areas of parks) are managed quite differently than wilderness areas. The National Park Service Organic Act of 1916 provides that "the fundamental purpose" of parks is twofold: "to conserve the scenery and the natural and historic objects and the wild life therein *and* to provide for the enjoyment of the same in such manner and by such means as will leave them unimpaired for the enjoyment of future generations" (National Park Service Organic Act of 1916, emphasis added). Congress began setting aside federal public lands as national parks for conservation and recreational purposes in 1872 with the establishment of Yellowstone National Park (Act of March 1, 1872). There is an elemental distinction between parks like Yellowstone and wilderness areas, however; there can be no permanent roads in wilderness areas, and motorized or mechanized means of transportation are generally prohibited in wilderness areas but are quite common—even prevalent—in national parks. The absence of roads and motors is the hallmark of wilderness, distinguishing wilderness areas from all other categories of federal as well as state and private land.

To ensure that natural conditions and wild characteristics are preserved, the Wilderness Act imposes a variety of management restrictions. Specifically, as noted above, the Act prohibits most roads, and it also forbids motor vehicles, motorized equipment, mechanical transport, aircraft landings, and structures or installations "except as necessary to meet minimum requirements for the administration of the area" (Wilderness Act § 1133(c)). Does this mean that land managers must stand back while wilderness areas "evolve in whatever direction nature chooses [be free-willed] . . . regardless of pre-existing condition or future consequences?" (Sydoriak et al. 2000). Not necessarily. As the U.S. Court of Appeals for the Ninth Circuit observed, "Congress did not mandate that the Service preserve the wilderness in a museum diorama, one that we might observe only from a safe distance, behind a brass railing and a thick glass window" (*Wilderness Watch et al. v. U.S. Fish and Wildlife Service et al.* 2010). Rather, wilderness is to be "made accessible to people, 'devoted to the public purposes of recreational, scenic, scientific, educational, conservation, and historical use'" (Wilderness Act § 1133(b)). In addition to limited use of motorized or mechanical measures "as necessary to meet minimum requirements for the administration of the area," the Act also authorizes "such measures . . . as may be necessary in the control of fire, insects, and diseases, subject to such conditions as the Secretary deems desirable" (Wilderness Act § 1133(d)(1)).

## Climate Threats to Naturalness and Wildness

In the mid-twentieth century, when the Wilderness Act was passed, preventing active manipulation of land and natural resources within this one special category of federal lands made good sense. The human population was growing, and Americans were becoming more affluent and had more free time and the means to travel to remote areas and to recreate with all sorts of mechanical or motorized devices. Meanwhile, industrialization—large-scale mining and pollution from a wide range of activities—was becoming more widespread and in many cases more destructive. In 1964 and the following few decades, creating and maintaining a system of untrammeled, natural preserves seemed attractive and even critical. In the twenty-first century, however, the changes wrought by climate change are making some question whether maintaining wilderness areas will be possible in the future, and whether devoting resources to such an effort makes any sense (Galatowitsch et al. 2009; Camacho 2011). Moreover, even if the effort is made, its not at all clear that it will be possible to keep something both wild—untrammeled and unmanipulated—and natural—exhibiting only those processes and functions that would be found in nature absent human influence (Cole 2001).

For some if not most areas, a dramatically warming climate creates a "no analog" future (Williams and Jackson 2007; Ruhl 2008). Although land managers might look to existing ecological conditions, processes, and functions in southern or low-elevation areas to predict future conditions, processes, and functions in northern or high-elevation areas and to plan future scenarios and management responses (Galatowitsch et al. 2009), bringing climate models down to the fine-scale level needed to make timely on-the-ground management decisions may seem little better than reading tea leaves. Precipitation patterns, vegetative shifts, species migration and invasion, wind, and soil composition are likely to change in unpredictable ways.

Temperature increases in the American West—where most wilderness areas exist—are likely to be even greater than the projected 3°F to 10°F worldwide increase by the end of the century (Saunders and Maxwell 2005). Storms, floods, drought, fire, disease, insect infestation, and species invasion are likely to become more severe and widespread. Some scientists believe that the effects will be most intense at higher elevations, including alpine and subalpine wilderness areas. As a result, the natural ecological characteristics that set an area apart and qualified it for wilderness designation will almost certainly change over time as glaciers melt and precipitation patterns change. Examples include the following:

*Diminished snowpack and earlier snowmelt.* More winter precipitation will fall as rain instead of snow, periods of snowpack accumulation will be shorter, and earlier springtime warming will melt snowpacks earlier in the year. Peak flows will occur sooner than the current situation of early to mid-summertime peak flows, and this may cause severe flooding and soil erosion downstream in the spring, as well as diminished water supplies later in the year (U.S. Global Change Research Program 2003; Saunders and Maxwell 2005).

*Increased evaporation, erosion, and dust.* Higher temperatures will cause greater evaporation from reservoirs, lakes, and streams, and will also cause soil dryness, loss of vegetation, and erosion. More dust and other airborne particulate matter will result in more air pollution and reduced visibility (Munson et al. 2011).

*Fire.* A warmer climate will lead to more frequent and more severe fires and a longer fire season in the West. Scientists with the U.S. Forest Service Climate Change Resource Center believe that relatively modest changes in mean climate will lead to substantial increases in area burned. For a mean temperature increase of 4°F, annual area burned by wildfire is expected to increase by as much as fivefold. Ponderosa pine forests at mid- to high elevations are already facing much harsher fire regimes due to fire suppression and drought. Crown fires in these forests will cause extensive tree mortality, severe soil erosion, and nutrient losses (McKenzie et al. 2004).

*Disease and infestation.* Plant diseases and insect infestations are strongly influenced by weather and climate. Heat and drought can stress and overwhelm the physiological capability and structural integrity of plants. Climate change, coupled with invasive species and fire suppression, creates conditions conducive for devastating forest diseases and infestations (Neilson 2008).

*Shifting ranges and extinctions.* Scientists have already begun to observe shifts in the ranges of plant and animal species in the last century. Some species have climbed upward in elevation or migrated toward the North or South Pole as they seek areas within their temperature tolerances. Cooler regions have been colonized by new species, including sea anemones in Monterey Bay and lichens and butterflies in northern Europe. Based on studies of over 1,700 species, Parmesan and Yohe found "highly significant, nonrandom patterns of change in accord with observed climate warming in the twentieth century, indicating a very high confidence (.95%) in a global climate change fingerprint" (Parmesan and Yohe 2003). But some species, such as the Arctic fox (*Alopex lagopus*), are occupying a smaller range—they have nowhere cooler to go (Parmesan 2005). In a 2004 paper in *Nature*, scientists concluded that climate change could shrink the ranges of 15 to 37 percent of all species so drastically that they would be "committed to extinction" (Thomas et al. 2004). It is not possible to place the blame solely on climate change, because other variables such as habitat destruction due to development also play a role, but it seems more likely than not that a warming climate is a substantial factor in these rapid changes (Ruhl 2008).

## Human Threats to Naturalness and Wilderness

Climate change is not only changing the composition of existing wilderness preserves, but it may also increase human pressure to intervene and alter ongoing processes in hopes of mitigating adverse effects or adapting to them. For example, there may be more pressure to log forests to contain fire, disease, and infestation; to eradicate invasive species with mechanical or chemical treatments; to provide artificial water supplies to imperiled species; to reintroduce native species into historic ranges they no longer occupy; and to translocate nonnative imperiled species to cooler, higher elevations in wilderness areas (Landres 2008). Some of these activities are under consideration, and in some regions they are already under way.

*Logging and other vegetation management.* Wilderness managers and owners or managers of adjacent lands may push for more logging or other measures to "fireproof" forests and to inhibit the spread of disease and insect infestation. Other proposed strategies to combat disease and infestation include planting disease-resistant trees (hybridized or genetically modified), using pesticides and fungicides, and introducing nonnative insect predators (Cole and Yung 2010).

*Artificial water deliveries.* As precipitation patterns change and droughts become more frequent and persistent, wilderness managers may resort to artificial delivery systems to provide water to imperiled species. When bighorn sheep (*Ovis canadensis*) populations began to decline in the Kofa National Wildlife Refuge and Wilderness of southwest Arizona, the U.S. Fish and Wildlife Service (FWS) built a number of water tanks, or guzzlers, within the wilderness area. FWS personnel, in partnership with the Arizona Game and Fish Department, maintain and monitor the tanks. Composed mostly of aerated PVC pipe buried underground and designed to catch rainwater and channel it into concrete weirs or troughs, each system is capable of holding approximately 13,000 gallons of water. During droughts, refuge personnel transport water to the structures using motorized vehicles and equipment (*Wilderness Watch et al. v. U.S. Fish and Wildlife Service et al.* 2010).

*Eradicating invasive species.* In wilderness areas, land management agencies sometimes engage in eradication programs involving shooting, trapping, poisoning, burning, and other measures. Nonnative fish species have been a recurring target for rotenone applications in wilderness streams and lakes, and these types of efforts are likely to increase as managers attempt to mitigate or adapt to the effects of climate change (U.S. Department of Agriculture [USDA] Forest Service 2010; U.S. Department of the Interior, FWS 2010).

*Reintroducing native species.* Reintroductions of species that historically occupied wilderness areas but that no longer persist in those areas have already occurred and may be expected to continue. Examples include aerial stocking of cutthroat trout (*Oncorhynchus clarkii*) in wilderness lakes and streams cleared of other species through the use of rotenone. One of the most controversial reintroductions involves the Rocky Mountain gray wolf (*Canis lupus*), which had been nearly extirpated throughout the Rockies by human depredation in the early twentieth century. Pursuant to an Endangered Species Act recovery plan, wolves were captured from viable populations in Canada and released into the Greater Yellowstone Ecosystem in the mid-1990s. Recently, a federal court in Idaho approved the use of intrusive monitoring techniques—helicopters—to inventory and track reintroduced wolves and their offspring in wilderness areas (*Wolf Recovery Foundation et al. v. USFS et al.* 2010).

*Assisted migration.* Climate-sensitive species may be subject to assisted migration or translocation proposals. Potential climate refugees include the American pika (*Ochotona princeps*), bighorn sheep, eastern red wolves (*Canis rufus*), San Bernardino flying squirrels (*Glaucomys sabrinus californicus*), white-tailed ptarmigans (*Lagopus leucura*), cold-water trout and other fish species, arroyo toads (*Bufo californicus*), checkerspot butterflies (*Euphydryas editha taylori*), and white bark pine (*Pinus albicaulis*). Pika, for example, historically resided at 5,700 feet elevation, but in recent decades they have crept uphill an additional 2,000 feet. In California and Nevada, they are running out of room to climb. The high peaks and cooler temperatures of wilderness areas in the northern Rockies of Colorado, Wyoming, Montana, and Canada may seem like an attractive new home, but the pika will need help getting there.

## Conservation Implications, Resilience, and Adaptive Management

Given the rapid changes occurring on the landscape, land managers and scholars alike have debated whether the idea of wilderness is an "anachronism" doomed to extinction (Landres 2008). Even in the 1960s, when the Wilderness Act was passed, wilderness designations were subject to criticism for "locking up" federal lands and making them off-limits to all but low-impact recreational uses. Other critics have focused on the lack of continuing ecological relevance. It is true that most wilderness areas were chosen for reasons other than their biological amenities. Unlike the National Wildlife Refuge System and some other types of preserves, the wilderness system was not designed to ensure that areas with the most biodiversity potential are included within the system; rather, Congress was more concerned with recreational and aesthetic values (Callicott 1998; Foreman 1998). "Consequently, the wilderness system generally protects scenic areas of 'rock and ice' rather than wetlands, grasslands, and other more biologically productive but less visually spectacular areas" (Zellmer 2004). Arguably, the failure to prioritize scientific criteria in designating wilderness areas has resulted in an "artificial human construct" that provides "a cursory snapshot of wild lands frozen in time" (Zellmer 2004). This circa-1960 mindset plays out in the management directives expressed in the Act, which assume that a preserved ecosystem will remain in a desired, steady-state condition. We have since learned that disturbance and change is not only inevitable, it is also elemental in maintaining ecological integrity. What, then, can a system of wilderness preserves do in terms of promoting biological diversity and ecological resilience? More to the point, should Congress revise the Wilderness Act to enable managers to employ active adaptive management to promote resilience rather than wild or natural, albeit historic, characteristics?

Scientists have begun to emphasize resilience—the capacity of an ecosystem to tolerate and adapt to disturbances without collapsing into a qualitatively different state—as a replacement for our present stationarity-based approaches, which assume that natural systems fluctuate in a predictable way and strive to keep ecosystems within the historic range of variability (Holling 1973; Folke et al. 2004). As a conservation strategy, resilience has gained some ground in Congress in recent years. The 2009 Waxman–Markey climate bill passed by the U.S. House of Representatives is a lead example. The bill highlights resilience as a key concept for managing natural resources. It defines resilience as the "ability to resist or recover from disturbance [in order to] preserve diversity, productivity, and sustainability" (American Clean Energy and Security Act of 2009). Although this bill has not been passed, it indicates that Congress is increasingly aware of the need for new conservation approaches.

In recognition of the complexity of ecosystem interactions, resilience theory emphasizes adaptation, flexibility, change, and transformation—concepts that seem antithetical to conservation strategies that insist on the ironclad preservation of areas perceived to be "wild" or "natural." It is not clear that the ecological and social values of wilderness can continue to be met if a greater degree of interactive management is proscribed.

On the scientific side of the ledger, wilderness matters, and the scientific values of many protected wilderness areas will remain intact despite climate change. According to scientists in the U.S. Forest Service—an agency that was once the most outspoken opponent of official wilderness designations—wilderness areas will play an even more critical role in the future. First, wilderness areas provide "baseline" places wherein ecological lessons can be learned and used to test more intensive adaptation strategies implemented in other areas. In addition, wilderness and other protected areas will continue to provide key ecosystem services, such as clean air and water. Roadless areas like wildernesses will also provide undisturbed migration corridors and large blocks of contiguous habitat for climate-threatened species. High-altitude wilderness areas also provide elevation gradients in landscapes that have become increasingly fragmented by roads and other development. Increasing connectivity by designing and protecting wildlife corridors and reducing human-made barriers such as roads and fences, and increasing the number of reserves, especially large protected areas connected by smaller reserves, are among the top climate change–adaptation priorities recommended in the scientific literature (Heller and Zavaleta 2009).

Moreover, our current track record for "ecosystem engineering" has been less than stellar. Even when decision makers have had the best of intentions and generous funding, their efforts to restore natural features and functions that were degraded or destroyed by development have been spotty. There have been at least as many missteps as successes in the Florida Everglades, the Missouri River, and the late successional reserves and key watersheds of the Pacific Northwest forests (Zellmer and Gunderson 2009). When it comes to translocating species into novel habitats through assisted migration efforts, ecosystem engineering is even trickier and even less likely to succeed. Selecting or designing new habitats that will be viable for communities of animal and plant species that have never lived together before and that have incredibly complex life cycle needs would seem to require godlike knowledge and foresight. Our record for ensuring that intentionally translocated species do not themselves become invasive nuisance species is at least as poor as our ecological restoration track record (Pimentel et al. 2000). Dramatic changes in climate will make our predictive challenges even greater. Active management interventions that upset natural functions and processes in wilderness areas might turn out to be catastrophic.

The social values of wilderness weigh against human-generated manipulations as well. The unique spiritual and symbolic attributes of leaving a few places on Earth wild and untouched are unparalleled. No other type of federal, state, or private land can provide the renewal, solitude, and peace found in an area of roadless, nonmotorized wilderness. These values are becoming ever more important as we become ever more technologically driven and connected to cellular towers and satellites and therefore less reflective and less humble in our everyday lives (Nagle 2005). As Howard Zahniser famously said, we should be guardians of wilderness, not gardeners (Zahniser 1963). A federal judge carried Zahniser's analogy forward into case law: "Nature may not always be as beautiful as a garden but producing gardens is not the aim of the Wilderness Act" (*Minnesota Public Interest Research Group v. Butz* 1975).

That said, there may be limited instances in which managers should actively intervene to protect or restore unique, irreplaceable wilderness characteristics that have been degraded by human activities. The challenge lies in determining such instances. There are at least three threshold inquiries to be answered in the affirmative before a wilderness restoration project is undertaken. First, is there sufficient understanding about reference conditions and processes, as well as the long-term effects of restoration actions? Second, is restoration even possible in a particular wilderness area, given the pervasiveness of ecological change? In other words, will intervention more likely than not improve the functioning and integrity of the ecosystem, including its biological, chemical, and geophysical characteristics (Landres 2004)? Third, can humans extricate themselves within some discrete period of time and let the ecological processes indicative of the pre-degraded characteristics of the wilderness area resume functioning? That is, is there a clear exit strategy—a viable end point based on identifiable and measurable benchmarks? If the answer to all of these questions is yes, then it may be acceptable to prioritize the need of the natural system for some sort of active intervention to restore ecological functions and processes over the social need to keep wilderness areas perfectly wild and untrammeled.

The National Park Service (NPS) has undertaken a potentially representative restoration project in Bandelier Wilderness in northern New Mexico. Although the area had been occupied and used for centuries by ancestral Pueblo people, historical data indicates that there was good grass cover and widely spaced, healthy woodlands of piñon and juniper trees. With Euro-American settlement in the nineteenth century, however, came fire suppression and heavy sheep and cattle grazing. President Woodrow Wilson established Bandelier National Monument in 1916 to preserve and protect the area, especially "prehistoric aboriginal ruins" of "unusual ethnologic, scientific, and educational interest" (Proclamation No. 1322 1916). When Congress took the additional step of designating 23,000 acres (about two-thirds) of the monument as the Bandelier Wilderness in 1976, the ecological characteristics of the area were by no means pristine, and various signs of human occupation remained (Public Law 94–567 1976). By the turn of the twenty-first century, overgrazing and fire suppression had caused "unprecedented change" in Bandelier's piñon–juniper woodlands. An ecological threshold had been crossed. Preventing grazing and other anthropocentric activities would not promote vegetative recovery or curtail soil erosion. Without active management intervention to restore understory plants and to stabilize soils, further deterioration of ecological function would be "highly persistent and irreversible." The Park Service predicted that some areas could lose *all* remaining soil within the next century (Sydoriak et al. 2000).

Studies in the late 1990s indicated that thinning some trees and using the cut branches as a slash "erosion blanket" on exposed soils generated a threefold increase in understory cover and a dramatic reduction in erosion. With these studies in mind, the Park Service prepared an environmental impact statement and adopted a restoration plan in 2007. The plan involves cutting small-diameter trees and scattering branches on bare soil on about 4,000 acres within the Bandelier Wilderness. The Park Service decided to use chain saws, stating that "treatment of such a large area would be infeasible without the use of motorized equipment, and that impacts to monument resources would be substantially reduced through the use of this equipment" (NPS 2007). The agency will use hand tools, however, near habitat that could be or is occupied by sensitive or federally listed species. To minimize impacts, the work will take place in the winter, when peregrine falcons, bald eagles, and Mexican spotted owls have not yet begun to nest; soils are drier; and fewer visitors are present. Work camps within a 3-hour walking distance from Bandelier headquarters will be supplied by mule pack trains. Those in more remote locations will get supplies via helicopter drops, but there will be no landings in the wilderness. For a period of time after the restoration work is completed, prescribed fire may be used to maintain mechanically thinned areas and to promote long-term recovery. Park Service staff will monitor each area's response to the restoration activities and will use the information gathered from the treated sites to modify future actions if warranted (NPS 2007).

The thinning/slash option selected by the Park Service was deemed the environmentally preferred alternative, that is, the alternative that "causes the least damage to the biological and physical environment; it also means the alternative which best protects, preserves, and enhances historic, cultural, and natural resources" (Council on Environmental Quality 1987). While the use of hand tools for all of the work would be less intrusive than chain saws, the Park Service found that it would take twenty times longer to accomplish the restoration and therefore result in substantially greater loss of soils, vegetation, and cultural resources. Likewise, under the no-action (no-intervention) alternative, ecological degradation would worsen, "with major adverse impacts to the naturalness aspect of wilderness character" (NPS 2007).

Why might active restoration be appropriate in Bandelier, while it may not be elsewhere? The Bandelier plan appears to meet the wilderness restoration criteria outlined above. First, there is sufficient understanding about the human impacts on reference conditions and processes, as well as the long-term effects of restoration actions. Second, experiments and studies had shown that restoration of vegetation and soils would not occur without active intervention, but restoration would be possible with relatively minimal intervention, and ecological functioning and integrity, including biological, chemical, and geophysical characteristics, would improve relatively quickly (Landres 2004). Finally, humans can disentangle themselves within a discrete period of time and let ecological processes function freely. In Bandelier, the restoration goal will be achieved when there is sufficient understory vegetation to carry naturally occurring fires (Sydoriak et al. 2000).

Sydoriak et al. (2000) provide detail:

> Since most of the soils of the park's piñon–juniper woodlands are over 100,000 years old . . . we can be sure that the natural range of variability in these ecosystems generally allowed for soil development and stability. Controlled, progressive experiments to restore vegetation and prevent soil erosion have taken place within and outside of Bandelier since 1992 and have proven successful. Treatment directly reduces tree competition with herbaceous plants for scarce water and nutrients, and the application of slash residues across the barren interspaces greatly reduces surface water runoff and ameliorates the harsh microclimate at the soil surface, immediately improving water availability for herbaceous plants. This restoration approach has produced a two- to seven-fold increase in total vegetation cover at three years post-treatment and reductions in erosion. Other experimental treatments, such as re-seeding and controlled burns, did not promote understory growth nearly as well. Moreover, evidence of management intervention (in the form of cut marks on small stumps and scattered slash mulch) superficially disappears within roughly ten years depending on site conditions.

In contrast, some strategies for active wilderness intervention are too uncertain or too likely to jeopardize complex relationships and processes (Landres 2008), while others have no discernible end point. For example, to diminish acidity caused by air pollution, Forest Service managers used helicopters to dump 140 tons of limestone into streams within the St. Mary's Wilderness in Virginia. The agency recognized that "The question is whether to allow continued loss of the aquatic biota while preserving the wilderness concept or ideal of 'untrammeled,' or compromise the wilderness ideal, to preserve the aquatic resource?" (USDA Forest Service 1998). The intervention worked—albeit briefly—to enhance the wilderness area's "outstanding aquatic resource." Within a few months, stream pH had returned to desirable levels and macroinvertebrate and fish populations began to improve. Within six years, however, the streams were once again experiencing high acidity and the limestone treatment was repeated (Cole and Yung 2010). The only human intervention that could provide long-term benefits for this area is to stop air pollution altogether. Attempting limited yet highly intrusive restoration activities on the ground is a mere Band-Aid.

Fish eradication and stocking represent another dubious type of restoration intervention. In both the Carson-Iceberg Wilderness of California and the Bob Marshall Wilderness of Montana, federal agencies, in cooperation with state fish and game managers, plan to eradicate introduced, nonnative trout by chemically treating wilderness streams with rotenone. In Montana, the agencies hope to clear the way for stocking westslope cutthroat trout—the official state fish—in over twenty high-elevation lakes, some of which were historically fishless. Managers plan to use outboard motors and aircraft to place their personnel and pumps to apply rotenone in their attempt to restock the lakes (USDA Flathead National Forest 2006). In California, rotenone would be applied through hand-spraying and drip stations. The end goal is to establish a genetically pure population of threatened Paiute cutthroat trout, a species that is experiencing hybridization with nonnative trout (U.S. Department of the Interior, FWS 2010). In addition to being treated with rotenone, which kills fish, amphibians, and everything else that absorbs oxygen through gills (Center for Biological Diversity 2003), the area downstream of the treated stream segment will be "neutralized" with potassium permanganate dispensed by a gas-powered generator and auger. The Forest Service determined that "chemical removal of hybridized trout with the piscicide rotenone and the use of motorized equipment (generator and auger) is the minimum activity within Wilderness needed to complete Paiute cutthroat trout restoration" (USDA Forest Service 2010). Chemical treatments were planned for three consecutive years, but were ultimately enjoined by a federal district court for violating the Wilderness Act (*Californians for Alternatives to Toxics et al. v. U.S. Fish and Wildlife Service et al.* 2011). Both of the proposed initiatives would involve significant trammeling of the wild with only questionable and likely short-lived benefits for natural processes and function. Neither includes a viable exit strategy. Neither should be undertaken in wilderness.

Similar examples of interventions that are too risky and too likely to harm essential wilderness values while providing little ecological benefit include translocating climate-sensitive species to high-elevation and/or northern wilderness areas, where existing communities and ecosystem processes are poorly understood, and the means of insuring successful translocation without unintended adverse consequences are limited or nonexistent. Additionally, the use of intrusive monitoring techniques, such as helicopters to inventory and track translocated or reintroduced species, adversely affects not only the species in question but also the surrounding ecosystem as a whole and is antithetical to the very concept of untrammeled, quiet wilderness (*Wolf Recovery Foundation et al. v. USFS et al.* 2010). Although monitoring is an essential element of learning and adapting our management approaches, in most cases there are less intrusive means of accomplishing those goals.

Delivering water to bighorn sheep in the desert presents one final example of a human intervention that makes little sense in a wilderness area (*Wilderness Watch et al. v. U.S. Fish and Wildlife Service et al.* 2010). No one wants to see members of an iconic species like bighorn sheep die of thirst. But human and wildlife depredation, disease, habitat degradation, and other stressors are likely contributing to the species' decline, and there do not appear to be effective, comprehensive recovery initiatives that address them in tandem with water shortages. Constructing tanks, pipes, and guzzlers and servicing them with motor vehicles may be the most popular and least costly means of assisting the sheep, but this approach is neither self-sustaining nor resilient. Moreover, like the limestone dumping and fish eradication and stocking programs, there is no viable exit strategy. In sum, water-delivery systems involve significant manipulation of wilderness characteristics while providing only short-term benefits for the sheep, with few if any long-term benefits for natural ecological processes and function.

## Conclusion: Adapting Human Interventions and Expectations

There are many compelling reasons why wilderness areas should continue to be protected from overt human-dominated manipulations in most circumstances. Intervention into degraded wilderness areas should only be an option if: (1) there is sufficient understanding about reference conditions and processes, as well as the long-term effects of restoration actions; (2) intervention will more likely than not improve the functioning and integrity of the ecosystem; and (3) humans can extricate themselves within some discrete period of time and let the area's ecological processes resume functioning. If these criteria are met and intervention is undertaken, it should be accomplished with the least intrusive means possible. In addition, secure financial resources must be dedicated to the project, to post-project monitoring, and to further adaptation of the restoration plan as necessary.

Unless we understand a system perfectly—an impossible task—interventions aimed at increasing the stability of the system in a particular historic state may, in fact, increase the fragility of the system and do more damage than the exogenous perturbations that caused the degradation in the first place. Interventions consistent with resilience theory would maintain not the historical state of the wilderness area but rather the essential ecosystem processes that structure the area and enable the wilderness ecosystem to self-organize into a sustainable and wild regime—a collection of mutually reinforcing ecological processes.

In the end, ensuring the resilience of wilderness areas will require humans to be more sophisticated and more adaptive than we have been in the past. Rather than acting as gardeners or, worse yet, curators of museum-like dioramas fighting to keep historic features in place, we can be humble yet strategic stewards—guardians—of resilient wilderness areas.

## References

Act of March 1, 1872, 16 U.S.C. § 21 (1872).

American Clean Energy and Security Act of 2009, H.R. 2454, 111th Cong. (2009–2010), § 453.

Californians for Alternatives to Toxics et al. v. U.S. Fish and Wildlife Service et al., 814 F.Supp.2d 992 (E. D. Cal. 2011).

Callicott, J. B. "Should Wilderness Areas Become Biodiversity Preserves?" In *The Great New Wilderness Debate,* eds. J. B. Callicott and M. P. Nelson, 12–13. Athens, Ga.: University of Georgia Press, 1998.

Camacho, A. "Transforming the Means and Ends of Natural Resources Management." *North Carolina Law Review* 89 (2011): 1405–1407.

Center for Biological Diversity. "Lawsuit Forces Reevaluation of Wilderness Stream Poisoning Project." Press Release, October 9, 2003. http://www.biologicaldiversity.org/news/press_releases/trout10-9-03.html.

Cole, D. N. "Management Dilemmas That Will Shape Wilderness in the 21st Century." *Journal of Forestry* 99 (2001): 4–8.

Cole, D. N., and L. Yung. "Park and Wilderness Stewardship: The Dilemma of Management Intervention." In *Beyond Naturalness: Rethinking Park and Wilderness Stewardship in an Era of Rapid Change,* eds. D. N. Cole and L. Yung, 1–11. Washington, D.C.: Island Press, 2010.

Council on Environmental Quality. *Memorandum to Agencies: Forty Most Asked Questions Concerning CEQ's National Environmental Policy Act Regulations 6a, 40 C.F.R. Parts 1500–08.* Washington, D.C.: CEQ, 1987.

Craig, R. K. "'Stationarity Is Dead'—Long Live Transformation: Five Principles for Climate Change Adaptation Law." *Harvard Environmental Law Review* 34 (2010): 9–37.

Folke, C., S. Carpenter, B. Walker, M. Scheffer, T. Elmqvist, L. Gunderson, and C. S. Holling. "Regime Shifts, Resilience and Biodiversity in Ecosystem Management." *Annual Review of Ecology, Evolution, and Systematics* 35 (2004): 557–581.

Foreman, D. "Wilderness: From Scenery to Nature." In *The Great New Wilderness Debate,* eds. J. B. Callicott and M. P. Nelson, 568–571. Athens, Ga.: University of Georgia Press, 1998.

Galatowitsch, S., L. Frelich, and L. Phillips-Mao. "Regional Climate Change Adaptation Strategies for Biodiversity Conservation in a Midcontinental Region of North America." *Biological Conservation* 142 (2009): 2012–2022.

Gorte, R. W. *Wilderness: Overview and Statistics.* Congressional Research Service Report No. RL31477 (2008).

Hays, S. P. *Conservation and the Gospel of Efficiency: The Progressive Conservation Movement, 1890–1920.* Cambridge, Mass.: Harvard University Press, 1959.

Heller, N. E., and E. S. Zavaleta. "Biodiversity Management in the Face of Climate Change: A Review of 22 Years of Recommendations." *Biological Conservation* 142 (2009): 14–32.

Holling, C. S. "Resilience and Stability of Ecological Systems." *Annual Review of Ecology and Systematics* 4 (1973): 1–23.

Intergovernmental Panel on Climate Change. "Summary for Policymakers." In *Contribution of Working Group II to the Fourth Assessment Report of the Intergovernmental Panel on Climate Change*, eds. M. L. Parry, O. F. Canziani, J. P. Palutikof, P. J. van der Linden, and C. E. Hanson, 7–22. Cambridge: Cambridge University Press, 2007.

Jantarasami, L. C., J. J. Lawler, and C. W. Thomas. "Institutional Barriers to Climate Change Adaptation in U.S. National Parks and Forests." *Ecology and Society* 15, no. 4 (2010): 33.

Landres, P. "Managing Wildness in Designated Wilderness." *Frontiers in Ecology and the Environment* 2 (2004): 498–499.

Landres, P. "Let It Be: A Hands-off Approach to Preserving Wildness in Protected Areas." In *Beyond Naturalness: Rethinking Park and Wilderness Stewardship in an Era of Rapid Change*, eds. D. N. Cole and L. Yung, 88–105. Washington D.C.: Island Press, 2008.

McKenzie, D., Z. Gedalof, D. Peterson, and P. Mote. "Climate Change, Wildfire, and Conservation." *Conservation Biology* 18 (2004): 890–902.

*Merriam-Webster's Collegiate Dictionary*, 11th ed. Springfield, Mass.: Merriam Webster, 2003, s.v. "natural," "wild."

Minnesota Public Interest Research Group v. Butz, 401 F.Supp. 1276, 1331 (D. Minn. 1975).

Munson, S. M., J. Belnap, and G. S. Okin. 2011. "Responses of Wind Erosion to Climate-Induced Vegetation Changes on the Colorado Plateau." *Proceedings of the National Academy of Science USA* 108 (2011): 3854–3859.

Nagle, J. C. "The Spiritual Values of Wilderness." *Environmental Law* 35 (2005): 955–1003.

National Park Service. "Bandelier Ecological Restoration Plan and EIS, Record of Decision, 1–13" (2007). http://parkplanning.nps.gov/document.cfm?parkID=27&projectID=10977&documentID=20655.

National Park Service Organic Act of 1916, 16 U.S.C. § 1.

Neilson, R. "Vegetation Distribution and Climate Change." USDA Forest Service, Climate Change Resource Center (2008). http://www.fs.fed.us/ccrc/topics/vegetation.shtml.

Parmesan, C. "Biotic Response: Range and Abundance Changes." In *Climate Change and Biodiversity*, eds. T. E. Lovejoy and L. Hannah, 41–55. New Haven, Conn.: Yale University Press (2005).

Parmesan, C., and G. Yohe. "A Globally Coherent Fingerprint of Climate Change Impacts across Natural Systems." *Nature* 421 (2003): 37–42.

Pimentel, D., L. Lach, R. Zuniga, and D. Morrison. "Environmental and Economic Costs of Nonindigenous Species in the United States." *BioScience* 50 (2000): 53–65.

Proclamation No. 1322, February 11, 1916, 39 Stat. 1764.

Public Law 94–567, § 1(a), October 20, 1976, 90 Stat. 2692.

Rodgers, W. H. "The Seven Statutory Wonders of U.S. Environmental Law: Origins and Morphology." *Loyola of Los Angeles Law Review* 27 (1994): 1009–1012.

Ruhl, J. B. "Climate Change and the Endangered Species Act: Building Bridges to the No-Analog Future." *Boston University Law Review* 88 (2008): 1–62.

Saunders, S. and M. Maxwell. "Less Snow, Less Water: Climate Disruption in the West." Rocky Mountain Climate Organization (2005). http://www.rockymountainclimate.org/website%20pictures/Less%20Snow%20Less%20Water.pdf.

Scott, D. W., ed. *Enduring Wilderness: Protecting Our Natural Heritage Through the Wilderness Act.* Golden, Colo.: Fulcrum, 2004.

Stegner, W. "Wilderness Letter to Outdoor Recreation Resources Review Commission." In *Enduring Wilderness: Protecting Our Natural Heritage Through the Wilderness Act*, ed. D. W. Scott, 11. Golden, Colo.: Fulcrum, 2004.

Sydoriak, C. A., C. D. Allen, and B. F. Jacobs. "Would Ecological Landscape Restoration Make the Bandelier Wilderness More or Less of a Wilderness?" *Wild Earth* 10 (2000): 83–90.

Thomas, C. D., et al. "Extinction Risk from Climate Change." *Nature* 427 (2004): 145–148.

USDA Flathead National Forest. "South Fork Flathead Watershed Westslope Cutthroat Trout Conservation Program, Record of Decision" (2006). http://efw.bpa.gov/environmental_services/Document_Library/South_Fork_Flathead/forest_service_rod2.pdf.

USDA Forest Service, Region 8. "Proposed St. Mary's Aquatic Restoration Project, Decision Notice and Finding of No Significant Impact" (1998). www.fs.usda.gov/Internet/FSE_DOCUMENTS/fsbdev3_000366.pdf.

USDA Forest Service. "Paiute Cutthroat Trout Restoration, Record of Decision" (2010). http://a123.g.akamai.net/7/123/11558/abc123/forestservic.download.akamai.com/11558/www/nepa/54914_FSPLT2_030233.pdf.

U.S. Department of the Interior, Fish and Wildlife Service. "Paiute Cutthroat Trout Restoration Project, Alpine County, CA, Notice of Availability: Final Environmental Impact Statement." Fed. Reg. 75, no. 68: 18235–18236 (April 9, 2010).

U.S. Global Change Research Program. "National Assessment of the Potential Consequences of Climate Variability and Change: Rocky Mountain/Great Basin Region" (2003). http://www.usgcrp.gov/usgcrp/nacc/education/rockies-greatbasin/rockiesandgreatbasin-edu-3.htm#Strategies.

Wilderness Act of 1964, 16 U.S.C. §§ 1131–1136.

Wilderness Society v. U.S. Fish & Wildlife Service, 353 F.3d 1051 (9th Cir. 2003).

Wilderness Society. "Frequently Asked Questions about Wilderness" (2011). http://wilderness.org/content/frequently-asked-questions-about-wilderness.

Wilderness Watch et al. v. U.S. Fish and Wildlife Service et al., 629 F.3d 1024 (9th Cir. 2010).

Williams, J. W., and S. T. Jackson. Novel climates, no-analog communities, and ecological surprises. *Frontiers in Ecology and the Environment* 5 (2007): 475–482.

Wolf Recovery Foundation et al. v. USFS et al., 692 F.Supp.2d 1264 (D. Id. 2010).
Zahniser, H. 1959. "Letter to C. Edwards Graves, April 25, 1959," cited in D. W. Scott, "'Wilderness Character,' and the Challenges of Wilderness Preservation," *Wild Earth* (2001–2002): 79.
Zahniser, H. "Guardians Not Gardeners." *Living Wilderness* 83 (1963): 2.
Zellmer, S. "Wilderness, Water, and Climate Change." *Environmental Law* 42 (2012): 313.
——. "A Preservation Paradox: Political Prestidigitation and an Enduring Resource of Wildness." *Environmental Law* 34 (2004): 1015–1089.
Zellmer, S., and L. Gunderson. "Why Resilience May Not Always Be a Good Thing: Lessons in Ecosystem Restoration." *Nebraska Law Review* 87 (2009): 893–949.

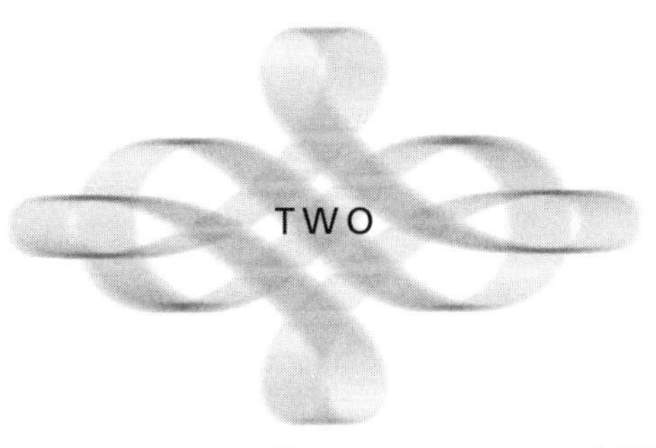


# Bringing Resilience to Wildlife Management and Biodiversity Protection

MELINDA HARM BENSON AND MATTHEW E. HOPTON

Biological diversity can be considered both temporally (i.e., evolutionary time) and/or spatially and reflects the number, variety, and variability of organisms. It includes diversity within species (i.e., genetic and morphological), between species (i.e., alpha and beta), and among ecosystems (i.e., beta and gamma). Over the past few hundred years, human activities have increased species extinction rates by as much as 1,000 times above the background rates that were typical over Earth's history (Figure 2.1) (Millennium Ecosystem Assessment 2005; but see He and Hubbell 2011). In the United States, there are approximately 1,900 species listed as threatened or endangered, with potentially thousands more at risk (U.S. Fish and Wildlife Service [USFWS] 2011a). The challenge of addressing biodiversity loss and the inevitable but largely unknown consequences associated with it presents a "wicked problem" characterized by complexity and high uncertainty (Farley 2007).

The current approach to wildlife management and the wicked problem of biodiversity loss in the United States is the subject of this chapter. We examine the nature in which existing legal frameworks and institutions address these issues and the extent to which they are compatible with a resilience-based approach. After providing a working definition of resilience, we then provide an overview of relevant state and federal

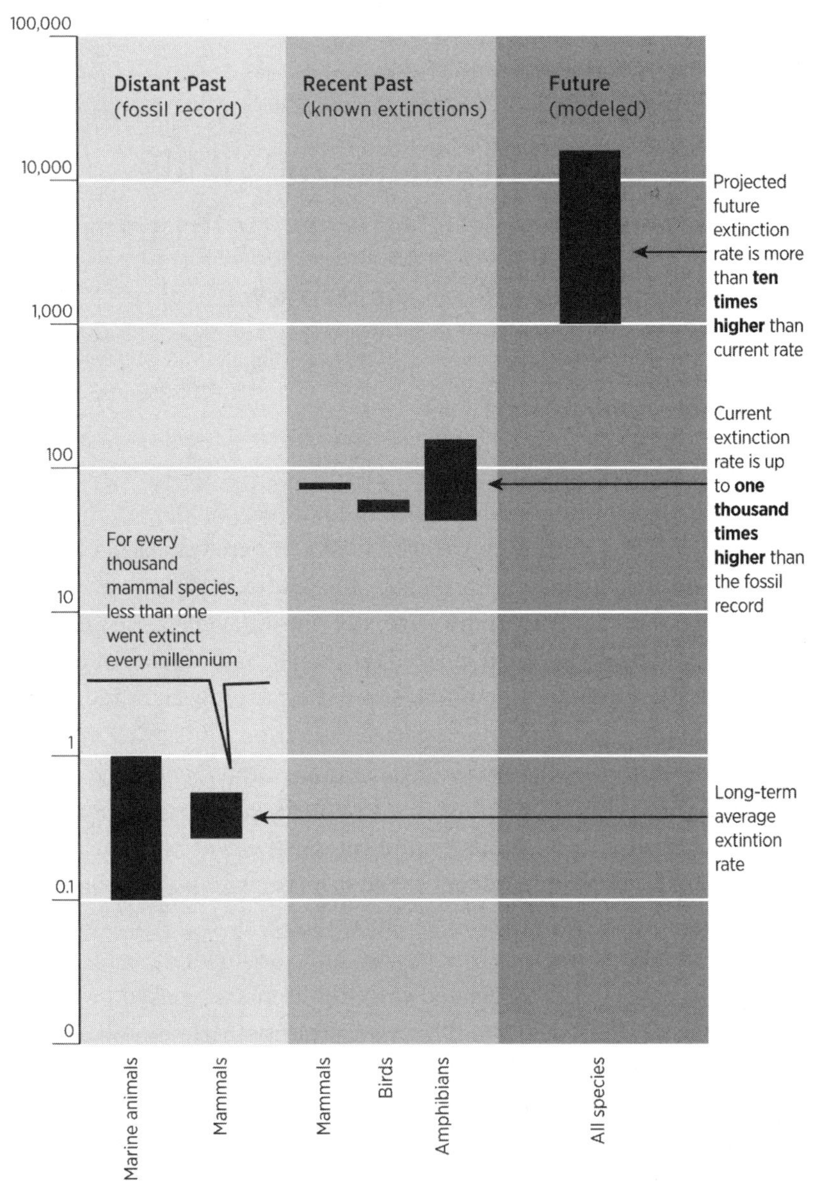


FIGURE 2.1. Millennium Ecosystem Assessment on extinction rates. *Source*: Millenium Ecosystem Assessment.

approaches to wildlife management and biodiversity protection in the United States. We place particular emphasis on the Endangered Species Act of 1973 (ESA), currently the strongest federal law capable of addressing biodiversity loss. We then explore the extent to which the ESA and other wildlife laws are compatible with resilience theory and provide some recommendations for legal and institutional reform based on a resilience-based perspective of social-ecological systems.

## Resilience, Biodiversity, and Management

Resilience, as used in this context, describes a suite of social or ecological system properties. Brand and Jax (2007) have acknowledged the differing definitions of resilience and point to the need for increased conceptual clarity to maintain the practical relevance of this concept. We use the description put forth by Carpenter et al. (2001), which characterizes resilience in three ways: (1) the amount of change the system can undergo and still retain the same controls on function and structure; (2) the degree to which the system is capable of self-organization; and (3) the ability to build and increase the capacity for learning and adaptation. Resilience is not only dependent on the functional diversity of a system, but also dependent on the range of responses (i.e., response diversity) within functional groups (Bellwood et al. 2004) and the redundancy of functions. Functional redundancy results when there are multiple species that can fill the same function or role in an ecosystem. In response diversity, individuals within a species or species within a functional group vary in their ability to deal with or respond to perturbations to system properties (Solbrig 1994), thereby enabling some species to survive and the system to continue operating. In cross-scale redundancy, members of the same functional group have influence at different scales. A functional group is a construct of organisms that typically perform the same function or processes (i.e., specific ecosystem-level biogeochemical processes [Vitousek and Hooper 1994]) in an ecosystem (Bellwood et al. 2004).

The difference between functional redundancy and response diversity is that functional redundancy will be ineffective if every species of the same functional group interacts in the same manner (Bellwood et al. 2004). For this reason, the value of high functional richness and redundancy is lost if redundant species do not "respond" differently to different stimuli (Bellwood et al. 2004). Response diversity is critical, as the interaction between species and stimuli is scale dependent (Nystrom 2006). Resilience results from species diversity, which increases the redundancy of species (Lawton and Brown 1993), and the increased biodiversity can increase the range of responses by individual species within such functional groups; the result is species may fill a function both within and across ecosystems (Solbrig 1994).

The multiple but distinct scales of self-organization and the distribution of function within and across scales also create resilient systems (Peterson et al. 1998). A system's resilience is dependent on the interactions between structure and dynamics at multiple scales (Garmestani et al. 2009). Because we cannot identify which species, functions, or responses are "key" to ecological resilience, caution dictates the wisdom of preserves of large regional species pools in heterogeneous locations to increase the chances that resilience will be maintained over time (Virah-Sawmy et al. 2009).

Resilience-based management focuses on specific attributes or drivers of complex ecological systems and crafts guiding principles for human intervention to improve long-term performance of the systems (Zellmer and Gunderson 2009). Within a social-ecological system, resilience reflects a system's capacity to absorb recurrent disturbances to retain essential structures, processes, and feedbacks. Resilience thinking acknowledges the potential for regime shifts, while providing a framework for building adaptive capacity within social-ecological systems.

## Wildlife Management and Biodiversity Protection Efforts in the United States

Most environmental and natural resource management approaches in the United States involve a mix of state and federal laws that came into being long before resilience became a theoretical orientation with regard to ecological systems and do not incorporate current scientific understanding or knowledge. The area of biodiversity and wildlife conservation is no exception. In general, individual states are recognized as having a legitimate sovereign interest in the task of regulating wildlife and the taking of wildlife within their borders. This approach dates back to English common law, which held that all valuable wild species were owned by the sovereignty and "Game could be hunted and harvested only with the Crown's permission" (Freyfogle and Goble 2009, 22). This state regulation of wildlife came with English colonialists to the New World, and the notion remains a part of the American legal system. Along with it came an important limit on this power; "the king was obligated to manage wildlife in the interests of the entire realm, not for personal benefit," and so contemporary wildlife management is subject to public trust principles (Freyfogle and Goble 2009, 23).

State control over wildlife has resulted in a fragmented system, in which each state creates its own rules and limits on the harvest of wildlife. It was almost immediately obvious that this was problematic, particularly with regard to migratory species. States had little incentive to limit harvest within their jurisdiction only to see the benefits obtained elsewhere. A primary example of this problem can be found in the extinction story of the American passenger pigeon (*Ectopistes migratorius*). Once one of the most abundant bird species in North America, the American passenger pigeon provides a cautionary tale of wildlife mismanagement (Forbush 1927). Its demise was the result of a combination of three factors relevant to biodiversity protection—a lethal combination of limited perception, fragmented management authority, and new technology. First and foremost, the species was perceived to be an inexhaustible resource. Pigeons were so numerous their flocks literally darkened the skies during their migration, and it was considered inconceivable they would ever be anything else. As a result, there were virtually no limits on their harvest. The second primary factor was increased technological capacity. While netting and gunshot were used to kill pigeons in great numbers, it was the advent of two new technologies that brought the species' rapid decline: the railroad and the telegraph. Combined, they allowed hunters to track the birds wherever they went, and soon entire flocks were captured and brought to market before they could reproduce and replenish their numbers. Third, state ownership of wildlife meant that each state was responsible for regulating harvest of the pigeon and other migratory birds only within their own borders, and states were reluctant to set limits on profit making in their own states when the benefits were sure to be reaped outside their borders. There were some last-ditch attempts at a multistate agreement placing limits on harvest, but these efforts came too late for the pigeon. The species went extinct in 1914 (Forbush 1927).

Jurisdictional limitations associated with state management have continued to relegate many state programs to a fairly limited focus in terms of wildlife management. Fontaine (2011) notes that in 2002, Congress created the State Wildlife Grant program to provide funding to state fish and wildlife management agencies, with the goal of maintaining biodiversity and avoiding the federal listing of species under the ESA. Still, he observes that state programs tend to focus on either regulating the harvest of game species or offsetting federal ESA listings, leaving the vast majority of species without any regulation or legal protection (Fontaine 2011).

Goble and Freyfogle (2002) place federal wildlife laws into three main categories (Table 2.1). The first category contains laws concerned with regulating and/or prohibiting the harvest or "take" of specific species. They include the Lacey Act (1900, 1981), the Migratory Bird Conservation Act of 1929, the Wild Free-Roaming Horses and Burros Act of 1971, and the Marine Mammal Protection Act of 1972. In the second category are those laws focused on funding, managing, and maintaining specific wildlife habitats. Funding-focused statutes in this category include the Migratory Bird Conservation Act of 1929, the Migratory Bird Stamp Act of 1934, and the Land and Water Conservation Fund Act of 1965. Refuge management statutes in this category include the National Marine Sanctuaries Act of 1972 and the National Wildlife Refuge System Administration Act of 1966 which was substantially amended in 1997 in a manner that provided a fourth unifying vision for the more than 545 national wildlife refuges across the United States. Next, there is the category of land and water management statutes that address wildlife and biodiversity issues as one aspect of a larger suite of management duties. These include the National Park Service Organic Act of 1916, the Wilderness Act of 1964, Wild and Scenic Rivers Act of 1968, the National Forest Management Act of 1976, and the Federal Land Policy and Management Act of 1976. Finally, there is the ESA, arguably the one law in the United States specifically and exclusively concerned with protecting

TABLE 2.1 Key Federal Wildlife Statutes

| Statutes | Purpose, policy, and key provisions |
|---|---|
| *1. Laws with a species-specific focus* | |
| Lacey Act of 1900 | Enacted in 1900, when illegal hunting threatened many game species; first federal wildlife statute with a national scope. It prohibits trade in wildlife, fish, and plants that have been illegally taken, transported, or sold. |
| Migratory Bird Treaty Act of 1918 | Implemented the 1916 Convention between the United States and Great Britain (for Canada) for the protection of migratory birds. Later amendments implemented similar treaties with Mexico, Japan, and Russia. |
| Wild Free-Roaming Horses and Burros Act of 1971 | "It is the policy of Congress that wild free-roaming horses and burros shall be protected from capture, branding, harassment, or death; and to accomplish this they are to be considered in the area where presently found, as an integral part of the natural system of the public lands." "All wild free-roaming horses and burros are hereby declared to be under the jurisdiction of the Secretary for the purpose of management and protection in accordance with the provisions of this Act." |
| Marine Mammal Protection Act of 1972 | Congress declared that all species and population stocks of marine mammals are or may be in danger of extinction or depletion due to human activities, and these mammals should not be permitted to diminish below their optimum sustainable population; prohibits the hunting, killing, capture, and/or harassment of marine mammals, and enacts a moratorium on the import, export, and sale of any marine mammal, along with any marine mammal part or product within the United States. First federal law with explicit ecosystem focus. |
| *2. Laws with a habitat acquisition and management focus* | |
| Migratory Bird Conservation Act of 1929 | This law established a Migratory Bird Conservation Commission, which was empowered to approve areas recommended by the secretary of the interior for acquisition with Migratory Bird Conservation Funds. These areas were selected specifically to maintain and develop refuges for North American birds. |
| Migratory Bird Stamp Act of 1934 | Requires use of a migratory bird stamp for hunting and raises funds for the conservation of migratory waterfowl by the 1916 migratory bird treaty between the United States and Great Britain (for Canada). All moneys received from stamp sales are used for migratory bird conservation. |
| Land and Water Conservation Fund Act of 1964 | Authorizes federal assistance to the states in planning, acquisition, and development of needed land and water areas and facilities, and provides funds for the federal acquisition and development of lands to enhance the quality and quantity of outdoor recreation. The primary sources of income to the fund are revenues from federal drilling offshore for oil and gas development. |

(*Continued*)

TABLE 2.1 Key Federal Wildlife Statutes (*continued*)

| Statutes | Purpose, policy, and key provisions |
|---|---|
| National Marine Sanctuaries Act of 1972 | Enacted to protect significant marine habitats and special ocean areas, such as national marine sanctuaries. It authorizes the designation and management of areas of the marine and Great Lakes environment that are considered to be nationally significant and that merit federal management. Designated areas are managed for multiple uses deemed compatible with resource protection. |
| National Wildlife Refuge System Administration Act of 1966, as amended in 1997 | Consolidated the various categories of lands into a single National Wildlife Refuge System. Amended in 1997, the National Wildlife Refuge System Improvement Act provides an organic act for the refuge system and sets forth unifying management principles for the more than 545 national wildlife refuges across the United States. |
| *3. Other laws with mandates that include a combined species and habitat focus* | |
| National Park Service Organic Act of 1916 | Authorizes the designation of national parks and monuments ". . . to conserve the scenery and the natural and historic objects and the wildlife therein and to provide for the enjoyment of the same in such manner and by such means as will leave them unimpaired for the enjoyment of future generations." While not directed at wildlife protection per se, the law's preservation mission creates opportunities for habitat conservation and species protection. Hunting is not generally allowed in national parks. |
| Wilderness Act of 1964 | Establishes the National Wilderness Preservation System, an area in which "the earth and its community of life are untrammeled by man, where man himself is a visitor who does not remain." Wilderness areas are generally roadless, undeveloped federal lands without permanent improvements or human occupation and are managed so as to preserve their natural conditions. |
| Wild and Scenic Rivers Act of 1968 | Allows for the designation as "wild and scenic" rivers that possess outstandingly remarkable scenic, recreational, geologic, fish and wildlife, historic, cultural, or other similar values. Wild and Scenic rivers are preserved in free-flowing conditions (i.e., no dams), and their immediate environments are protected for the benefit and enjoyment of present and future generations. |
| National Forest Management Act of 1976 | Amending the Forest and Rangeland Renewable Resources Planning Act of 1974 and in accordance with the Multiple-Use, Sustained-Yield Act of 1960, this law authorizes the secretary of agriculture to develop a program for lands managed by the U.S. Forest Service. The law directs management of renewable resources on these lands based on multiple-use, sustained-yield principles, as well as the development and implementation of resource management plans for each unit of the National Forest System. It is the primary statute governing the administration of national forests. The law includes a specific requirement related to management of a diversity of plant and animal communities. |

| | |
|---|---|
| Federal Land Policy and Management Act of 1976 | Constitutes the organic act for the Bureau of Land Management and directs its lands to be managed based generally on multiple-use and sustained-yield principles. This law specifically directs public lands be managed to protect the quality of ecological resources and to provide food and habitat for fish and wildlife. It also provides for the protection of public lands of critical environmental concern. |
| *4. Law with a biodiversity focus* | |
| Endangered Species Act of 1973 | Provides for the protection of species threatened with or in danger of extinction. This law explicitly references the need for conservation of not only species but also the ecosystems on which they depend. It authorizes a process for the listing of species under the act and for the designation of critical habitat necessary for species recovery. Details regarding statutory provisions are found in Table 2.2. |

*Note*: This list is not comprehensive. See Goble and Freyfogle (2002) for a complete review of applicable statutes.

species and the ecosystems on which they depend, which will be discussed extensively below.

As can be seen from this brief summary and the outline of statutory purposes provided for each of these laws, federal wildlife protection in the United States reflects a piecemeal approach. Early on, federal efforts attempted to address the limits on the ability of individual states to conserve wildlife and began with efforts to protect migratory birds. The funding and acquisition of wildlife habitat complemented these initial efforts to protect migratory species and coincided with early decisions to reserve forest lands under federal protection and designate national parks beginning in the late 1800s and continuing through the Great Depression. The focus of these laws was largely human-centered enjoyment of hunting and other recreational values related to wildlife. It was not until the 1960s and 1970s that the rise of the ecology movement set in motion the enactment of a suite of natural resource protection laws with a more ambitious and slightly less anthropogenic focus that mixed both individual species and habitat and land management concerns. For the most part, these laws reflect the "multiple use-sustained yield" principles of the era that continue to be a primary governing principle to this day (Nie 2009). Only one law sought to protect species directly and exclusively, and thereby indirectly can protect biodiversity: the ESA.

## *The Endangered Species Act: Our Current Approach to Biodiversity Protection*

The ESA is the strongest federal protection against species loss (Salzman and Thompson 2010). It was passed into law in 1973 during a time of unprecedented optimism in the United States with regard to our ability to have both a healthy environment and unlimited economic prosperity (Shellenberger and Nordhaus 2007). It embodies a view of species as parts—components of a larger ecosystem that are worth saving, even when there is little understanding regarding why they might be important. The ESA has become the major driver of most large-scale biodiversity protection and habitat restoration efforts in the United States (Benson 2012). To understand why the ESA has been so influential, it is important to understand the basic requirements of ESA and how they operate. The key provisions of the ESA are summarized in table 2.2.

TABLE 2.2 Key Provisions of the Endangered Species Act

| Provision | Statutory section | Description and enforceability |
|---|---|---|
| Purpose and policy; ESA section 2 | 16 U.S.C. § 1531 | To provide: "a means whereby the ecosystems upon which endangered species and threatened species depend may be conserved, to provide a program for the conservation of such endangered species and threatened species." |
| Definition of "take; ESA section 3 | 16 U.S.C. § 1532 | "Take" is broadly defined to include any actions that harm species, including "habitat modification or degradation where it actually kills or injures wildlife by significantly impairing essential behavior patterns, including breeding, feeding, or sheltering." |
| Listing; ESA section 4 | 16 U.S.C. § 1533(a)(1) | The listing is deemed appropriate if the continued existence of the species is jeopardized by one or more of the following factors: (1) the present or threatened destruction, modification, curtailment of the species' habitat or range; (2) overutilization for commercial, recreational, scientific, or educational purposes; (3) disease or predation; (4) inadequacy of existing regulatory mechanisms; or (5) "other factors" affecting the species' continued existence (16 U.S.C. § 1533(a)(1)). Several hard deadlines for acting on listing decisions are given. |

| | | |
|---|---|---|
| Critical habitat designation; ESA sections 3 and 4 | 16 U.S.C. §§ 1532 (5)(A), 1533(a)(3)(A) | "Critical habitat" is defined as: (1) specific areas within the geographical area occupied by the species at the time of listing, if they contain physical or biological features essential to conservation, and those features may require special management considerations or protection; and (2) specific areas outside the geographical area occupied by the species if the agency determines the area itself is essential for conservation. Economic factors are considered in designation. Alteration of critical habitat triggers the consultation requirement. Critical habitat must be designated concurrently or within one year of decision to list. |
| Recovery plans; ESA section 4(f) | 16 U.S.C. § 1533(f) | Recovery plans must contain: (1) objective measurable criteria for delisting the species; (2) site-specific actions; and (3) estimates of the time and cost for implementing the recovery plan. This is a guidance document that is not independently enforceable under ESA section 11. |
| Consultation process; ESA section 7 | 16 U.S.C. § 1536 | Requires all federal agencies to consult with the appropriate wildlife agency to ensure their actions are not likely to jeopardize the continued existence of listed species or result in destruction or adverse modification of critical habitat. Compliance is mandatory; failure to engage in consultation legally enforceable. |
| Prohibition against "take"; ESA section 9 | 16 U.S.C. § 1540 | It is illegal to "take" a listed species (see definition above) without a permit under sections 7 or 10. Seldom enforced against private parties due to burden of proof issues—must show "actual injury" to listed species. |
| Habitat conservation planning and nonessential/ experimental populations; ESA section 10 | 16 U.S.C. § 1539 | Exceptions to "take" prohibition. Allows for permits for incidental take to be granted in association with the development of habitat conservation plans on private lands and the establishment of and maintenance of experimental populations to facilitate recovery. |
| Citizen enforcement mechanism; ESA section 11 | 16 U.S.C. § 1540 (g) | Section 11 provides for civil and criminal penalties for ESA enforcement. Subsection (g) allows any citizen to petition for listing of a species and/ or to compel the government to perform nondiscretionary duties under the law (e.g., engage in consultation under section 7). |

When Congress passed the ESA, it recognized that depleted species are of "esthetic, ecological, educational, recreational, and scientific value to our Nation and its people" and expressed concern that many native plants and animals were in danger of becoming extinct (USFWS 2011b). The ESA was therefore passed into law with the ambitious purpose of providing "a means whereby the ecosystems upon which endangered species and threatened species depend may be conserved, to provide a program for the conservation of such endangered species and threatened species" (ESA § 1531).

The main mechanism for biodiversity protection under the ESA is the placement of individual species on the official list of species that are either in danger of extinction or under the threat of becoming endangered (i.e., "threatened or endangered"). The list is created and maintained by the two federal agencies with the primary responsibility for ESA implementation and enforcement: the USFWS for land and freshwater species and National Marine Fisheries Service for marine and anadromous species. The decision regarding whether to list a species as threatened or endangered is supposed to be based on "the best scientific and commercial data available" (ESA § 1533(b)(1)(A)). Economic or other social considerations are not considered. Listing is deemed appropriate if the continued existence of the species is threatened by one or more of the following factors: (1) the present or threatened destruction, modification, curtailment of the species habitat or range; (2) overutilization for commercial, recreational, scientific, or educational purposes; (3) disease or predation; (4) inadequacy of existing regulatory mechanisms; or (5) "other factors" affecting the species' continued existence. These listing factors provide the government with a great amount of discretion regarding the decision whether to list a species.

Any citizen or group of individuals can petition the government to list a species under the ESA. In fact, most of the species currently listed receive protection as a direct result of citizen petitions. Once a petition is received, the government has 90 days to make a finding regarding whether the petition presents sufficient information to require further review. If a determination is made that the listing of a species may be warranted, the government announces that it will conduct a review of the relevant scientific information. This is often referred to as a 12-month status review, and as the name suggests, the agency has one year to make a decision. At the end of the 12 months, the agency can do one of three things: (1) it can reject the listing petition, (2) it can determine that listing is "warranted but precluded" by other petitions, or (3) it can propose to list the species. If the government decides to propose the listing of a species, it then has one additional year to issue a final listing decision. The final listing decision then triggers a number of other obligations under the ESA, including the development of a recovery plan, the designation of critical habitat for the species, and a review of the listing decision every five years. Not only does a citizen petition force consideration of a particular species, it sets off a cascade of duties and deadlines. At each stage in the process, the government may be challenged in court. Many of the deadlines are "hard" (the 12-month status review, the final listing decision, and the critical habitat designation) in the sense that, as soon as the government misses the deadline, it is in violation of the ESA and vulnerable to a lawsuit. This process is one reason why ESA enforcement results in so much litigation, which in turn makes it one of the more controversial environmental statutes of its era.

Once a species is listed, a number of protections immediately fall into place. It becomes illegal to "take" a listed species, with limited exceptions. The term "take" is broadly defined to include any actions that harm the species, including "habitat modification or degradation where it actually kills or injures wildlife by significantly impairing essential behavior patterns, including breeding, feeding, or sheltering" (50 CFR § 17.3). Listing triggers the requirement that the appropriate wildlife agency designate critical habitat for the species either concurrently with the listing of the species or within one year of listing. Though the critical habitat designation includes all habitat area considered essential to the conservation of a listed species, this is one area in which economic considerations can come into play, and the agency is allowed to consider the economic or other impact of the designation on humans.

Critical habitat becomes important with regard to the activities of federal agencies, because the proposed adverse modification of critical habitat triggers the ESA's consultation requirement. Under Section 7 of the ESA, all federal agencies are required to consult with the appropriate wildlife agency to ensure their actions are not likely to jeopardize the continued existence of listed species or result in destruction or adverse modification of critical habitat. The consultation process applies to all federal "actions," a term broadly defined by the courts to include not only direct construction projects but also the granting of licenses and contracts and the promulgation of regulations (Sullins 2001). For example, the annual delivery of water under Federal Bureau of Reclamation water service projects has been a source of much ESA litigation and is the main reason why the ESA has been described "as the major federal environmental constraint on water" (Tarlock 2004, 2008).

Once an action agency determines that its proposed activity may affect a listed species or its critical habitat, it proceeds in one of two directions (Figure 2.2). If the activity "may affect, but is not likely to affect" the species, the consultation required is "informal." Informal consultation involves the action agency and the appropriate wildlife agency to share information regarding the proposed activity to assist the action agency in determining whether formal consultation is necessary. If the agencies conduct a "Biological Assessment" and determine the proposed activity is not likely to adversely affect the species or its critical habitat, no further consultation is required.

If, however, either through informal consultation or because the action agency is already certain the action "may affect and is likely to affect" the species, formal consultation is required. Formal consultation is a comprehensive process that results in the issuance of a Biological Opinion, an analysis of whether the proposed action would be "likely to jeopardize the continued existence of the species or adversely modify designated critical habitat" (ESA § 7). If a jeopardy determination is made, the Biological Opinion identifies any "reasonable and prudent alternatives" that would allow the action agency to move forward with the proposed activity. Biological Opinions include an Incidental Take Statement, anticipating that some "take" of species may result from the proposed project. The Incidental Take Statement includes terms and conditions designed to reduce the impact of the anticipated "take" that are binding on the action agency (USFWS 2007).

Finally, the ESA provides for the development of recovery plans for listed species. Recovery plans must contain objective measurable criteria for delisting, the species site-specific actions, and estimates of the time and cost for implementing the recovery plan. In theory, the recovery plan functions as the central organizing tool for guiding the recovery process for each species and for implementing the ESA as a whole.

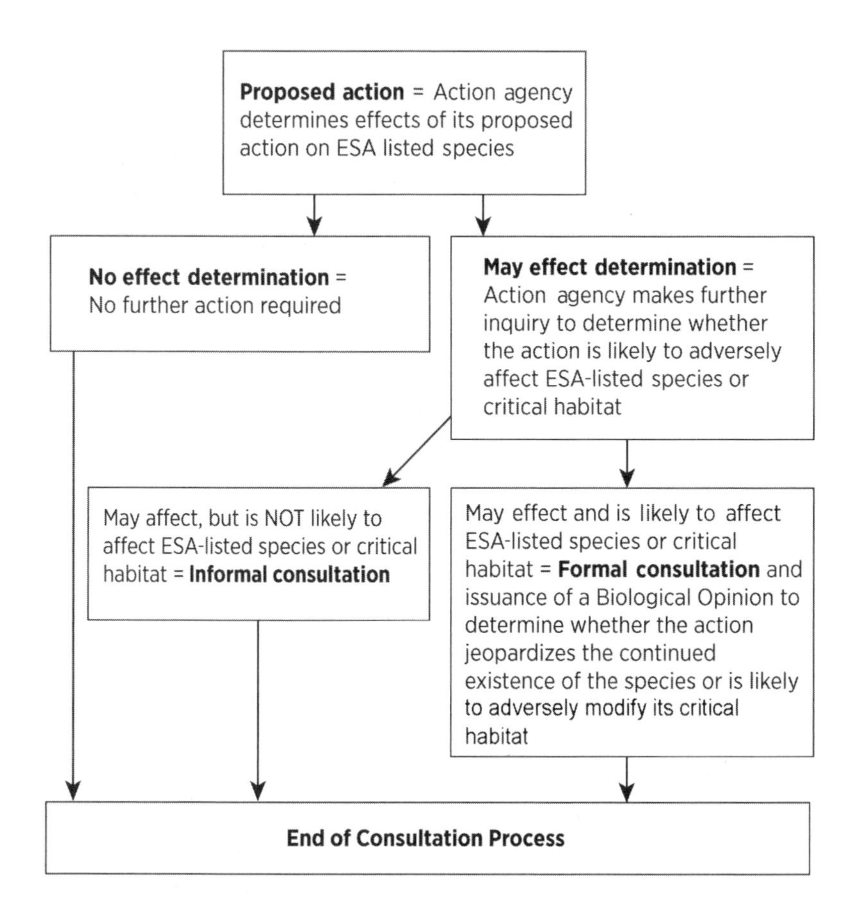


FIGURE 2.2. The ESA consultation process.

In actual practice, however, the lack of easily enforceable deadlines for developing and updating recovery plans, combined with the fact they are guidance documents and not independently enforceable, means they are often less influential than perhaps originally intended.

In summation, the ESA is a relatively simple and straightforward statute. Its uncompromising attitude toward species recovery and its many legally enforceable deadlines has made it one of the more controversial and highly litigated environmental laws in the United States. First and foremost, the ESA is helpful, because it gives biodiversity a seat at the table. There is little question the ESA has had enormous benefits related to the protection of biodiversity (Scott et al. 2005). Although some critics claim the ESA has been unsuccessful in achieving species recovery, it depends on which half of the glass you choose to focus. Since its enactment, only forty-two species have been delisted—and of those fifteen were delisted because they recovered. On the other hand, the ESA is estimated to have prevented 2,227 species from going extinct (Salzman and Thompson 2010). Moreover, the mere possibility of listing a species under the ESA often inspires efforts to protect the species, pre-empting the need for listing.

Without some legal mechanism for actually valuing biodiversity and other ecosystem services more broadly, the ESA has become a primary tool in the area of wildlife management and biodiversity protection. There are many other biodiversity protection tools provided for in the ESA, including programs designed to create incentives for owners of private lands, among them are habitat conservation plans and safe harbor agreements not discussed here. It is important to acknowledge the ESA has resulted in an unprecedented level of effort for the protection of species (Goble 2005). However, as will be discussed in the next section, the general framework and approach to wildlife protection reflected in and operationalized by the ESA has its limits.

### *Limitations of the Endangered Species Act and Other Current Wildlife Laws from a Resilience-Based Perspective*

Although the law as currently enforced focuses on individual species, the overarching purpose of the ESA is to protect and recover imperiled species and the ecosystems on which they depend. Legislative history indicates that Congress is concerned with ecosystems, as demonstrated "in the legislative history's proclamation that the ESA's 'essential purpose' is 'to protect the ecosystems upon which we and other species depend' " (Blumm and Kimbrell 2004, 351). Viewed from the perspective of resilience theory, the ESA has several limitations. The first point is perhaps the most obvious. The ESA focuses on the well-being of individual species rather than that of ecosystems, and it has long been recognized that a better approach would be to protect large areas of land as well as interacting components of ecosystems (function, processes, communities, etc.; Franklin 1993). Whereas the law itself acknowledges the importance of the "ecosystems upon which endangered species and threatened species depend," the actual implementation of the law is currently geared toward the listing and eventual recovery of individual species. If the goal is to protect biodiversity, the number of species needing protection for system resilience is too great to approach protection for each one on an individual basis (e.g., Franklin 1993). While enforcement agencies occasionally list several species conjunctively, success or failure of management efforts is tied to the recovery of individual species and not ecological resilience. Even the critical habitat designations for species center on the needs of individual species rather than the ecosystem. One unintended consequence of this approach is that there are often conflicts over the competing needs of species. A recent example can be found in the needs of the Steller sea lion (*Eumetopias jubatus*)—a species protected under the Marine Mammal Protection Act—and endangered Chinook salmon (*Oncorhynchus tshawytscha*) in the Pacific Northwest. Litigation has ensued over attempts to control sea lions as they encroach further into the Columbia River system in search of food, thereby inhibiting salmon recovery efforts (Milliman 2011). Aldo Leopold famously wrote "To keep every cog and wheel is the first precaution of intelligent tinkering" (1949, 190). One of the unintended consequences of the ESA is that it has concentrated management efforts on "parts" at the expense of a more complex assessment of the resilience of social-ecological systems.

The second limitation of the ESA relates to the fact the law only begins to protect species when they are threatened with or in danger of extinction. For this reason, the ESA has been referred to as an "emergency room" approach to protect species (Salzman and Thompson 2010, 277). It is initiated only when there is specific scientific information supporting the decision that a species is on the brink of extinction. This is problematic from a resilience standpoint for two reasons. First, the emphasis of resilience-based management is on the capacity of a system to maintain essential processes and feedbacks based on self-organization. The system is functional to the extent that it does not require external intervention by humans to maintain itself. By its very nature, an ESA listing of a species and the associated protections the species receives do not kick in until management efforts to save the species are required. The management efforts can range from relatively straightforward (i.e., banning DDT to save peregrine falcons [*Falco peregrinus*]) to much more complex (recovery needs for anadromous fish species, e.g., salmon [*Oncorhynchus* spp.], in the Columbia River Basin and the associated harvest issues, operation of dams, etc.). Second, because species only receive protection when they are in precipitous decline, there is limited capacity for the level of experimentation often necessary to better understand the needs of the endangered species or its role in the ecosystem. Although the ESA does provide for the designation of nonessential/experimental populations of imperiled species under Section 10(j), reintroduction efforts, hatcheries, and other processes that take place under this provision are highly regulated and limited in their flexibility.

The third major limitation of the ESA from a resilience perspective is the ESA's main enforcement focus on federal agency actions. Much of the ESA's implementation has focused on litigation-driven listing decisions, including actions related to USFWS's ability to meet the various deadlines associated with the listing process. As a result, the focus regarding ESA implementation is often on this listing process itself. While listing is important, it is only the beginning of a long process. Furthermore, the focus on federal agency action is very limiting, given that private land comprises approximately 60 percent of land in the United States, and the percentage of important remaining habitat (e.g., forest lands at 72 percent) is often much higher (USFWS 2001; U.S. Department of Agriculture 2002). As a result of the ESA enforcement and design, many of the habitat alterations that threaten species, such as land use planning decisions, fall outside federal jurisdiction. Most species protection on private lands occurs through voluntary agreements under the safe harbor and habitat conservation planning provisions of the law (Sullins 2001). The consultation process is mandatory. It is also time-consuming. Although the consultation process addresses the adverse modification of critical habitat, the focus is only on the need of the listed species and the ESA's prohibition against the alteration of critical habitat provisions only applies to federal actions.

Finally, and perhaps most importantly, the ESA struggles to accommodate a resilience perspective, because the law itself is based on outdated assumptions regarding ecological equilibrium and stationarity and it builds in an assumption that biological systems are basically static systems composed of a suite of unchanging variables. The policies and goals of the law focus on restoration and recovery, as opposed to resilience and adaptive capacity (Thrower 2006). Therefore, while legal scholars such as Robin Craig (2010) are starting to realize that "stationarity is dead," assumptions about stationarity are alive and well in our environmental laws. Failure to embrace the complexities associated with ecological systems, including their capacity for regime change, makes the ESA limited in terms of its capacity to address the challenges ahead—most notably the "no analog" future brought by global climate change (Ruhl 2008a). The use of a list as the major vehicle for species protection reduces the focus to one of parts rather than systems, even though the threats to those species are almost always systemic.

## Conclusion

For wildlife management in the United States to be more integrative of resilience theory, several changes must be made. The recommendations below may seem dramatic, but they are achievable without congressional amendment of the ESA. Whereas the ESA as currently enforced focuses on individual species, its overarching purpose is to protect not only imperiled species but also their associated habitats: "the ESA's 'essential purpose' is to protect 'the ecosystems upon which we and other species depend' " (Blumm and Kimbrell 2004, 351). First and foremost, there is a need to shift our management strategies from a species-centered to a systems-based approach. Moving from a focus on specific species or even particular habitats to one that seeks to understand system dynamics will allow our efforts to better capture the complexity associated with the challenges of biodiversity loss. Although ecology, by its very nature, takes a systems approach, conservation biology is increasingly focused on systems (e.g., Nassauer 2006), but the laws that currently govern state action (as described in Table 2.1) do not reflect these approaches and this understanding of the natural world.

A systems-based approach will require several basic shifts in our current thinking and management approaches (Benson and Garmestani 2011). Chief among the shifts required will be a more integrated approach to governance that includes a willingness to reassess demands placed on ecological systems by our social systems (e.g., water allocation priorities and land use decisions) in recognition of the interconnectedness of social-ecological systems (e.g., Nassauer 2006; Benson and Garmestani 2011). New approaches need to allow for the formulation of meaningful responses that foster biodiversity while increasing our understanding of the systems involved (Nassauer 2006). Adaptive management is a vehicle for achieving the integration of resilience theory into decision making to protect biodiversity (Holling 1978). A central tenet of adaptive management is the embrace of uncertainty and complexity in natural systems and recognition that "management involves a continual learning process that cannot conveniently be separated into functions like 'research' and ongoing 'regulatory activities,' and probably never converges to a state of blissful equilibrium involving full knowledge and optimum productivity" (Walters 1998).

Although adaptive management is increasingly embraced by federal agencies and is now a major organizing principle for many ESA efforts (Williams et al. 2009), the underlying theory of resilience is not necessarily being embraced, because the law itself does not easily accommodate a resilience perspective. There are exciting developments on the integration of adaptive management for species conservation from outside the ESA context that may provide important guideposts for the future. Examples include the Conservation Measures Partnership's development of a set of "Open Standards for the Practice of Conservation" designed to facilitate conservation project design, management, and monitoring to help practitioners improve the practice of conservation (Grantham et al. 2010) and the Draft Strategic Plan for Responding to Accelerating Climate Change for the National Wildlife Refuge System (2010), which embraces ecological resilience as an organizing principle.

The second and related issue is that managing for resilience will require us to be more proactive in our management efforts and to support the functioning of system feedbacks and processes before they are endangered and on the brink of regime change. This will require a better understanding of how systems function and what properties (i.e., species and their interactions) impart resilience to a given system. As discussed above, the ESA emergency room approach has a limited capacity to meet the challenges ahead. Just as an emergency room is no place for the treatment of chronic conditions, biodiversity protection requires a more preventative strategy. Furthermore, the ESA listing process, which takes a linear approach to species listing and management, does not accommodate the reality that many species are and will continue to be "conservation-reliant," in the sense that maintenance of viable populations of many species will require continuing, species-specific intervention (Scott et al. 2005, 2010). In their coining of the term *conservation-reliant*, Scott et al. (2005, 2010) argue for a new relationship to the concept of recovery under the ESA. They suggest that viewing "recovery" as a continuum of states rather than a simple "recovered/not recovered" dichotomy may enhance our ability to manage such species within the framework of the ESA. A resilience-based framework would also strengthen the role and authority of recovery planning as an organizational tool.

Next, there is a need to embrace more intentionally polycentric approaches to governance, including approaches that go beyond traditional jurisdictional boundaries and authorities. A systems-based approach will need to take on the challenges and opportunities associated with both management gaps and the overlapping nature of jurisdictional issues related to wildlife management, including nongovernmental efforts to protect species such as the Nature Conservancy's conservation inventories and Conservation by Design strategy (Noss 1987; Nature Conservancy 2000). As Cosens (2010) recently observed, "adaptive governance moves from a focus on efficiency and lack of overlap among jurisdictions to a focus on diversity, redundancy, and multiple levels of management that include a role for local knowledge and local action" (239). To date, the ESA and its enforcement have focused mainly on federal actions, whereas state wildlife management focuses on game species. More ambitious approaches are needed that take advantage of current state, federal, and private authorities in a coordinated way. Fortunately, our capacity to generate, coordinate, and share data and other information related to biodiversity has never been greater. Seemingly ever-increasing technological capacities related to geographic information science, remote sensing, and other tools allow for a much more integrated approach to experimentation, decision making, and governance.

Finally, a resilience perspective will move management to a much-needed focus beyond notions of preservation, restoration, and optimization.

Virtually all of our current management approaches are based on assumptions of system equilibrium and are made at spatial and temporal scales that do not reflect the current understanding in ecology and conservation biology. They also tend to focus on anthropocentric notions of optimization and efficiency. The early environmental history of wildlife protection and migratory birds provides an excellent example. However, as Walker and Salt (2006, 141) note, "optimization (in the sense of maximizing efficiency through tight control) is a large part of the problem, not the solution. . . . When the aim is to increase the efficiency by trying to tightly control it, we usually do so at the cost of the system's resilience." Similarly, Grantham and others (2010) observe that "conservation planning is a dynamic process, the science of which has generally focused on one-time-only assessments of optimal protected area configuration" (436). They suggest a shift to a more adaptive approach to the conservation planning process that more deliberately includes incorporating learning back into conservation design.

The challenge of moving away from optimization and preservation focuses is made more necessary when we are confronted with the realities of climate change. The future will require us to be more intentional in assessing the role of management in climate change mitigation and adaptation. For the most part, state and federal management approaches still fail to face this reality. ESA is perhaps the predominant example, in that it assumes recovery efforts based on historic ranges for species and does not easily accommodate the notion of shifting habitats, migration patterns, and the need for ongoing management. And yet, the ESA has been used by environmentalists to drive climate change action at the federal level, successfully moving for the listing and critical habitat designation of the polar bear (*Ursus maritimus*) as the first mammal listed with climate change as the primary threat to its recovery (Ruhl 2008).

The challenges associated with current and projected rates of biodiversity loss are great. Resilience thinking has the capacity to allow for a more meaningful response to these challenges, providing a conceptual basis for understanding regime change and providing a framework for maintaining or improving adaptive capacity. From a governance perspective, any real integration of resilience theory will require a number of changes in our approach to governance, including new laws and institutions that better equip us to face and acknowledge that regime shifts are occurring, can be natural or human-induced processes, and will continue to occur. Questions surrounding the appropriate roles of mitigation and adaptation responses will frame the future of biodiversity protection. To date, the ESA has dominated in the federal approach to biodiversity protection. However, for the reasons outlined in this chapter, it is now time to address wicked problems that are beyond the capacity of ESA's statutory framework. By moving to a more complex, systems-based approach, we can craft biodiversity protection approaches that are more reflective of our increasing understanding of the complexity of both social and ecological systems.

## References

Bellwood, D. R., T. P. Hughes, C. Folke, and M. Nystrom. "Confronting the Coral Reef Crisis." *Nature* 429 (2004): 827–833.

Benson, M. H. "Intelligent Tinkering: The Endangered Species Act and Resilience." *Ecology and Society* 17, no. 4 (2012): 28. http://dx.doi.org/10.5751/ES-05116–170428.

Benson, M. H., and A. S. Garmestani. "Can We Manage for Resilience? The Integration of Resilience Thinking into Natural Resource Management in the United States." *Environmental Management* 48 (2011): 392–399.

Blumm, M. C., and G. Kimbrell. "Flies, Spiders, Toads, Wolves and the Constitutionality of the Endangered Species Act's Take Provision." *Environmental Law* 34 (2004): 309–361.

Brand, F. S., and K. Jax. "Focusing the Meaning(S) of Resilience: Resilience as a Descriptive Concept and a Boundary Object." *Ecology and Society* 12, no. 1 (2007): 23. http://www.ecologyandsociety.org/vol12/iss1/art23.

Carpenter, S., B. Walker, J. M. Anderies, and N. Abel. "From Metaphor to Measurement: Resilience of What to What?" *Ecosystems* 4 (2001): 765–781.

Cosens, B. "Transboundary River Governance in the Face of Uncertainty: Resilience Theory and the Columbia River Treaty." *Journal of Land, Resources and Environmental Law* 30 (2010): 229–267.

Craig, R. K. "'Stationarity Is Dead'—Long Live Transformation. Five Principles for Climate Change Adaptation Law." *Harvard Environmental Law Review* 34 (2010): 10–31.

Endangered Species Act of 1973, 16 U.S.C. §§ 1531–1544.

Farley, J. "Wicked Problems." *BioScience* 57 (2007): 797–798.

Federal Land Policy and Management Act of 1976, 43 U.S.C. §§ 1701–1782.

Fontaine, J. J. "Improving Our Legacy: Incorporation of Adaptive Management into State Wildlife Action Plans." *Journal of Environmental Management* 92 (2011): 1403–1408.

Forbush, E. H. "Passenger Pigeon." In *Birds of Massachusetts and Other New England States*, vol. 2, ed. E. H. Forbush, 54–82. Boston: Massachusetts Board of Agriculture, 1927.

Franklin, J. F. "Preserving Biodiversity: Species, Ecosystems, or Landscapes?" *Ecological Applications* 3 (1993): 202–205.

Freyfogle, E. T., and D. D. Goble. *Wildlife Law: A Primer*. Washington, D.C.: Island Press, 2009.

Garmestani, A. S., C. R. Allen, and H. Cabezas. "Panarchy, Adaptive Management and Governance: Policy Options for Building Resilience." *Nebraska Law Review* 87 (2009): 1036–1054.

Goble, D. D. "By the Numbers." In *The Endangered Species Act at Thirty: Renewing the Conservation Promise*, eds. D. D. Goble, J. M. Scott, L. K. Svancara, and A. Pidgorna, 16–35. Washington, D.C: Island Press, 2005.

Goble. D. D., and E. T. Freyfogle. *Federal Wildlife Statutes: Texts and Contexts*. New York: Foundation, 2002.

Grantham, H. S., M. Bode, E. McDonald-Madden, E. T. Game, A. T. Knight, and H. P. Possingham. "Effective Conservation Planning Requires Learning and Adaptation." *Frontiers in Ecology and the Environment* 8 (2010): 431–437.

He, F., and S. P. Hubbell. "Species-Area Relationships Always Overestimate Extinction Rates from Habitat Loss." *Nature* 473 (2011): 368–371.

Holling, C. S. *Adaptive Environmental Assessment and Management*. London: Wiley, 1978.

Lacey Act of 1900, 16 U.S.C. §§ 3371–3378.

Land and Water Conservation Fund Act of 1965, 16 U.S.C. §§ 460l–4604.

Lawton, J. H., and V. K. Brown. "Redundancy in Ecosystems." In *Biodiversity and Ecosystem Function*, eds. E. D. Schulze and H. A. Mooney, 255–270. Berlin: Springer-Verlag, 1993.

Leopold, A. *A Sand County Almanac and Sketches Here and There*. Oxford: Oxford University Press, 1949.

Marine Mammal Protection Act of 1972, 16 U.S.C. §§ 1361–1407.

Migratory Bird Conservation Act of 1929, 16 U.S.C. §§ 715–715d, 715e, 715f–715r.

Migratory Bird Stamp Act of 1934, 16 U.S.C. §§ 718–718j.

Migratory Bird Treaty Act of 1918, 16 U.S.C. §§ 703–712.

Millennium Ecosystem Assessment. *Ecosystems and Human Well-Being: Biodiversity Synthesis*. Washington, D.C.: World Resources Institute, 2005.

Milliman, J. "Sea Lions Turn Salmon Run into Buffet: Feeding Frenzy at Columbia River's Bonneville Dam Becomes Federal Case Weighing the Fate of Two Protected Species." *Wall Street Journal* May 23, 2011. http://online.wsj.com/article/SB10001424052702304066504576339311088801864.html.

Nassauer, J. I. "Landscape Planning and Conservation Biology: Systems Thinking Revisited." *Conservation Biology* 20 (2006): 677–678.

National Forest Management Act of 1976, 16 U.S.C. §§ 1600–1614.

National Marine Sanctuaries Act of 1972, 16 U.S.C. §§ 1431–1445b.

National Park Service Organic Act of 1916, 16 U.S.C. §§ 1–4.

National Wildlife Refuge System Administration Act of 1966, as amended in 1997, 16 U.S.C. §§ 668dd–ee.

Nature Conservancy. *Conservation by Design: A Framework for Mission Success.* Arlington, Va.: Nature Conservancy, 2000.

Nie, M. *The Governance of Western Public Lands: Mapping Its Present and Future.* Lawrence: University Press of Kansas, 2009.

Noss, R. "From Plant Communities to Landscapes in Conservation Inventories: A Look at the Nature Conservancy (USA)." *Biological Conservation* 41 (1987): 11–37.

Nystrom, M. "Redundancy and Response Diversity of Functional Groups: Implications for the Resilience of Coral Reefs." *Ambio* 35 (2006): 30–35.

Peterson, G. D., C. R. Allen and C. S. Holling. "Ecological Resilience, Biodiversity and Scale." *Ecosystems* 1 (1998): 6–18.

Ruhl, J. B. "Climate Change and the Endangered Species Act: Building Bridges to the No-Analog Future." *Boston University Law Review* 88 (2008): 1–62.

Salzman, J., and B. Thompson. *Environmental Law and Policy.* Eagan, Minn.: Foundation, 2010.

Scott, M. J., D. D. Goble, J. A. Wiens, D. S. Wilcove, M. Bean, and T. Male. "Recovery of Imperiled Species Under the Endangered Species Act: The Need for a New Approach." *Frontiers in Ecology and the Environment* 3 (2005): 383–389.

Scott, M. J., D. D. Goble, A. M. Haines, J. A. Wiens, and M. C. Neel. "Conservation-Reliant Species and the Future of Conservation." *Conservation Letters* 3 (2010): 91–97.

Shellenberger M. and T. Nordhaus. *Breakthrough: From the Death of Environmentalism to the Politics of Possibility.* New York: Houghton Mifflin, 2007.

Solbrig, O. T. "Plant Traits and Adaptive Strategies: Their Role in Ecosystem Function." In *Biodiversity and Ecosystem Function,* eds. E. D. Schulze and H. A. Mooney, 97–116. Berlin: Springer-Verlag, 1994.

Sullins, T. *ESA: Endangered Species Act.* Chicago: American Bar Association Section of Environment Energy and Resources Press, 2001.

Tarlock, A. D. "Water Law Reform in West Virginia: The Broader Context." *West Virginia Law Review* 106 (2004): 495–538.

Tarlock, A. D. *Law of Water Rights and Resources.* Eagan, Minn.: West, 2008.

Thrower, J. "Adaptive Management and NEPA: How a Nonequilibrium View of Ecosystems Mandates Flexible Regulation." *Ecology Law Quarterly* 33 (2006): 871–895.

U.S. Department of Agriculture. *Major Uses of Land in the United States, 2002.* Economic Research Service Bulletin No. EIB-14, 2002. http://www.ers.usda.gov/publications/EIB14/eib14j.pdf.

U.S. Fish and Wildlife Service. *U.S. Forest Facts and Historical Trends.* Washington, D.C.: USFWS, 2001. http://fia.fs.fed.us/library/briefings-summaries-overviews/docs/ForestFactsMetric.pdf.

U.S. Fish and Wildlife Service. *Formal Consultation on the Proposed Relicensing of the Klamath Hydroelectric Project, FERC Project No. 2082, Klamath River, Klamath County, Oregon, and Siskiyou County, California*. Prepared by United States Department of the Interior, Fish and Wildlife Service, Yreka Fish and Wildlife Office, Yreka, California. 2007.

U.S. Fish and Wildlife Service. Rising to the Challenge: Draft Strategic Plan for Responding to Accelerating Climate Change. Arlington, Va.: USFWS, 2010.

U.S. Fish and Wildlife Service. *Species Reports*. Arlington, Va.: USFWS, 2011a. http://ecos.fws.gov/tess_public.

U.S. Fish and Wildlife Service. *ESA Basics: More than 30 Years of Conserving Endangered Species*. Arlington, Va.: USFWS, 2011b. http://www.fws.gov/endangered/esa-library/pdf/ESA_basics.pdf.

Virah-Sawmy, M., L. Gillson, and K. J. Willis. "How Does Spatial Heterogeneity Influence Resilience to Climatic Changes? Ecological Dynamics in Southeast Madagascar." *Ecological Monographs* 79 (2009): 557–574.

Vitousek, P. M., and D. U. Hooper. "Biological Diversity and Terrestrial Ecosystem Biogeochemistry." In *Biodiversity and Ecosystem Function*, eds. E. D. Schulze and H. A. Mooney, 3–14. Berlin: Springer-Verlag, 1994.

Walker, B., and D. Salt. *Resilience Thinking: Sustaining Ecosystems and People in a Changing World*. Washington, D.C.: Island Press, 2006.

Walters, C. J. *Adaptive Management of Renewable Resources*. New York: Macmillan, 1998.

Wilderness Act of 1964, 16 U.S.C. §§ 1131–1136.

Wild Free-Roaming Horses and Burros Act of 1971, 16 U.S.C. §§ 1331–1340.

Wild and Scenic Rivers Act of 1968, 16 U.S.C. §§ 1271–1287.

Williams, B. K., R. C. Szaro, and C. D. Shapiro. *Adaptive Management: The U.S. Department of the Interior Technical Guide*. Washington, D.C.: Adaptive Management Working Group, U.S. Department of the Interior, 2009. http://www.doi.gov/initiatives/AdaptiveManagement/TechGuide.pdf.

Zellmer, S., and L. Gunderson. "Why Resilience May Not Always Be a Good Thing: Lessons in Ecosystem Restoration from Glen Canyon and the Everglades." *Nebraska Law Review* 87 (2009): 894–947.

# Landscape-Level Management of Parks, Refuges, and Preserves for Ecosystem Resilience

ROBERT L. GLICKSMAN AND GRAEME S. CUMMING

The laws that govern the location and management of national parks, wildlife refuges, and other federal land preserves were designed to protect and "conserve" natural resources such as fish, wildlife, and other "natural objects" found in some of the nation's most highly treasured locations. It is clear, however, that the nature of conservation has changed substantially since Congress first enacted laws such as those from which the National Park Service (NPS) and the U.S. Forest Service first derived their authority. Recent years have seen an increasing awareness that ecosystems depend on other systems and may be influenced by them from afar (Polis et al. 2004). Contemporary conservation strategies are often preoccupied with events outside protected areas, such as the maintenance of connectivity and the creation of buffer zones (McCook et al. 2009; Raatikainen et al. 2009). As the area of wild land outside protected areas declines, human-dominated ecosystems, such as farms, forestry plantations, and riparian zones, are becoming essential for the long-term sustainability of many animal and plant populations and communities and entire ecosystems (Cumming and Spiesman 2006).

Adding further complexity to the problem, the effects of climate change and other broad-scale anthropogenic influences (e.g., nitrogen deposition, introduction of invasive species) are anticipated to increase (Millennium Ecosystem Assessment 2005). These influences are expected to alter both the structure and functioning of natural systems, such that it will be impossible to keep them looking and functioning as they have in the past. As J. B. Ruhl has argued, "environmental law is going to have to give up on the preservation strategy" that has sought for much of the last century to either protect nature as it exists in the parks and refuges from human-caused impairment or to restore conditions as they existed during some historical baseline reference point (Ruhl 2010, 394). Instead, there is broad consensus that resource management and the laws that govern it should shift toward protecting ecosystem resilience, focusing on external influences and connections important for system persistence (although that consensus disappears if managing for resilience is interpreted as striving to maintain systems in a relatively static condition in the face of disturbances; Doremus 2010).

Given that management for resilience appears set to become a new paradigm for the role of protected areas in conservation (Walker et al. 2002), it is important to define what management for resilience means and to determine how best to implement such a management strategy. The starting point for definitions of resilience in this context is the notion of a social-ecological system: a fully integrated system of people and nature that behaves differently than either a social system or an ecosystem alone (Westley et al. 2002). Recognition of the social-ecological nature of conservation (conservation is done by people, for people; without people, there would be no need for conservation) represents a significant shift away from earlier notions of people as somehow independent of ecosystems and/or in a detached external role as "managers" (Cundill et al. 2012).

Carpenter et al. (2001) provide a comprehensive definition of the resilience of a social-ecological system as (1) the amount of disturbance that a system can absorb while still remaining within the same state or domain of attraction; (2) the degree to which the system is capable of self-organization (versus lack of organization or organization forced by external factors); and (3) the degree to which the system can build and increase its capacity for learning and adaptation. Resilience can also be thought of as the ability of a system to maintain its identity (Cumming and Collier 2005). Thus, the resilience of biodiversity in a protected area encompasses not only the ecosystem's ability to respond to physical changes, but also its ability to continue to function despite failed management or regulatory actions (Angelo 2009) and the ability of the social elements of the system to withstand political and economic pressures (for instance, a demand to open a park to drilling for oil).

Despite their acknowledged weaknesses, such as their potential to negatively impact the resource access rights of marginalized human communities, the creation and maintenance of protected areas remains one of the most important and widely applied strategies for biodiversity conservation. If the primary goal of protected areas is to conserve critical elements of biodiversity for the future, while continuing to provide ecosystem services to contemporary human communities (Bengtsson et al. 2003), then it is important that protected areas are made resilient (as far as possible) to whatever the future may bring. Analysis of the social-ecological resilience of protected areas from a legal perspective thus focuses on the questions of: (1) What kinds of protected areas are best suited to protecting the resilience of natural systems? (2) Do adequate legal structures exist to allow for the creation and management of such areas?

In the rest of this chapter, we first explain why the development of policy and management strategies at broad scales is critical for the existence of resilient social-ecological systems. We then summarize the legal regimes under which the agencies responsible for managing the national parks, wildlife refuges, and similar land designation categories in the United States operate. We also explore the challenges and opportunities those regimes present for effective broad-scale management of protected areas, focusing on national parks and wildlife refuges as examples. We conclude that existing laws governing the national parks and wildlife refuges provide considerable flexibility for agencies to anticipate and react to climate change and other disruptions of natural systems through landscape-level management mechanisms, and that the responsible agencies have begun to exercise their discretion in ways consistent with the promotion of ecosystem resilience. At the same time, the laws governing management of the parks and refuges were largely enacted to fit a different conception of scientific reality from the one that now exists, and entrenched policies adopted under those laws may no longer fit that reality. As a result, there is an increasing mismatch between legal mandates and protected area resilience. Both statutory and administrative changes are needed if national parks and wildlife refuges are to maintain social-ecological resilience.

## Broad-Scale Planning and Management Strategies for Protected Areas

Resilience is an emergent, whole-system property that arises from the interactions between different system components and their environment (Levin 2005). The focal scale of analysis that is adopted in addressing any scientific problem is dictated by the phenomena under study. Explanations for phenomena usually come from patterns and processes at lower hierarchical levels than the focal scale (e.g., the outcome of an election depends on the behaviors of individual voters), whereas patterns and processes at higher hierarchical levels impose constraints on outcomes at the focal scale (e.g., state laws are subordinate to federal laws and the national constitution; Allen and Starr 1982). To understand the resilience of protected areas therefore requires a multiscale, hierarchical approach that considers both the broader constraints on protected areas and the finer-scale mechanisms that drive their emergent dynamics.

Current approaches to the design of conservation networks emphasize the need to conserve representative examples of ecosystems and ecological communities within a given area, as well as focal or "target" populations and species (e.g., valuable, rare, charismatic, endemic, and/or endangered species and populations [e.g., Margules and Pressey 2000; Lindenmayer et al. 2002; Baskett et al. 2007; Diniz et al. 2008]). Multiscale conservation planning efforts in the United States, such as the Nature Conservancy's Conservation by Design framework (Nature Conservancy 2003), begin at a national level with coarse ecoregions and then zoom in hierarchically to finer grains of analysis and increasingly smaller expanses (Poiani et al. 2000). Although Conservation by Design deliberately considers a set of species that use the landscape at different scales, it does not directly incorporate a scale-based awareness of the accompanying human elements of the problem (and their feedbacks and interactions with ecosystems) and pays little attention to institutions.

As part of broad-scale conservation planning initiatives, a range of quantitative approaches (e.g., MARXAN and C-Plan: Pressey et al. 1996; Ball and Possingham 2000; Possingham et al. 2002) have been developed to facilitate the objective selection of new areas to prioritize for addition to existing reserve networks. While such approaches can be extremely useful in phrasing the problem and aiding decision makers, the outcomes are strongly dependent on the spatial data sets (e.g., breeding bird survey data, national land cover maps, and digital elevation models) used as inputs to the reserve selection algorithms (Geselbracht et al. 2009). Most conservation-related data sets are static, in the sense that they consist of aggregations of data that often come from multiple years of field observations or one-off mapping exercises. While reserve selection approaches are spatially extensive, therefore, they often lack temporal depth and detail. This in turn means that they have little to say about the resilience of the broader landscape and can easily be derailed by the emergent dynamics that result from interactions between different landscape constituents (particularly people and ecosystems). For example, the expansion of a reserve network may increase local land prices, making the plan harder to implement; fire management to maintain critical habitats on small land parcels may become increasingly difficult as a landscape becomes increasingly populated; and unanticipated changes in resource use (e.g., increased water extraction or greater demand for timber) can modify conservation priorities. Although spatially explicit conservation planning in its current form captures many important elements of the problem of reserve design for resilience, it does not provide a rigorous approach to designing reserve networks for resilience (see the example presented by Leroux et al. 2007).

In a changing world, the traditional approach to the creation and management of parks, refuges, and reserves (i.e., treating them as static and discrete entities that should remain essentially the same with reference to a historical state) is clearly inadequate. As argued by Camacho (2010), "Reserves—the fundamental strategy of conventional natural resources management—embody this model of ecology that emphasizes stasis and natural stability" (221). Climate change, for example, is likely to fundamentally alter—and in some cases destroy (as in the Arctic)—existing ecological systems (Craig 2010). If reserve networks are to persist, they must be built with system resilience to future perturbations in general and climate change in particular in mind. This in turn means understanding the nature of broad-scale threats and developing approaches to maintain system attributes at multiple scales that have been demonstrated to lead to greater resilience in other contexts.

At local scales, system attributes that influence resilience include (among other things) key components (e.g., organisms, actors, soils, infrastructure, etc.), interactions, and feedbacks; diversity, in both the social and the ecological systems; and system elements that contribute to system continuity and memory (Cumming et al. 2005). The resilience of a given reserve will also be influenced by a number of spatial attributes, including, for example, its morphology (e.g., reserve size and shape, including perimeter-to-area ratio), the nature of its boundaries (e.g., hard or soft, permeable or not by both people [e.g., poachers] and organisms [e.g., cattle, dispersing wild ungulates]), the arrangement of its parts (e.g., location of catchments and water resources, position along an elevational gradient), and its proximity to sources of propagules and perturbations (Cumming 2011a, b). At a regional scale, the spatial properties of the reserve network most relevant to resilience include not only the connectivity of reserves to one another and to other ecological sources and sinks, but also their socioeconomic context; the degree to which they are influenced by spatial feedbacks and subsidies (spatial feedbacks occur when A influences B, B influences A, and A and B are in different locations, as with rural supply and urban demand); and regional perturbations and drivers, such as the economics of resource-related supply and demand, pest outbreaks, and extreme weather events (Cumming 2011a).

Not all problems will require management at regional scales (Cosens 2011). Species losses in the national parks in the United States have, however, often resulted from the unavailability of sufficient lands to provide habitat and meet other species needs (Keiter 2010, 92). For example, species such as the Florida panther require large areas for dispersal, both to escape parental territories and find mates. Because it is often impossible to exclude large disturbances from small protected areas, mosaics of intensively managed landscapes interspersed with smaller "pristine" reserves often lack resilience. Even the larger national parks are not isolated entities, or islands, that are somehow separate from the larger world (Keiter 2010, 91).

These considerations, taken together, imply there is a need for a system of nature protection that better matches the dynamism and the multiscale nature of the real world. In essence, this means creating institutions that will facilitate the existence of dynamic reserves with different zones that may be exploited in different ways at different times (Carpenter and Brock 2004). For example, sections of a river may be closed to fishing during the fish-breeding season; timber extraction may be restricted to periods of the year when it has lower impacts on breeding or migratory birds; and a no-tolerance zone for campfires may be extended during particularly dry times of year. Such regulations may be applied at different scales (e.g., local, district, or regional), depending on the spatial and temporal scales of the process that is being managed. Dynamic reserves will need to extend beyond the parameters of current publicly owned protected areas, such as national parks and wildlife refuges, and hence will affect lands owned by states and private individuals. The multiplicity of tenure regimes occurring in the vicinity of protected areas will present significant management challenges, if, for example, private owners favor development that governments managing nearby reserves believe will cause detrimental spillover effects (Berkes et al. 2006). In particular, effective management of these extended reserves will require appropriate incentive structures, such as payments to private landowners who grant conservation easements or otherwise agree not to develop, for the coordination of a diversity of stakeholders (Bengtsson et al. 2003).

## Agency Management Authority

As Alejandro Camacho has recognized, the existing legal framework for management by federal agencies of public land systems such as parks and wildlife refuges "was not designed to facilitate a wide-scale ecosystem-based or landscape-based approach that blurs the distinctions between legally separate public lands. . . . [F]requent, wide-scale, and synchronized interaction between agencies was simply not contemplated by existing natural resource management laws, so that coordinated ecosystem-based management is still the exception in natural resource governance" (Camacho 2010, 208). Figure 3.1 illustrates the allocation of authority

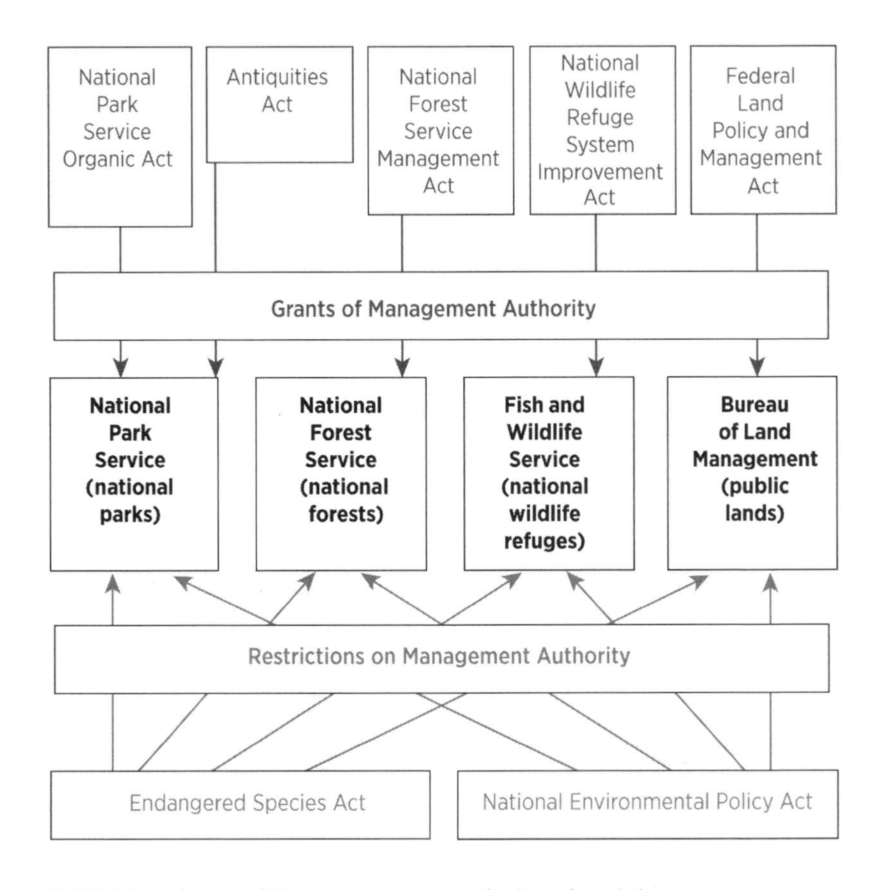


FIGURE 3.1. Federal Land Management Agency authority and restrictions.

among the federal land management agencies, as well as the source of some statutory restrictions on that authority or on private activity on lands managed by those agencies. A key question is whether, despite the absence of such an ecosystem-based focus, agencies such as the NPS and the U.S. Fish and Wildlife Service (USFWS) can nevertheless manage the resources over which they have jurisdiction in combination with lands owned or managed by other public and private entities and in ways that enhance the resilience of parks, refuges, and preserves in the face of the unprecedented disruptions expected to result from climate change and the continued encroachments caused by development.

The more than 500 million acres of land owned by the federal government is divided into multiple systems administered by different federal agencies. Some of these systems, such as the National Park System, the National Wildlife Refuge System, and the National Wilderness Preservation System, are "dominant-use" systems devoted to a limited number of uses that include recreation, wildlife protection, and resource preservation. The NPS administers the 80 million acres of national parks and national monuments under the National Park Service Organic Act of 1916. That act declares the purpose of these areas to be to conserve "the scenery and the natural and historic objects and the wild life therein and to provide for the enjoyment of the same in such manner and by such means as will leave them unimpaired for the enjoyment of future generations" (National Park Service Organic Act of 1916 § 1). The Antiquities Act of 1906 authorizes the President to designate as national monuments "historic landmarks, historic structures, and other objects of historic or scientific interest" situated on federal lands (Antiquities Act § 431). The NPS must prepare management plans for the national parks and monuments that, among other things, include measures for the preservation of the area's resources (National Park Service Organic Act § 3). Commodity uses such as logging, grazing, and mining are either prohibited or tightly restricted in the national parks and monuments.

The National Wildlife Refuge System, which comprises about 90 million acres, is administered by the USFWS. Under the National Wildlife Refuge System Improvement Act of 1997 (NWRSIA), the mission of the USFWS is to administer a network of lands and waters for the conservation, management, and restoration of fish, wildlife, and plant resources and their habitats for the benefit of present and future generations of Americans. The statute affords priority to the conservation of wildlife, plants, and their habitats, defining conservation to mean sustaining, restoring, and enhancing healthy populations (NWRSIA § 668ee(4)). The Act allows compatible wildlife-dependent recreational uses, including hunting and fishing, in the refuges. Recreational uses are compatible if they will not materially interfere with or detract from the mission of the Refuge System or the purposes of individual refuges (NWRSIA § 668ee(1)).

Other systems, such as the National Forest System and the public lands administered by the Bureau of Land Management (BLM), are governed by statutes allowing a range of "multiple uses" that include not only recreation and natural resource protection, but also timber production, range management, and mineral production. Even these systems include areas set aside for preservation purposes, however. Congress may designate areas of the national forests and the BLM public lands as wilderness, just as it may do in the national parks, monuments, and wildlife refuges (Glicksman and Coggins 1999). The National Wilderness Preservation System, which comprises more than 100 million acres, is the subject of chapter 1 in this book and will not be discussed further here. The National Forest Service has designated nearly 60 million acres of land as roadless areas in which timber harvesting and road construction are tightly restricted (Glicksman 2004). The BLM's organic act, the Federal Land Policy and Management Act (FLPMA), authorizes the agency to identify "areas of critical environmental concern." These are areas "where special management attention is required . . . to protect and prevent irreparable damage to important historic, cultural, or scenic values, fish and wildlife resources or other natural systems or processes." Land use plans must give priority to the protection of these areas, and they are not available for the full array of multiple uses (FLPMA §§ 1702(a), 1711, 1712(c)(3)). The reference to protection of natural processes seems to be consistent with an ecosystem management focus. The Wild and Scenic Rivers Act provides protection for congressionally designated rivers. The areas affected by the designation include both private and public lands (Wild and Scenic Rivers Act of 1968). Use restrictions vary, depending on whether a river has been designated as wild, scenic, or recreational.

There are U.S. laws that allow for the protection of marine protected areas (MPAs), but there is no single law that governs the creation of MPAs. Moreover, the range of permissible uses varies considerably, from multiple-use management similar to that which exists in the national forests and BLM lands to smaller "no-take" areas governed by a preservation mandate similar to the ones that govern national parks and wildlife refuges. One observer has concluded that "MPAs will not be effective in preserving marine mammals if the MPAs are surrounded by uncontrolled areas where pollution, habitat destruction, and overfishing exist" (Abate 2009, 274).

One example of spatial planning to protect MPAs might provide a model for protection in the United States of not only MPAs but of reserves more generally. The Great Barrier Reef Marine Park Act, adopted by the Australian legislature in 1975, established the Great Barrier Reef Marine Park Authority. The Authority manages the reef primarily through a zoning system that imposes different levels of restrictions based on the zoning designation (including zones for general use, habitat protection, conservation park, buffer, scientific research, marine national park, and preservation). Over the years, the percentage of reef waters subject to the highest levels of protection has increased significantly in response to monitoring that revealed a failure to achieve ecosystem protection at lower protection levels (Sivas and Caldwell 2008, 247–248). By at least one account, the park is the "foremost example of a zoning system that has been successfully applied in the context of a large, nationally significant MPA" (Baur et al. 2004, 565).

## Climate Change as a Spur to Increased Emphasis on Resilience: Opportunities and Challenges

Until fairly recently, the NPS has not incorporated the need to manage parks to enhance ecosystem resilience, particularly in the face of climate change, into its management directives in any significant way. In its 2006 *Management Policies*, an internal document designed to guide NPS employees fulfilling their statutory duties, including implementation of management plans, the NPS mentioned the concept of resilience only once (in connection with recreational boating campgrounds; NPS 2006). The policies referred to climate change only twice. First, the NPS recognized that "accelerated climate change may significantly alter park ecosystems" (NPS 2006, 53). But it required only that the parks gather and maintain baseline climatological data for reference, not that they use the data in any particular way to enhance resilience in the face of climate change. Second, it encouraged park managers to educate visitors about the influence of climate change, again without mandating that managers prepare for or react to climate change threats or events in any particular way that would enhance the agency's ability to achieve its statutory mission.

In 2010, however, the agency issued its *Climate Change Response Strategy*. The strategy includes six principles to guide the development and implementation of an integrated science-management response capacity for the NPS and its partners. Among these principles are several that comport with the concepts of resilience thinking and management at a landscape scale, as discussed above. The agency committed itself, for example, to building connections at local to regional scales within the agency and with other agencies and stakeholders. It also committed to the use of adaptive management (Walters 1986; McLain and Lee 1996; Gunderson 1999), recognizing the need to be "structured for adaptability, grounded in experience," and to rely on "approaches to learning and management that evolve as knowledge increases" (NPS 2010, 9). It relied on Secretarial Order 3289, which established a department-wide approach for increasing science-based understanding of climate change and coordinating an effective response to the resources it manages, as the supporting authority for this approach (Secretarial Order 3289). The strategy identified increasing the resilience of natural systems as an important approach to climate change adaptation. One of its objectives is to collaborate with federal, state, and local partners and programs to develop tools, such as vulnerability assessments and scenario planning (Peterson et al. 2003), to help develop adaptation plans at appropriate scales. Another goal is to collaborate to develop "cross-jurisdictional conservation plans to protect and restore connectivity and other landscape-scale components of resilience" (NPS 2010, 16). The agency recognized that protecting and restoring corridors and connectivity across landscapes "will require strong collaboration with partners and programs to share knowledge, develop repositories of genetic resources, and, where appropriate, develop cross-jurisdictional conservation" (NPS 2010, 16).

Announcing resource management goals that are consistent with enhancing the resilience of park and refuge ecosystems is one thing. Implementing those strategies is another (see Hillborn 1992; Walters 1997). As the NPS has recognized, current laws may present challenges to the goal of achieving resilient ecosystems in the face of threats such as climate change. The 2010 NPS Climate Change Response Strategy noted that "[m]ost resource protection laws with which the NPS must comply were not written considering a changing climate" (NPS 2010, 23). As a result, it is necessary to reestablish service-wide consistency in interpreting the NPS mission and mandates so that climate change adaptation efforts can be fashioned in ways that comply with laws governing management of the National Park System.

The questions the agency must face in doing so are difficult. One question is how to reconcile the need for human intervention to adapt to climate change with the core statutory mandate to conserve the "natural objects" within the parks and to "leave them unimpaired for the enjoyment of future generations" (NPS Organic Act § 1). One study found that NPS staffers regard the NPS Organic Act as a barrier to the agency's efforts to adapt to climate change as a result of its emphasis on preservation and naturalness, which has led to agency policies that minimize intervention and management in the parks (Jantarasami et al. 2010). This same legal focus may hinder efforts to promote resilience in park ecosystems more generally.

The legislation that governs administration of the national wildlife refuges and provides the USFWS with its management authority accommodates resilience thinking more easily. The USFWS must ensure maintenance of the "biological integrity, diversity, and environmental health" of the refuge system and plan and direct the growth of the system "to contribute to the conservation of the ecosystems of the United States" (NWRSIA § 668dd(a)(4)(B)–(C)). According to a leading expert on refuge management, this mandate "introduces broader, synthetic, ecological process concepts to the management objectives' of the national wildlife refuge system, and advances a holistic, ecological standard for evaluating conservation" (Fischman and Adamcik 2011, 23–24). Administration of the national wildlife refuge system on a landscape scale would be fully consistent with these prescriptions (figure 3.2).

Indeed, the USFWS has taken steps to manage ecosystems rather than individual wildlife species or individual refuges. The agency has described the goals of some refuges in terms of broad ecosystem-based goals. Its policies implementing the 1997 NWRSIA's mandate to maintain the biological integrity, diversity, and environmental health of the refuge system proclaim that fragmentation of the system's wildlife habitats is "a direct threat to the integrity" of the system. As a result, uses reasonably anticipated to reduce the quality or quantity of habitats, or fragment habitats on a refuge, will not be deemed a compatible wildlife-dependent use (U.S. Fish and Wildlife Service 2011b).

FIGURE 3.2. Migratory bird hunting & conservation stamp for wildlife habitat preservation.

Even more significantly, the USFWS's 2010 strategic plan lists as one of its foundational principles an "emphasis on conservation of habitats within sustainable landscapes." According to the plan, the USFWS will apply strategic habitat conservation as its framework for landscape conservation. This approach aims to conserve terrestrial, freshwater, and marine habitats within sustainable landscapes to conserve target populations of species or suites of species and the ecological functions that sustain them. The plan commits the USFWS to managing toward future landscape conditions by predicting and working with the effects of climate change. The goal is to facilitate the transition of ecosystems from current, natural states to new conditions resulting from climate change. These include mimicking or assisting natural adaptive processes such as species dispersal or migration to avoid catastrophic conversions that might occur otherwise. Realignment may be appropriate for ecosystems that have already been significantly disturbed (U.S. Fish and Wildlife Service 2010).

## Available Legal Tools for Managing Resilient Landscape-Level Parks and Refuges

### *Cooperative Resource Management to Achieve Resilience*

Notwithstanding the potential difficulties in implementing landscape-level management programs to achieve ecosystem resilience posed by the century-old NPS Organic Act, that statute, the NWRSIA, and other federal natural resource management laws provide the NPS and the USFWS with a variety of tools to implement conservation strategies that are built on the formation of dynamic reserves. The broad discretion the NPS has to regulate use of the national parks and monuments "by such means and measures as conform to the fundamental purpose" of those places should facilitate the NPS's ability to take actions that enhance the resilience of park ecosystems in the face of climate change and other threats. The interior secretary's authority to issue "such rules and regulations as he may deem necessary or proper for the use and management of the parks, monuments, and reservations under the jurisdiction of the National Park Service" reinforces this discretion (16 U.S.C. § 3).

Indeed, these laws and others described below provide the NPS with surprisingly broad explicit authority to engage in the kind of landscape-scale management efforts that will be necessary to protect ecosystem resilience in the parks. The agency has acted on that authority. NPS management policies recognize that:

> Cooperative conservation beyond park boundaries is necessary as the National Park Service strives to fulfill its mandate to preserve the natural and cultural resources of parks unimpaired for future generations. Ecological processes cross park boundaries, and park boundaries may not incorporate all of the natural resources . . . that relate to park resources. . . . Therefore, activities proposed for adjacent lands may significantly affect park programs, resources, and values. Conversely, NPS activities may have impacts outside park boundaries. . . . [P]arks are integral parts of larger regional environments. (NPS 2006, § 1.6)

The agency's management policies therefore commit the NPS to cooperative endeavors with other federal, regional, state, and local agencies, with tribes, and with neighboring landowners and nongovernmental organizations (NGOs).

The agency has explicit authority to enter cooperative management agreements in which a unit of the NPS is located adjacent to or near a state or local park area to ensure effective and efficient management of the parks or for the purpose of protecting natural resources of park units through collaborative efforts on land inside and outside of National Park System units (NPS Organic Act §§ 1a–2(*l*)(1), 1j). These powers should assist the NPS in fulfilling its commitment to collaborate with other public and private landowners in devising management approaches that enhance resilience on a landscape scale. Such cooperative agreements may address problems likely to result from climate change, such as invasive species within or adjacent to park system units (Crowl et al. 2008), and may provide for restoration of natural resources, including native wildlife habitat or ecosystems (NPS Organic Act § 1j–(b)). Similarly, the Organic Act authorizes the NPS, in managing national monuments contiguous to national forests, to cooperate with the National Forest Service (NPS Organic Act § 2).

Likewise, the NWRSIA makes available to the USFWS tools that enable the service to enlarge the scope of the resources it protects beyond the boundaries of the refuges themselves. The Act requires the USFWS, for example, to "ensure effective coordination, interaction, and cooperation with owners of land adjoining refuges and the fish and wildlife agency of the States in which units of the System are located." In addition, the statute mandates that the USFWS "ensure timely and effective cooperation and collaboration with federal agencies and state fish and wildlife agencies during the course of acquiring and managing refuges" (NWRSIA § 668dd(a)(4)(E), (M)). Importantly, none of these provisions dictates the form of the required cooperative efforts.

These statutory provisions authorize the NPS and the USFWS to engage in cooperative management efforts but do not require the agencies' desired partners to respond favorably when the NPS makes cooperative overtures, nor do they otherwise assure the success of such cooperative ventures. The obstacles to cooperative management among federal agencies are especially problematic when the desired partner of the NPS or USFWS operates under a statutory mandate or is pursuing a mission whose priorities conflict with the steps the NPS or USFWS deem necessary to enhance resilience and protect ecosystem integrity. The National Forest Service and the BLM, both of which operate under multiple-use mandates, may have no discretion to deny activities with the potential to disrupt ecosystem resilience, such as the issuance of hardrock mining activities. Even if these agencies have such discretion, they may not regard the approaches of the two dominant-use agencies to be consistent with their multiple-use mandates. NPS officials have cited differences in agency missions and the absence of an agency with the authority to resolve disputes among the land management agencies as reasons why interagency cooperation has failed in some instances to protect park resources (Shafer 2010).

The federal land management agencies have embarked on some cooperative ventures that aim to achieve landscape-level ecosystem management. Joseph Sax and Robert Keiter, for example, have described the NPS's embrace of regionalism as:

> the primary long-term strategy for protecting [Glacier National Park's] ecological integrity. The overall goal is to knit the entire Glacier region together as an entity with the park at the core of the larger ecosystem, primarily by creating transboundary management forums, institutions, or incentives consistent with the park's conservation objectives. (Sax and Keiter 2006, 300–301)

A web of natural resource management laws played a role in the NPS's effort to fashion a plan for protecting the integrity of the regional landscape surrounding Glacier National Park. Sax and Keiter identify the Endangered Species Act (ESA), National Environmental Policy Act of 1969 (NEPA), and the Wilderness Act of 1964 as imposing "managerial consistency across boundaries" (Sax and Keiter 2006, 307). The regulations issued by the Council on Environmental Quality (CEQ) to implement NEPA, for example, require agencies to consider the cumulative effects of an agency proposal and "other past, present, and reasonably foreseeable future actions regardless of what agency (Federal or non-Federal) or person undertakes such other actions" (CEQ regulations § 1508.7). The ESA requires agencies to consult with the USFWS to insure that their actions do not jeopardize listed species or adversely affect critical habitat (ESA § 7(a)(2)). Under the USFWS's regulations, an agency considering whether the statute requires consultation must consider "the effects of the action as whole," including indirect effects (USFWS regulations § 402.14(c)). Indirect effects include issues such as private land development that a federal action makes possible. An agency that fails to properly take those effects into account risks the issuance of an injunction that blocks its proposed action. As a result, agencies have an incentive to consider the consequences not only of their own actions but of related or resulting actions by others, including other federal agencies, states, tribes, and private landowners. That incentive may foster agency consideration of and coordination with others with the ability to control activities that are beyond the jurisdiction of the proposing agency. Sax and Keiter have surmised in connection with the presence of grizzly bears in and around Glacier National Park that the ESA has "forced other federal agencies to cooperate in management planning that crosses boundary lines, and provoked the development of a program of cumulative effects modeling that encourages forest service managers to adopt an expansive, long-range perspective in their planning processes" (Sax and Keiter 1987, 248).

Similar initiatives have begun elsewhere, such as in the northern Rockies. Efforts such as the Yellowstone to Yukon Initiative have sought "to break down the traditional jurisdictional boundaries that impede rational management of wildlife and water systems" (Sax and Keiter 2006, 301). More specifically, one of the initiative's goals has been "to create a linear network of interconnected nature reserves extending from the Yukon Territory to Yellowstone National Park to protect the grizzly bear, wolf, and other wide ranging wildlife species" (Sax and Keiter 2006, 302). According to Sax and Keiter, however, the initiative has not yet succeeded in fully achieving regional managerial integration, in part due to community resistance stemming from the initiative's impact on pre-existing mineral development projects.

Sax and Keiter detect within the last twenty years or so "a pronounced shift in thinking and management away from formal enclaves (such as the park and the [adjacent national] forest) and toward the region seen as an integral ecological unit, in particular as the habitat needs of target wildlife populations" (Sax and Keiter 2006, 306). They explain this shift as the product of several extralegal factors, including the increased emphasis among national forest users on recreation as opposed to commodity production, a reduced dependence among surrounding communities on commodity production as the driver of the local economy, and communities that increasingly regard themselves as inhabiting a regional environment. Where these features exist in other locales in which disturbances threaten resilience, federal land managers may be able to fashion similar landscape-level ecosystem management strategies. Education can assist in fostering among residents in a region containing a critical and threatened ecosystem a sense of belonging to a regional environment, and in convincing them that preservation of a fully functioning ecosystem will be to their economic benefit because of the ecosystem services it provides and the tourist revenues it can help generate. These kinds of endeavors have succeeded in other contexts, such as in efforts by the International Gorilla Conservation Project and the Institute for Tropical Forest Conservation to protect African gorillas (Sawyer and Sawyer 2011, 390–391). Similarly, education and outreach apparently have helped local governments to persuade residents of the Anacostia River watershed to buy into restoration and enhancement initiatives (Arnold 2011, 852–853). The USFWS has used education and outreach to supplement its efforts under the ESA to protect the Mexican wolf (Greaves 2009). Efforts to convince farmers and ranchers to take voluntary measures to protect lesser prairie chickens in New Mexico had mixed success (Paez 2009).

The USFWS also has made efforts to integrate protection of resources within and outside refuge boundaries. The USFWS Manual, an internal agency guidance document whose legal status is variable and subject to dispute, provides:

> Events occurring off refuge lands or waters may injure or destroy the biological integrity, diversity, and environmental health of a refuge. Given their responsibility to the public resources with which they have been entrusted, refuge managers should address these problems. It is critical that they pursue resolution fully cognizant and respectful of legitimate private property rights, seeking a balance between such rights and the refuge manager's own responsibility to the public trust. (U.S. Fish and Wildlife Service 2011b)

The agency's preference is to reach a solution through direct negotiations with the private landowner, agency, or other entity that is the source of the problem. For example, the USFWS has purchased private land outside the Lower Rio Grande Valley National Wildlife Refuge along the river to create a wildlife corridor that extends to the Gulf of Mexico, and has worked with state wildlife agencies and NGOs to assist in habitat restoration in the south Texas area (U.S. Fish and Wildlife Service 2012).

If those kinds of efforts fail, the USFWS may engage in collaborative discussions with state or local authorities or other organizations that can help in cooperative resolution of the problem. The next step typically will be to work with state or local regulators. As a last resort, the USFWS will pursue available legal remedies "with full respect to private property rights." The agency also has created a series of Landscape Conservation Cooperatives, which it describes as public–private partnerships that recognize that climate change challenges "transcend political and jurisdictional boundaries and require a more networked approach to conservation—holistic, collaborative, adaptive and grounded in science to ensure the sustainability of America's land, water, wildlife and cultural resources" (U.S. Fish and Wildlife Service 2011a).

Still, a clash of agency cultures and missions stands as a formidable obstacle to these cooperative efforts and suggests the need for a mediating force. The Interagency Climate Change Adaptation Task Force, created by President Obama in 2009 and co-chaired by the CEQ, the Office of Science and Technology Policy, and the National Oceanic and Atmospheric Administration, may be capable of resolving disputes among the federal land management agencies that relate to the impacts of climate change on ecosystem resilience. President Obama also issued an executive order requiring federal agencies to "participate actively" in the task force's development of a national strategy for adaptation and to develop policies and practices compatible with that strategy (Executive Order 13514, 2009). Among the task force's guiding principles is inducing agencies to implement adaptation approaches that seek "to increase ecosystem resilience and protect critical ecosystem services on which humans depend to reduce vulnerability of human and natural systems to climate change." The task force also recognized the need for a national approach to problems that cut across sectors and agencies, including strengthening coastal and ocean resilience and protecting fish, wildlife, plant resources, and their habitats (CEQ 2010). In 2011, the CEQ issued instructions for implementing adaptation planning in conformity with the adaptation executive order. The instructions require agencies to assess the vulnerability of the resources they manage to climate change, identify priority adaptation actions, and submit to the CEQ adaptation plans for implementation (CEQ 2011).

A quarter of a century ago, George Coggins rendered this pessimistic assessment: "History . . . demonstrates that the prospects for preventing and abating threats through intergovernmental coordination alone are bleak" (Coggins 1987). It remains to be seen whether, notwithstanding that dubious track record, the task force and CEQ directives will induce agencies, individually and through coordinated initiatives, to take effective steps to foster ecosystem resilience in the face of climate change.

### *Boundary Revisions*

Changing the boundaries of the national parks and wildlife refuges, or creating a spatiotemporally variable buffer zone around protected areas (e.g., to extend the area of strongly protected wetlands and/or no-go areas during periods when migratory or breeding birds are present), may offer another approach to managing reserves at the scale and in the locations necessary to promote ecosystem resilience. Spatially variable boundaries have been used in other environmental protection contexts. Louisiana, for example, has restricted the application of pesticides in buffer zones between inhabited residences and targeted crops that vary in breadth depending on factors such as wind speed and method of application (ground or aerial; Feitshans 1999).

The NPS has the power to make minor boundary revisions that it deems necessary for the proper preservation, protection, or management of an area of the national park system (Land and Water Conservation Fund Act of 1965 § 460*l*–9(c)(1)). The NPS Organic Act also vests in the NPS the responsibility to develop criteria to evaluate more significant proposed changes to the existing boundaries of individual park units if those boundaries do not adequately allow for the protection and preservation of the natural resources integral to the unit (NPS Organic

Act § 1a–12). In proposing specific boundary changes to Congress, the NPS must consult with affected state and local agencies, surrounding communities, affected landowners, and private organizations before applying the criteria (NPS Organic Act § 1a–13). The NPS's criteria support making boundary adjustments to prevent harm caused by activities on adjacent lands (e.g., pesticide applications or inappropriate use of fire) that pose a direct and substantial threat to the continued existence of the park's primary resources and values.

The NPS has stated, however, that adjustments to address external threats "should be considered a last resort when cooperative efforts have been fully explored and found to be inadequate" and that it does not endorse the creation of buffer zones. Moreover, the NPS has cautioned that boundary adjustments to protect wildlife populations whose habitat requirements or migratory patterns extend over large areas "are likely to be seriously limited by feasibility considerations." As a result, it has urged addressing regional natural resource issues through cooperation with other landowners (such as through the acquisition of fee interests or easements by nonprofit organizations for conservation purposes) and other governments (such as through local zoning constraints) as an alternative to expanding park boundaries (National Park Service 1991). Notwithstanding concerns about excessive expansion of the parks estate, Congress has moved lands administered by the National Forest Service and the BLM into the national parks to enable the NPS to fulfill its preservation mandate (Shafer 2010).

### *Congressional or Presidential Actions to Link Lands in Larger Landscapes*

The existence of protected areas, such as wilderness areas, in federal land systems under the management of agencies other than the NPS and the USFWS provides opportunities "to link protected areas together to protect much larger and more ecologically intact landscapes" (Keiter 2010, 82). The Antiquities Act of 1906 vests in the President the unilateral authority to create national monuments from less protected land systems, to protect "objects of historic or scientific interest" (Antiquities Act § 431). Courts have upheld presidential designations of places that were later transformed into national parks, such as the Grand Canyon and Jackson Hole. A dramatic example of the Act's potential for sweeping large tracts of federal land into a protective regime is President Jimmy Carter's reservation of 56 million acres in Alaska as a series of national monuments (Coggins and Glicksman 2007). Presidents have exercised their authority under the Antiquities Act to create national monuments on lands administered by the multiple-use agencies as a means of protecting adjacent national parks, monuments, and recreation areas from development that threatened resource degradation. Interior Secretary Bruce Babbitt, for example, justified President Clinton's designation of the Grand Staircase-Escalante National Monument as a way to protect Bryce Canyon and Capital Reef National Parks and the Glen Canyon National Recreation Area (Shafer 2010).

Congress has expanded existing parks such as Denali, Death Valley, and the Joshua Tree National Park to reconnect fragmented landscapes. These expansions become more viable if they have the support of surrounding communities who regard national parks and similar preserves as sources of economic opportunity (Keiter 2010, 82, 95–96). Researchers have concluded that protected areas such as parks and refuges succeed in protecting biodiversity and resilience only if local communities understand and embrace protective efforts, which requires education to build social and political support (Moral and Sale 2011).

Another potentially promising tool for achieving landscape-scale management efforts that enhance resilience is the National Heritage Area (NHA) program. As the NPS has explained, "National Heritage Areas (NHAs) are designated by Congress as places where natural, cultural, and historic resources combine to form a cohesive, nationally important landscape" (NPS 2011). The federal government does not own lands designated as NHAs. Rather, the NPS provides technical, planning, and matching financial assistance to communities in which these areas are located. The NPS approves management plans for NHAs, but the plans are largely crafted and are implemented by partnerships of local governments, private landowners, and nongovernmental organizations. As of September 2012, Congress had designated about fifty NHAs in different kinds of landscapes spread across the country. Some have succeeded in restoring dysfunctional ecosystems. The Yuma Crossing NHA in southwestern Arizona, for example, helped to restore wetlands through the removal of nonnative vegetation and planting of trees. Previous efforts at wetlands restoration in the area had foundered as a result of patchwork land-ownership patterns and conflicting land use goals. Moreover, NHAs can be used in conjunction with national park management to provide the kind of buffers needed to promote ecosystem resilience. As an NPS report recognized, "[w]hen a heritage area is adjacent to a national park or a national park is within a heritage area, both parties benefit from the expanded opportunity to interpret and protect resources over a larger landscape" (National Park System Advisory Board 2006, 3, 14).

### *Condemnation or Acquisition of Private Lands*

Another way to expand the impact or influence of management of current federal parks and refuges to encompass landscapes large enough to provide ecological memory is to arrange voluntary sales of private lands to the government or use the power of eminent domain to purchase conservation easements or fee title to private lands adjacent to preserves. Acquisition efforts should be directed toward resources essential to the preservation or restoration of ecosystem functions, such as lands that reconnect federal lands in headwaters with downstream areas. In the past, Congress has provided the NPS with condemnation authority to assist in the preservation of landscapes adjacent to national parks such as the Cape Cod National Seashore. The USFWS may be able to use funds generated by the sale of migratory bird-hunting stamps to help acquire new lands for the National Wildlife Refuge System (Fischman and Adamcik 2011).

The ESA, discussed in Chapter 2 of this book, also authorizes the secretary of the interior to use land acquisition authority under statutes such as the Fish and Wildlife Act of 1956, the Fish and Wildlife Coordination Act, and the Migratory Bird Treaty Act, and using funds available under the Land and Water Conservation Fund Act of 1965, to acquire "by purchase, donation, or otherwise" lands or interests in lands to assist in conserving fish, wildlife, and plants (Endangered Species Act § 1534).

Federal land exchanges (either with private parties or between agencies) may serve the same functions as acquisitions, and sales of federal lands that are not essential for the conservation of ecosystems may help finance some of the acquisitions (Leshy 2010). Trades between the NPS and the National Forest Service have sought to create more ecologically appropriate boundaries for Rocky Mountain National Park, for example (Shafer 2010). The USFWS has the authority to acquire lands by exchange under the NWRSIA (NWRSIA § 668dd(b)(3)). This authority enables the agency to integrate management of the lands found within the refuges themselves with private landowners and with management of state parks and preserves, and to acquire lands adjacent to current refuge lands if doing so will enhance the Act's ecosystem conservation purposes.

The states can play a useful role in contributing lands that can be used to aggregate parks and other preserves large enough to assist in protecting ecosystem resilience. In the past, state acquisitions have assisted in the aggregation of lands that were eventually included in national parks such as Shenandoah and the Great Smoky Mountains. As Robert Keiter has recognized, nonfederal options for resource protection have grown with the expansion of state and local park systems (Keiter 2010, 81). The federal government can finance state acquisitions of lands for outdoor recreational purposes under the Land and Water Conservation Fund Act. The federal government provides funds from sources such as offshore oil and gas leasing to the states to purchase lands for these purposes, although this mechanism could be made more powerful by dedicating money from the fund to acquisition of lands for conservation purposes, rather than requiring annual appropriations by Congress (Leshy 2010). The statute restricts the ability of states to convert lands acquired with federal money to nonrecreational purposes without the approval of the interior secretary (Coggins and Glicksman 1984, 2007). The states have the capacity to assist in federal ecosystem protection efforts in other ways. In recent years, groups of western states have created processes for providing coordinated protection of wildlife corridors. Tax breaks could induce private landowners to participate as well (Keiter 2010, 102).

### *Regulatory and Related Actions to Protect Resilience*

If it is not possible to aggregate sufficiently large tracts of land in the hands of sympathetic public or private owners to provide landscape-level resource management for resilience, the NPS and the USFWS may be able to exercise their regulatory authority to abate external threats. The Property Clause of the Constitution vests in the federal government the authority to regulate conduct on nonfederal lands that threatens to harm federal holdings. Despite the Supreme Court's broad interpretation of that power (Appel 2001; *Camfield v. United States*; *United States v. Alford*; Property Clause), the scope of the federal land management agencies' authority to abate threats originating off federal lands (so-called external threats) is unclear. "Congress seldom exercises [its power to control external threats], the land agencies seldom claim it, and the Department of Justice seldom asserts it in federal land litigation" (Coggins and Glicksman 2007; *Stupak-Thrall v. United States*).

Coordination among federal agencies in addressing external threats, including the appearance of invasive species or the spread of insect infestations or pathogens that threaten forests, tends to be especially difficult, legally and practically (Sax and Keiter 2006; Glicksman 2009). The NPS and other agencies nevertheless have successfully invoked on occasion their authority to abate activities, such as spraying pesticides (*United States v. South Florida Water Management District*) and forcing reductions in air pollution that would have affected the Grand Canyon (*Central Arizona Water Conservation District v. EPA*), that created threats to federal lands and resources. The USFWS has also claimed authority to address external threats such as invasive species and habitat destruction (USFWS 2000).

Lessons from the work of Lin Ostrom and her collaborators (e.g., Ostrom et al. 1999; Ostrom 2003, 2007, 2009) suggest that the protection of common property resources is facilitated by the existence of clearly defined, graduated sanctions. Another option in abating external threats may thus be to define spatially explicit thresholds of potential concern (TPCs; Biggs and Rogers 2003) that use science to identify a clear, explicit set of criteria indicating a looming decline in system resilience (or a potential loss of identity). When a TPC is crossed, legislation would trigger the imposition of restrictions on the intensity or nature of activities that could be undertaken in a given area. For example, if forest loss in the broader landscape is a concern for the protection of forest-dependent animals, a reduction in tree cover below a certain percentage of its historical extent might trigger restrictions on the licensing of timber companies to harvest trees beyond a certain volume. Once the system has revegetated, such restrictions could be relaxed.

Examples of this kind of approach already exist under U.S. environmental laws. Under the ESA, for example, incidental take statements in a USFWS biological opinion must define the extent of permissible incidental take. If the level is exceeded, the agency sponsoring the project must reinitiate consultation with the USFWS, and the take statement must specify monitoring or reporting requirements that enable the USFWS to determine whether the trigger for further consultation has been crossed (*Wild Fish Conservancy v. Salazar*). In other contexts, the exceedance of preset thresholds automatically requires management actions, either by agencies or regulated entities. Under the Clean Water Act, for example, Oregon has issued permits to the operators of storm sewer systems that contain benchmark numbers based on state water quality standards that, if exceeded, trigger an adaptive management process for the imposition of more effective, best management practices (Dunn and Burchmore 2007). If it could be implemented in a way that was politically viable, legislation that explicitly includes graded restrictions or sanctions for different zones around protected areas might create a strong incentive for responsible environmental management by all parties concerned.

## Conclusion

Ultimately, efforts such as those described above may be inadequate to allow management of parks, refuges, and other preserves at a sufficient scale to provide the regional resilience that will be needed to withstand severe perturbations. The combination of development that continues to reduce the size of spaces devoted to conservation and the emergence of more dramatic climatic change at some point may outstrip the ability of resource managers, such as the NPS and the USFWS, to promote resilient ecosystems at broad extents under the current legal regime (Rockström et al. 2009).

If physical realities preclude even well-informed agency efforts to manage for resilience, Congress should consider revising the legal framework that has governed federal land management for most of the past century by realigning agency jurisdiction to facilitate cohesive and coordinated management of ecosystems, rather than continuing to rely on the somewhat artificial boundary lines between parks, refuges, forests, and other federal land systems. As at least one scholar has recognized, this kind of revision would not require a rejection of all boundary lines. Instead, "it demands redrawing spatial boundaries and time frames to suit the particular purposes of the ecological analysis" (Breckenridge 1995, 374).

Consolidation of federal land systems is one option. Some have proposed combining the land systems managed by the two dominant-use agencies, the NPS and the USFWS (Keiter 2005). Others have argued, however, that consolidation might reduce opportunities for innovation and adaptive management and that past agency consolidations have produced unwieldy and complacent organizations (Fischman 2002; Zellmer 2004). At a minimum, it would be helpful if the organic statutes for the NPS and the multiple-use agencies contained explicit authority to manage for ecosystem resilience or at least contained provisions like the one in the NWRSIA establishing a policy in favor of maintaining biological integrity.

One way to provide parks and preserves of a scale compatible with natural landscape dynamics would be to add significant new lands to the parks and refuge systems. Political support for a significant expansion of the federal land base does not exist and is unlikely to develop, however, if for no other reason than the displacement of private ownership it would entail. It is unlikely that the federal land base could be expanded enough to achieve effective government control to ensure landscape-level resilience in any event. Nor is extensive federal regulation of private land use viable. Land use regulation has long been largely, though not exclusively, the prerogative of local governments. The existing federal environmental regulatory programs that tend to generate the most heated opposition are those, such as the ESA's taking prohibition and the Clean Water Act's dredge and fill permit program, that look like land use control mechanisms. As John Leshy has argued, federal regulatory initiatives designed to make private land bear much of the burden of biodiversity protection would be "controversial [and] fiercely resisted" (Leshy 2010, 124).

Accordingly, efforts at statutory reform will likely need to focus on providing enhanced authority for federal resource managers to coordinate their efforts with state and private landowners. In particular, one attractive option may be for the federal government to provide monetary incentives for private landowners, similar to the payments made under the Conservation Reserve Program to farmers who take wetlands out of production (Coggins and Glicksman 2007, § 19:26), to refrain from engaging in the kinds of development that threaten the goal of management for ecosystem resilience.

Finally, Congress may need to amend the laws that govern the national parks, wildlife refuges, and other preserves to deal with intractable questions that federal land managers will likely face if climate change results in the loss of resources that induced Congress to create a particular preserve in the first place. In such cases, how can the management agency comply with a mandate to leave natural resources unimpaired if climate change results in plant or animal range shifts that make a park unsuitable for species that previously thrived there (NPS 2010, 23)? Explicit statutory authorization to manage for the maintenance of ecological processes is desirable, even if the components of the ecosystems in question differ from those traditionally found there. The NPS insists that "[p]resent NPS management policies are probably sufficient to guide many potential climate change response actions," but Congress may need to either amend the enabling legislation for specific parks if they contain detailed prescriptions that no longer reflect the on-the-ground realities, or provide general authority to manage for resilience notwithstanding dramatic physical changes that frustrate the ability to retain the iconic features for which a preserve is known.

## References

Abate, R. S. "Marine Protected Areas as a Mechanism to Promote Marine Mammal Conservation: International and Comparative Law Lessons for the United States." *Oregon Law Review* 88 (2009): 255–309.

Allen, T. F. H., and T. B. Starr. *Hierarchy: Perspectives for Ecological Complexity.* Chicago: University of Chicago Press, 1982.

Angelo, M. J. "Stumbling Toward Success: A Story of Adaptive Law and Ecological Resilience." *Nebraska Law Review* 87 (2009): 950–1007.

Antiquities Act of 1906, 16 U.S.C. §§ 431–433.

Appel, P. "The Power of Congress 'Without Limitation': The Property Clause and Federal Regulation of Private Property." *Minnesota Law Review* 86 (2001): 1–130.

Arnold, C. A. "Fourth Generation Environmental Law: Integrationist and Multimodal." *William & Mary Environmental Law & Policy Review* 35 (2011): 771–884.

Ball, I., and H. Possingham. *MARXAN (V1.8.2): Marine Reserve Design Using Spatially Explicit Annealing, a Manual.* Brisbane, Australia: Ecology Centre, University of Queensland, 2000.

Baskett, M. L., F. Micheli, and S. A. Levin. "Designing Marine Reserves for Interacting Species: Insights from Theory." *Biological Conservation* 137 (2007): 163–179.

Baur, D.C., W. R. Irvin, and D. R. Misenko. "Putting 'Protection' into Marine Protected Areas." *Vermont Law Review* (2004) 497–577.

Bengtsson, J., P. Angelstam, T. Elmqvist, U. Emanuelsson, C. Folke, M. Ihse, F. Moberg, and M. Nystrom. "Reserves, Resilience and Dynamic Landscapes." *Ambio* 32 (2003): 389–396.

Berkes, F., et al. "Globalization, Roving Bandits, and Marine Resources." *Science* 311 (2006): 1557–1558.

Biggs, H. C., and K. H. Rogers. "An Adaptive System to Link Science, Monitoring, and Management in Practice." In *The Kruger Experience: Ecology and Management of Savanna Heterogeneity*, eds. J. T. DuToit, K. H. Rogers, and H. C. Biggs, 59–80. Washington, D.C.: Island Press, 2003.

Breckenridge, L. P. "Reweaving the Landscape: The Institutional Challenges of Ecosystem Management for Lands in Private Ownership." *Vermont Law Review* 19 (1995): 363–422.

Camacho, A. E. "Assisted Migration: Redefining Nature and Natural Resource Law Under Climate Change." *Yale Journal on Regulation* 27 (2010): 171–255.

Camfield v. United States, 167 U.S. 518 (1897).

Carpenter, S., B. Walker, J. M. Anderies, and N. Abel. "From Metaphor to Measurement: Resilience of What to What?" *Ecosystems* 4 (2001): 765–781.

Carpenter, S. R., and W. A. Brock. Spatial complexity, resilience and policy diversity: fishing on lake-rich landscapes. Ecology and Society 9 no. 1, ( 2004): 8. http://www.ecologyandsociety.org/vol9/iss1/art8.

Central Arizona Water Conservation District v. EPA, 990 F.2d 1531 (9th Cir. 1993).

Coggins, G. C. "Protecting the Wildlife Resources of National Parks from External Threats." *Land and Water Law Review* 22 (1987): 1–27.

Coggins, G. C., and R. L. Glicksman. "Federal Recreational Land Policy: The Rise and Decline of the Land and Water Conservation Fund." *Columbia Journal of Environmental Law* 9 (1984): 125–236.

Coggins, G. C., and R. L. Glicksman. *Public Natural Resources Law*. Eagan, Minn.: Thomson Reuters/West, 2007.

Cosens, B. 2011. "Legitimacy, Adaptation and Resilience in Ecosystem Management." Working Papers Series, University of Idaho, College of Law, Moscow, Idaho, 2011. http://papers.ssrn.com/sol3/papers.cfm?abstract_id=1942875.

Council on Environmental Quality. *Progress Report of the Interagency Climate Change Task Force: Recommended Actions in Support of a National Climate Change Adaptation Strategy*. Washington, D.C.: CEQ, 2010.

Council on Environmental Quality. *Federal Agency Climate Change Adaptation Planning: Implementing Instructions*. Washington, D.C.: CEQ, 2011.

Council on Environmental Quality. Regulations for Implementing NEPA, 40 C.F.R. Part 1500.

Craig, R. K. "'Stationarity Is Dead'—Long Live Transformation: Five Principles for Climate Change Adaptation Law." *Harvard Environmental Law Review* 34 (2010): 9–73.

Crowl, T. A., T. O. Crist, R. R. Parmenter, G. Belovsky, and A. E. Lugo. "The Spread of Invasive Species and Infectious Disease as Drivers of Ecosystem Change." *Frontiers in Ecology and the Environment* 6 (2008): 238–246.

Cumming, G. S. *Spatial Resilience in Social-Ecological Systems*. Dordrecht, Netherlands: Springer, 2011a.

Cumming, G. S. "Spatial Resilience: Integrating Landscape Ecology, Resilience, and Sustainability." *Landscape Ecology* 26 (2011b): 899–909.

Cumming, G. S., G. Barnes, S. Perz, M. Schmink, K. E. Sieving, J. Southworth, M. Binford, R. D. Holt, C. Stickler, and T. Van Holt. "An Exploratory Framework for the Empirical Measurement of Resilience." *Ecosystems* 8 (2005): 975–987.

Cumming, G. S., and J. Collier. "Change and Identity in Complex Systems." *Ecology and Society* 10 (2005): 29. http://www.ecologyandsociety.org/vol10/iss21/art29.

Cumming, G. S., and B. J. Spiesman. "Regional Problems Need Integrated Solutions: Pest Management and Conservation Biology in Agroecosystems." *Biological Conservation* 131 (2006): 533–543.

Cundill, G., G. S. Cumming, D. Biggs, and C. Fabricius. "Soft Systems Thinking and Social Learning for Adaptive Management." *Conservation Biology* 26 (2012): 13–20.

Diniz, J. A. F., et al. "Conservation Planning: A Macroecological Approach Using the Endemic Terrestrial Vertebrates of the Brazilian Cerrado." *Oryx* 42 (2008): 567–577.

Doremus, H. "Adapting to Climate Change with Law That Bends Without Breaking." San Diego *Journal of Climate & Energy Law* 2 (2010): 46–85.

Dunn, A.D., and D. W. Burchmore. "Regulating Municipal Separate Storm Sewer Systems." *Natural Resources and Environment* 21 (2007): 3–6.

Endangered Species Act of 1973, 16 U.S.C. §§ 1531–1544.

Executive Order No. 13514, Federal Leadership in Environmental, Energy, and Economic Performance, § 16, 74 Fed. Reg. 52,117 (Oct. 5, 2009).

Federal Land Management and Policy Act of 1976, 43 U.S.C. §§ 1701–1782.

Feitshans, T. A. "An Analysis of State Pesticide Drift Laws." *San Joaquin Agricultural Law Review* 9 (1999): 37–93.

Fischman, R. L. "The National Wildlife Refuge System and the Hallmarks of Modern Organic Legislation." *Ecology Law Quarterly* 29 (2002): 457–622.

Fischman, R. L., and R. S. Adamcik. "Beyond Trust Species: The Conservation Potential of the National Wildlife Refuge System in the Wake of Climate Change." *Natural Resources Journal* 51 (2011): 1–33.

Geselbracht, L., R. Torres, G. S. Cumming, D. Dorfman, M. Beck, and D. Shaw. "Identification of a Spatially Efficient Portfolio of Priority Conservation Sites in Marine and Estuarine Areas of Florida." *Aquatic Conservation—Marine and Freshwater Ecosystems* 19 (2009): 408–420.

Glicksman, R. L. "Traveling in Opposite Directions: Roadless Area Management Under the Clinton and Bush Administrations." *Environmental Law* 34 (2004): 1143–1208.

Glicksman, R. L. "Ecosystem Resilience to Disruptions Linked to Global Climate Change: An Adaptive Approach to Federal Land Management." *Nebraska Law Review* 87 (2009): 833–892.

Glicksman, R. L., and G. C. Coggins. "Wilderness in Context." *Denver University Law Review* 76 (1999): 383–411.

Great Barrier Reef Marine Park Act 1975. http://www.comlaw.gov.au/Details/C2011C00149.

Greaves, N. "Unlucky Number 13: The Endangered Species Act, Subdelegation, and How Standard Operating Procedure Jeopardized Mexican Gray Wolf Reintroduction." *Arizona State Law Journal* 41 (2009): 905–931.

Gunderson, L. Resilience, flexibility and adaptive management—antidotes for spurious certitude? *Conservation Ecology* 3, no. 1 (1999): 7. http://www.consecol.org/vol3/iss1/art7.

Hillborn, R. "Can Fisheries Agencies Learn from Experience?" *Fisheries* 17 (1992): 6–14.

Jantarasami, L. C., J. J. Lawler, and C. W. Thomas. "Institutional Barriers to Climate Change Adaptation in National Parks and Forests of Washington State." *Ecology and Society* 15 (2010):33. http://www.ecologyandsociety.org/vol15/iss4/art33.

Keiter, R. B. "Public Lands and Law Reform: Putting Theory, Policy, and Practice in Perspective." *Utah Law Review* (2005): 1127–1226.

Keiter, R. B. "The National Park System: Visions for Tomorrow." *Natural Resources Journal* 50 (2010): 71–110.

Land and Water Conservation Fund Act of 1965, 16 U.S.C. §§ 460l–4604.

Leroux, S. J., F. K. A. Schmiegelow, G. S. Cumming, R. B. Lessard, and J. Nagy. "Accounting for System Dynamics in Reserve Design." *Ecological Applications* 17 (2007): 1954–1966.

Leshy, J. D. "Federal Lands in the Twenty-First Century." *Natural Resources Journal* 50 (2010): 111–137.

Levin, S. A. "Self-Organization and the Emergence of Complexity in Ecological Systems." *BioScience* 55 (2005): 1075–1079.

Lindenmayer, D. B., A. D. Manning, P. L. Smith, H. P. Possingham, J. Fischer, I. Oliver, and M. A. McCarthy. "The Focal-Species Approach and Landscape Restoration: A Critique." *Conservation Biology* 16 (2002): 338–345.

Margules, C. R., and R. L. Pressey. "Systematic Conservation Planning." *Nature* 405 (2000): 243–253.

McCook, L. J., et al. "Management Under Uncertainty: Guidelines for Incorporating Connectivity into the Protection of Coral Reefs." *Coral Reefs* 28 (2009): 353–366.

McLain, R. J., and R. G. Lee. "Adaptive Management: Promises and Pitfalls." *Environmental Management* 20 (1996): 437–448.

Millennium Ecosystem Assessment. *Ecosystems and Human Well-Being: Scenarios. Findings of the Scenarios Working Group.* Washington, D.C.: Island Press, 2005.

Moral, C., and P. F. Sale. "Ongoing Global Biodiversity Loss and the Need to Move Beyond Protected Areas: A Review of the Technical and Practical Shortcomings of Protected Areas on Land and Sea." *Marine Ecology Progress Series* 434 (2011): 251–266.

National Environmental Policy Act of 1969, 42 U.S.C. §§ 4321–4375.

National Park Service. *Criteria for Boundary Adjustments, Supplement to Planning Process Guidelines,* NPS-2. Washington, D.C.: NPS, 1991. http://planning.nps.gov/document/boundary%20criteria.pdf.

National Park Service. *Management Policies 2006.* Washington, D.C.: NPS, 2006. http://www.nps.gov/policy/mp2006.pdf.

National Park Service. *Climate Change Response Strategy, September 2010.* Washington, D.C.: NPS, 2010. http://www.nature.nps.gov/climatechange/docs/NPS_CCRS.pdf.

National Park Service. *What Are National Heritage Areas?* (2011). http://www.nps.gov/history/heritageareas/FAQ.

Natural Park Service Organic Act of 1916, 16 U.S.C. §§ 1–4.

National Park System Advisory Board. *Charting a Future for National Heritage Areas.* Washington, D.C.: NPS Advisory Board, 2006. http://www.nps.gov/resources/upload/NHAreport.pdf.

National Wildlife Refuge System Improvement Act of 1966, as amended in 1997, 16 U.S.C. §§ 668dd–668ee.

Nature Conservancy. *The Five-S Framework for Site Conservation: A Practitioner's Handbook for Site Conservation Planning and Measuring Conservation Success.* Washington, D.C.: Nature Conservancy, 2003.

Ostrom, E. "How Types of Goods and Property Rights Jointly Affect Collective Action." *Journal of Theoretical Politics* 15 (2003): 239–270.

Ostrom, E. "A Diagnostic Approach for Going Beyond Panaceas." *Proceedings of the National Academy of Sciences USA* 104 (2007): 15181–15187.

Ostrom, E. "A General Framework for Analyzing Sustainability of Social-Ecological Systems." *Science* 352 (2009): 419–422.

Ostrom, E., J. Burger, C. B. Field, R. B. Norgaard, and D. Policansky. "Revisiting the Commons: Local Lessons, Global Challenges." *Science* 284 (1999): 278–282.

Paez, S. A. "Preventing the Extinction of Candidate Species: The Lesser Prairie Chicken in New Mexico." *Natural Resources Journal* 49 (2009): 525–582.

Peterson, G. D., G. S. Cumming, and S. R. Carpenter. "Scenario Planning: A Tool for Conservation in an Uncertain Future." *Conservation Biology* 17 (2003): 358–366.

Poiani, K. A., B. D. Richter, M. G. Anderson, and H. E. Richter. "Biodiversity Conservation at Multiple Scales: Functional Sites, Landscapes, and Networks." *BioScience* 50 (2000): 133–146.

Polis, G. A., M. E. Power, and G. R. Huxel. *Food Webs at the Landscape Level.* Chicago: University of Chicago Press, 2004.

Possingham, H., I. Ball, and S. Andelman. "Mathematical Methods for Identifying Representative Reserve Networks." In *Quantitative Methods for Conservation Biology*, eds. S. Ferson and M. Burgman, 232–287. New York: Springer, 2002.

Pressey, R. L., H. P. Possingham, and C. R. Margules. "Optimality in Reserve Selection Algorithms: When Does It Matter and How Much?" *Biological Conservation* 76 (1996): 259–267.

Property Clause, U.S. Constitution art. IV, § 3, cl. 2.

Raatikainen, K. M., R. K. Heikkinen, and M. Luoto. "Relative Importance of Habitat Area, Connectivity, Management and Local Factors for Vascular Plants: Spring Ephemerals in Boreal Semi-Natural Grasslands." *Biodiversity and Conservation* 18 (2009): 1067–1085.

Rockström, J., et al. "A Safe Operating Space for Humanity." *Nature* 461 (2009): 472–475.

Ruhl, J. B. "Climate Change Adaptation and the Structural Transformation of Environmental Law." *Environmental Law* 40 (2010): 363–435.

Sawyer, J. M., and S. C. Sawyer. "Lessons from the Mist: What Can International Environmental Law Learn from Gorilla Conservation Efforts?" *Georgetown International Environmental Law Review* 23 (2011): 365–396.

Sax, J., and R. B. Keiter. "Glacier National Park and Its Neighbors: A Study of federal Interagency Relations." *Ecology Law Quarterly* 14 (1987): 207–263.

Sax, J., and R. B. Keiter. "The Realities of Regional Resource Management: Glacier National Park and Its Neighbors Revisited." *Ecology Law Quarterly* 33 (2006): 233–311.

Secretarial Order No. 3289, Addressing the Impacts of Climate Change on America's Water, Land, and Other Natural and Cultural Resources, § 1 (Sept. 14, 2009). http://www.fws.gov/home/climatechange/pdf/SecOrder3289.pdf.

Shafer, C. L. "The Unspoken Option to Help Safeguard America's National Parks: An Examination of Expanding U.S. National Park Boundaries by Annexing Adjacent Federal Lands." *Columbia Journal of Environmental Law* 35 (2010): 57–125.

Sivas, D. A., and M. R. Caldwell. "A New Vision for California Ocean Governance: Comprehensive Ecosystem-based Marine Zoning." *Stanford Environmental Law Journal* 28 (2008): 209–270.

Stupak-Thrall v. United States, 70 F.3d 881 (6th Cir. 1995), aff'd by an equally divided court, 89 F.3d 1269 (8th Cir. 1996).

U.S. Fish and Wildlife Service, Endangered Species Act Regulations, 50 C.F.R. § 402.14.

U.S. Fish and Wildlife Service. Refuge Planning Policy Pursuant to the National Wildlife Refuge System Administration Act as Amended by the National Wildlife Refuge System Improvement Act of 1997, 65 Fed. Reg. 33892 (2000).

U.S. Fish and Wildlife Service. *Rising to the Urgent Challenge: Strategic Plan for Responding to Accelerating Climate Change.* Washington, D.C.: USFWS, 2010. http://www.fws.gov/home/climatechange/pdf/ccstrategicplan.pdf.

U.S. Fish and Wildlife Service. "Landscape Conservation Cooperatives" (2011a). http://www.fws.gov/science/shc/lcc.html.

U.S. Fish and Wildlife Service. "The Fish and Wildlife Service Manual," part 601, § 3.20; part 603, § 2.5A (2011b). http://www.fws.gov/policy/manuals.

U.S. Fish and Wildlife Service. "Lower Rio Grande Valley: Creating a Wildlife Corridor" (2012). http://www.fws.gov/refuge/Lower_Rio_Grande_Valley/resource_management/wildlife_corridor.html.

United States v. Alford, 274 U.S. 264 (1927).

United States v. South Florida Water Management District, 922 F.2d 704 (11th Cir. 1991).

Walker, B., S. Carpenter, J. Anderies, N. Abel, G. Cumming, M. Janssen, L. Lebel, J. Norberg, G. D. Peterson, and R. Pritchard. "Resilience Management in Social-Ecological Systems: A Working Hypothesis for a Participatory Approach." *Conservation Ecology* 6, no. 1 (2002): 14. http://www.consecol.org/vol6/iss1/art14.

Walters, C. "Challenges in adaptive management of riparian and coastal ecosystems. *Conservation Ecology* 1, no. 2 (1997): 1. http://www.consecol.org/vol1/iss2/art1.

Walters, C. J. *Adaptive Management of Renewable Resources.* New York: McGraw-Hill, 1986.

Westley, F., S. R. Carpenter, W. A. Brock, C. S. Holling, and L. H. Gunderson. "Why Systems of People and Nature Are Not Just Social and Ecological Systems." In *Panarchy: Understanding Transformations in Human and Natural Systems,* eds. L. H. Gunderson and C. S. Holling, 103–199. Washington, D.C.: Island Press, 2002.

Wild and Scenic Rivers Act of 1968, 16 U.S.C. §§ 1271–1287.

Wild Fish Conservancy v. Salazar, 628 F.3d 513, 531–32 (9th Cir. 2010).

Wilderness Act of 1964, 16 U.S.C. §§ 1131-1136.

Zellmer, S. "A Preservation Paradox: Political Prestidigitation and an Enduring Resource of Wilderness." *Environmental Law* 34 (2004): 1015–1089.

FOUR

# Marine Protected Areas, Marine Spatial Planning, and the Resilience of Marine Ecosystems

ROBIN KUNDIS CRAIG AND TERRY P. HUGHES

At first blush, a concern for improving ocean resilience—or, more properly, the resilience of marine ecosystems—might seem misdirected. Oceans cover 71 percent of the Earth's surface and, because of their depth, provide 99 percent of the habitat available for life (Ogden 2001). Biological diversity in the oceans exceeds that on land (Craig 2005). In addition, the seas moderate and buffer the most fundamental physical and chemical processes of the planet, including temperature regulation, the hydrological cycle, and carbon sequestration. Changes in ocean temperature and ocean currents in one part of the world affect weather over a much greater area, as the La Niña/El Niño oscillation, or ENSO, demonstrates through its three- to seven-year cycles, driven by temperature and current changes in the eastern Pacific Ocean off the coast of South America. Barriers to dispersal are less prevalent in the sea than on land, promoting larval connectivity and migration over very large scales.

The oceans, therefore, maintain world-spanning, interconnected, physical, chemical and biological processes that seem far too large and complex for mere humans to damage. Indeed, in terms of both effective governance and scientific research, "marine systems have been relatively neglected because they are 'out of sight, out of mind' to most people, including most scientists" (Ray and Grassle 1991, 453). Until recently, a "paradigm of inexhaustibility" prevailed, a mindset that human managers did not need to worry about ocean health, because marine ecosystems would always be resilient enough to absorb and recover from the multiple and interactive stresses—overfishing, pollution, and now climate change—that humans impose on them (Connor 1999; Ogden 2001; Craig 2005).

Unfortunately, we now know that marine ecosystems often *cannot* in fact absorb the multitude of anthropogenic stressors imposed upon them, even before the accelerating impacts of climate change become more severe and add to existing drivers of change, such as overfishing (Laffoley et al. 2008; Agardy 2010). Many marine ecosystems have lost their resilience to recurrent natural and man-made disturbances and have undergone long-term shifts to new, degraded regimes (Hughes et al. 2005). In coastal regions in particular, fishing has substantially altered marine ecosystems for centuries (Jackson et al. 2001). For example, many coral reefs have undergone a regime shift to macroalgae following the overexploitation of herbivores and the addition of land-based nutrients. A study published in the journal *Science* in 2008 concluded that no area of the world's oceans is completely unaffected by human impacts, and 41 percent of the oceans are *strongly* affected by multiple human impacts (Halpern et al. 2008). In the face of additional climate change–induced stresses, marine governance systems and marine managers need to find mechanisms for increasing the resilience of ocean ecosystems.

This chapter explores one set of those mechanisms—place-based marine management, especially marine protected areas (MPAs)—and the various legal regimes that encourage use of these tools in pursuit of increased marine ecosystem resilience. We begin by reviewing the variety of threats to marine ecosystems, then turn to a basic discussion of the tools of marine spatial planning—MPAs, marine reserves, and marine zoning. Although marine spatial planning cannot address all threats to the oceans, it can quite effectively address the most pernicious governance challenge for sustaining or rebuilding marine resilience in most coastal nations: fragmentation of regulatory authority. We conclude that most coastal nations' marine governance regimes would more effectively protect the resilience of both marine ecosystems and the societies that depend on them by explicitly incorporating place-based marine management.

## Threats to Marine Ecosystem Resilience

In discussing requisite changes to marine governance regimes, it is first important to understand the range of threats to marine ecosystems. The main threat to marine ecosystem resilience arises from a loss of the ecosystem functions that marine biodiversity provides, leading to undesirable regime shifts. From a resilience perspective, the number of species is less important than the functions that they perform (Bellwood et al. 2004). Depauperate regions are more vulnerable to human impacts, because critical ecosystem functions may be performed by only one or two species. In that case, the loss of a top predator, the only major herbivore, or a key structural, habitat-forming species can have profound ecological impacts. Biodiversity hot spots, on the other hand, typically have many species performing similar ecological roles, providing some insurance against modest losses of biodiversity (Bellwood et al. 2004). According to the United Nations Education, Scientific, and Cultural Organization (UNESCO), at least forty-three of the seventy phyla of life—the second-most general classification of life on Earth, signaling broad genetic diversity—are found in the oceans, and 45 percent of known phyla exist only in the oceans; UNESCO also suggests that at least half, and probably more, of all species live in the seas (UNESCO 1996). Canadian scientists estimated in 2005 that there may be as many as 10 million undescribed species living in the seas (Center for Marine Biodiversity 2011), although the recently completed international Census of Marine Life is starting to fill in some of those holes. By 2010, the census had increased the number of known marine species from approximately 230,000 to approximately 250,000 (Ausubel et al. 2010). However, the ecological roles of each of those species and their contributions to ecosystem resilience are very poorly understood.

Biodiversity contributes to the resilience of a given marine ecosystem by increasing the redundancy of structure and function within that ecosystem (Craig 2005). As a result, marine biodiversity loss undermines marine ecosystem resilience. For example, the gradual depletion by overfishing of a suite of herbivore species in the Caribbean over the past century has eroded the resilience of coral reefs, leading to persistent regime shifts from corals to ecosystems dominated by blooms of seaweed (Hughes 1994). Nevertheless, significant concerns about loss of marine biodiversity did not emerge until the 1990s (Craig 2005). As ecological historian and scientist Jeremy B. C. Jackson commented in 2001, "The persistent myth of oceans as wilderness blinded ecologists to the massive loss of marine ecological diversity caused by overfishing and human inputs from the land over the past centuries" (Jackson 2001, 5411).

The three principal drivers of change in marine ecosystems are land-based marine pollution, overfishing, and climate change (Jackson et al. 2001; Hughes et al. 2003). Marine pollution takes many forms, but it is, critically, predominantly a land-based governance problem that cannot be addressed through direct regulation of the marine environment, including marine spatial planning. Because of fairly effective international treaties limiting direct pollution of the oceans from ocean dumping and from ships, most (e.g., around 80 percent in the United States) marine pollution comes from land, the result of direct and controlled discharge of pollutants from coastal facilities; indirect and largely uncontrolled runoff from agricultural and urban areas; the legacy of prior ocean dumping; and atmospheric deposition of pollutants initially emitted into the air (Craig 2012). Deforestation, intensification of agriculture, urban sprawl, industrialization, population growth, and migration to the coast all contribute to increased nearshore pollution (Craig 2012).

Offshore ocean dumping is now curtailed in the eighty or so countries that are parties to the 1972 Convention on the Prevention of Marine Pollution by Dumping of Wastes and Other Matter (the London Convention; Craig 2012). Nevertheless, for decades, nations and individuals intentionally dumped wastes at sea, including nuclear and toxic chemical waste, and this legacy of pollution from ocean dumping remains a concern for marine ecosystem health. As the Millennium Ecosystem Assessment (MEA) noted in 2005, for example, "the estimated 313,000 containers of low-intermediate emission radioactive waste dumped in the Atlantic and Pacific Oceans since the 1970s pose a significant threat to deep sea ecosystems should the containers leak, which seems likely over the long term" (MEA 2005, 483).

Coastal development directly spurs marine pollution problems, for instance, by increasing the amount of polluted runoff reaching the oceans and by promoting the discharge of sewage into the sea. According to the UN Environment Programme (UNEP), it would cost US$56 billion per year to adequately address discharges of untreated sewage from coastal communities. Depending on the social and economic contexts, different nations vary greatly in their capacity to address this problem. For example, "[a]round 60% of the wastewater discharged into the Caspian Sea is untreated, in Latin America and the Caribbean the figure is close to 80%, and in large parts of Africa and the Indo-Pacific the proportion is as high as 80–90%" (UNEP 2011). Certain kinds of land-based marine pollution, especially nutrient pollution from sewage, runoff of agricultural fertilizers, and atmospheric deposition of nitrogen and phosphorus compounds, are also linked to the proliferation of harmful algal blooms (HABs)—the rapid reproduction and multiplication of small marine plants (phytoplankton and dinoflagellates) that then cause harmful effects on the environment. HABs, for example, can lead to "red tides," the release of neurotoxins and the contamination of shellfish, and eutrophication and coastal "dead zones."

The various forms of marine pollution can, unquestionably, impair the resilience of marine ecosystems, especially coastal ecosystems. Sedimentation from muddy runoff can smother and kill marine organisms, particularly juveniles, leading to recruitment failure. Without recruitment, ecosystems lose their capacity to absorb and recover from recurrent shocks such as hurricanes, leading to regime shifts and long-term degradation. Turbid waters also have lower light penetration, affecting light-dependent species such as kelp and corals. Added nutrients from terrestrial sources, in combination with depleted herbivores from overfishing, promote the establishment and growth of phytoplankton and macroalgae. The gradual buildup of pollution often goes unnoticed as a threshold level is approached, leading eventually to a regime shift to a new, degraded system that is difficult to reverse. Nevertheless, resolving these pollution problems is predominantly a terrestrial rather than a marine governance issue, requiring improved management of land use practices, protective control of air emissions, water discharges, plastic and other waste disposal, and better management of polluted runoff.

The newest major stressor to ocean ecosystems—climate change—is likewise predominantly a terrestrial governance issue: the ultimate "solution" to climate change is to reduce greenhouse gas concentrations in the global atmosphere, which in turn requires controlling greenhouse gas emissions from the various anthropogenic sources that are predominantly land-based. Nevertheless, until such controls are implemented and take effect, it is worth emphasizing that climate change poses new challenges to marine ecosystem resilience, making effective ocean governance that much more important. Climate change is affecting atmospheric, land, freshwater, and ocean temperatures—but not uniformly (Intergovernmental Panel on Climate Change [IPCC] 2007). Temperatures toward the poles are increasing faster than temperatures near the equator, and land temperatures are rising faster than temperatures in the ocean (IPCC 2007; Craig 2012). Many of these climate change–driven ecological changes are likely to become worse in the coming decades, because global average temperature increases of at least 0.1°C to 0.2°C *per decade* are expected for the rest of this century (IPCC 2007).

The oceans play a significant role in moderating climate change impacts, and understanding the interaction of the atmosphere and the oceans is widely acknowledged to be critical to understanding and modeling climate change. For the purposes of this chapter, however, the main point is that climate change alters the oceans themselves, decreasing the resilience of marine ecosystems and threatening to push those ecosystems over critical thresholds into new ecological regimes that function very differently. Generally, these alternate regimes are undesirable, because they deliver fewer ecosystem goods and services, such as income from fisheries and coastal tourism. Three climate change–related effects are particularly important for marine ecosystems: increasing ocean temperatures and consequent changes in ocean dynamics (e.g., current strengths and storms); sea level rise; and ocean acidification.

Increasing sea surface temperatures (SSTs) and ocean heat content are two of the most direct impacts of increasing global average atmospheric temperatures. Increasing ocean temperatures have already changed important ecosystem functions and services. The IPCC in 2007 expressed "high confidence . . . that observed changes in marine . . . biological systems are associated with rising sea temperatures, as well as related changes in ice cover, salinity, oxygen levels, and circulation," including "shifts in ranges and changes in algal, plankton, and fish abundance in high-latitude oceans" (IPCC 2007, 3). In November 2009, researchers at the U.S. National Oceanic and Atmospheric

Administration (NOAA) reported that about half of the commercially important fish stocks in the western North Atlantic Ocean, such as cod and haddock, had been shifting poleward in response to rising sea temperatures. Similar shifts are underway in the southern hemisphere. Changes in ocean currents are also occurring, with the potential for major impacts on marine ecosystems. For example, "[e]vidence is starting to accumulate that global warming may contribute to—or even trigger—troubling ecological changes taking place in . . . key regions of coastal upwelling"—including areas off the coasts of northern California and Oregon—"where some of the world's richest fisheries exist," collectively accounting for 20 percent of the world's fish catch (Spotts 2007). In addition, the IPCC projected increasing coral-bleaching events as current levels of SST increase, and widespread coral mortality is likely to occur if SSTs increase by approximately 2.5–3.0°C (IPCC 2007). The largest coral-bleaching event to date occurred in 1998 during one of the warmest years on record, killing approximately 16 percent of the world's corals (Hughes et al. 2003).

Increases in ocean temperature also contribute to increasingly violent storm events and to sea level rise, which in turn is expected to lead to coastal erosion and inundation of coastal wetlands (IPCC 2007). According to a 2010 report from NOAA, "[t]he rate of global mean sea level (GMSL) rise is estimated currently to be 3.1 ± 0.4 mm yr-1 (3.4 mm yr-1 with correction for global isostatic adjustment)" (Levy 2009). In the United States, the Environmental Protection Agency (EPA) estimated in a report to Congress that a 2-foot (60 cm) increase in sea level could destroy 17 to 43 percent of the United States' coastal wetlands, with over 50 percent of that destruction occurring in Louisiana alone (U.S. EPA 2011), and the IPCC has "suggest[ed] that by 2080, sea level rise could convert as much as 33 percent of the world's coastal wetlands to open water" (IPCC 2007).

Finally, the most insidious side effect of climate change—"side effect" because it is not the result of atmospheric warming, but rather of the chemistry of increased atmospheric carbon dioxide concentrations—is ocean acidification. As the U.K. Royal Society reported in June 2005, increased carbon dioxide levels in the atmosphere are reducing the pH of the oceans, especially at high latitudes (Royal Society 2005). This acidification most directly affects species that calcify carbonate skeletons or shells, a process with which acidification interferes (Royal Society 2005). The IPCC similarly concluded in 2007 that "[t]he progressive acidification of oceans due to increasing atmospheric carbon dioxide is expected to have negative impacts on marine shell forming organisms (e.g., corals) and their dependent species" (IPCC 2007, 8). As such, ocean acidification increasingly threatens the biodiversity and ecological functions that oceanic and coastal ecosystems support and maintain. As one example of the human impacts beginning to occur from ocean acidification, shellfish farmers in Puget Sound, Washington, in the United States must start their mussels and other shellfish in Hawaii, because Puget Sound waters are now too acidic to foster their early growth.

Land-based pollution and climate change both demonstrate that governance for marine resilience requires all aspects of environmental governance regimes to pay attention to the oceans—something that these governance regimes have tended not to do. These two stressors to the marine environment thus underscore the primary *governance* threat to marine ecosystem resilience: regulatory fragmentation (Craig 2012). In both the United States and Australia, for example, land-based diffuse or nonpoint source pollution is primarily the governance province of the states and territories, whereas ocean ecosystems are generally subject to the federal or commonwealth government's authority. This fragmentation of both regulatory authority and regulatory goals has led in the United States to, among other problems, a large "dead zone" (hypoxic zone) in the Gulf of Mexico caused primarily by runoff from thousands of farms in the thirty-three states of the Mississippi River Basin (National Research Council [NRC] 2008); agricultural operations in Queensland, Australia, contribute pollution stressors to the Great Barrier Reef with inadequate state oversight and limited commonwealth authority to act (Craig 2012). Coal mining, port dredging, and poorly regulated coastal development in Queensland have compromised the outstanding universal values of the Great Barrier Reef world heritage area (Douvere and Badman 2012), which lies overwhelmingly in commonwealth waters:

> The establishment of the Great Barrier Reef Marine Park by the (Australian) Commonwealth in 1976, then the world's largest marine protected area, was a visionary step. It created an agency

> that had jurisdiction and responsibility for the whole ecosystem (GBRMPA—the Great Barrier Reef Marine Park Authority). The State of Queensland opposed the Park, took the Commonwealth to the High Court, and lost. But GBRMPA has almost no capacity to influence two major drivers of change that are increasingly affecting the Great Barrier Reef—activities on land and in Queensland coastal waters, and climate change. (Hughes 2012)

Comprehensive ocean governance to achieve marine ecosystem resilience, therefore, requires comprehensive governance coordination across land and sea and across governmental units at a scale that most coastal nations are unwilling or unlikely to achieve in the near future. Nevertheless, marine governance could be significantly improved in many coastal nations through a more limited focus on the marine environment itself and overfishing's effects on marine biodiversity, ecosystem functions, and resilience. In contrast to both climate change and marine pollution, overfishing is decisively a *marine* governance challenge—and, arguably, the most important marine governance issue for improving marine ecosystem resilience (Jackson et al. 2001). Fishing threats to marine biodiversity and ecosystem resilience include the impacts on target species, the phenomenon of bycatch, and the destruction of important habitat (Jackson et al. 2001, 2011). Harvesting of targeted species and bycatch changes the structure of food webs and can tip a marine ecosystem into a new regime. For example, in the Gulf of Maine, overharvesting of groundfishes such as cod has led to population explosions of lobsters and sea urchins. In turn, hyperabundances of herbivorous sea urchins have destroyed kelp beds, creating "urchin barrens"—low-diversity ecosystems in which recovery of kelp is prevented by continuous grazing, the indirect result of overfishing (Steneck et al. 2011).

Despite greatly increased fishing effort worldwide, the annual catch of wild fish has leveled off, suggesting that humans have reached—and probably exceeded—the sustainable yield of the world's fish stocks. As the UN Food and Agriculture Organization (UNFAO) reported in 2011, "Global capture fisheries production in 2008 was about 90 million tonnes, with an estimated first-sale value of US$93.9 billion, comprising about 80 million tonnes from marine waters and a record 10 million tonnes from inland waters. . . . World capture fisheries production has been relatively stable in the past decade" (UNFAO 2011: 5).

Nevertheless, even the normally taciturn UNFAO found the most recent marine fish stock statuses—and overall trends in stock status—to be cause for concern. Its 2011 summary of the status of world fish stocks is worth presenting in its entirety:

> The proportion of marine fish stocks estimated to be underexploited or moderately exploited declined from 40 percent in the mid-1970s to 15 percent in 2008, whereas the proportion of overexploited, depleted or recovering stocks increased from 10 percent in 1974 to 32 percent in 2008. The proportion of fully exploited stocks has remained relatively stable at about 50 percent since the 1970s. In 2008, 15 percent of the stock groups monitored by FAO were estimated to be underexploited (3 percent) or moderately exploited (12 percent) and able to produce more than their current catches. This is the lowest percentage recorded since the mid-1970s. Slightly more than half of the stocks (53 percent) were estimated to be fully exploited and, therefore, their current catches are at or close to their maximum sustainable productions, with no room for further expansion. The remaining 32 percent were estimated to be either overexploited (28 percent), depleted (3 percent) or recovering from depletion (1 percent) and, thus, yielding less than their maximum potential production owing to excess fishing pressure, with a need for rebuilding plans. This combined percentage is the highest in the time series. The increasing trend in the percentage of overexploited, depleted and recovering stocks and the decreasing trend in underexploited and moderately exploited stocks give cause for concern.
>
> Most of the stocks of the top ten species, which account in total for about 30 percent of the world marine capture fisheries production in terms of quantity, are fully exploited. The two main stocks of anchoveta (*Engraulis ringens*) in the Southeast Pacific and those of Alaska pollock (*Theragra chalcogramma*) in the North Pacific and blue whiting (*Micromesistius poutassou*) in the Atlantic are fully exploited. Several Atlantic herring (*Clupea harengus*) stocks are fully exploited, but some are depleted. Japanese anchovy (*Engraulis japonicus*) in the Northwest Pacific and Chilean jack mackerel (*Trachurus*

> *murphyi*) in the Southeast Pacific are considered to be fully exploited. Some limited possibilities for expansion may exist for a few stocks of chub mackerel (*Scomber japonicus*), which are moderately exploited in the Eastern Pacific, while the stock in the Northwest Pacific was estimated to be recovering. In 2008, the largehead hairtail (*Trichiurus lepturus*) was estimated to be overexploited in the main fishing area in the Northwest Pacific. Of the 23 tuna stocks, most are more or less fully exploited (possibly up to 60 percent), some are overexploited or depleted (possibly up to 35 percent) and only a few appear to be underexploited (mainly skipjack). In the long term, because of the substantial demand for tuna and the significant overcapacity of tuna fishing fleets, the status of tuna stocks may deteriorate further if there is no improvement in their management. Concern about the poor status of some bluefin stocks and the difficulties in managing them led to a proposal to the Convention on International Trade in Endangered Species of Wild Fauna and Flora (CITES) in 2010 to ban the international trade of Atlantic bluefin. Although it was hardly in dispute that the stock status of this high-value food fish met the biological criteria for listing on CITES Appendix I, the proposal was ultimately rejected. Many parties that opposed the listing stated that in their view the International Commission for the Conservation of Atlantic Tunas (ICCAT) was the appropriate body for the management of such an important commercially exploited aquatic species. Despite continued reasons for concern in the overall situation, it is encouraging to note that good progress is being made in reducing exploitation rates and restoring overfished fish stocks and marine ecosystems through effective management actions in some areas such as off Australia, on the Newfoundland–Labrador Shelf, the Northeast United States Shelf, the Southern Australian Shelf, and in the California Current ecosystems. (UNFAO 2011, 8)

Somewhat ironically, moreover, although the UNFAO characterized capture fishing overall as "stable," it noted that total catch has declined in most oceans of the world and that several fish stocks have significantly decreasing yields (UNFAO 2011). Most particularly, "[t]he decline of the gadiformes ('cods, hakes, haddocks' . . . ) seems relentless. In 2008, catches of this species group as a whole did not total 8 million tonnes, a level that had been until then consistently exceeded since 1967 and that reached a peak of almost 14 million tonnes in 1987. In the last decade, catches of Atlantic cod, the iconic species of this group, have been somewhat stable in the Northwest Atlantic at about 50,000 tonnes (very low by historical standards), but in the Northeast Atlantic catches have further decreased by 30 percent" (UNFAO 2011, 15). Total tonnage of marine fisheries captures peaked in 1996 (UNFAO 2011).

Commercial and artisanal fishing can undermine marine ecosystem resilience (Bellwood et al. 2004; Hughes et al. 2005; Steneck et al. 2011), but so too can high levels of recreational fishing. A 2004 study in the journal *Science*, for example, concluded that:

> Recreational landings in 2002 account for 4% of total marine fish landed in the United States. With large industrial fisheries excluded (e.g., menhaden and pollock), the recreational component rises to 10%. Among populations of concern, recreational landings in 2002 account for 23% of the total nationwide, rising to 38% in the South Atlantic and 64% in the Gulf of Mexico. Moreover, it affects many of the most-valued overfished species—including red drum, bocaccio, and red snapper—all of which are taken primarily in the recreational fishery. (Coleman et al. 2004, 1958)

Moreover, the study emphasized that both commercial and recreational fishing "can cause cascading trophic effects that alter the structure, function, and productivity of marine ecosystems," and where recreational fishing actually outstrips commercial fishing, it "can have equally serious ecological and economic consequences on fished populations" (Coleman et al. 2004, 1959).

Fishers, quite naturally, prefer to catch the largest fish of the species they are targeting, and they often prefer to target large apex predators, such as tuna, swordfish, and sharks (Coleman et al. 2004). Both preferences have consequences for marine ecosystem function. First, by targeting large predators, fishers can wipe out (or nearly so) an entire trophic level of a food web, changing the abundances of smaller predators and prey species at lower trophic levels and eroding the resilience of the ecosystem. In the heavily fished main Hawaiian Islands, for example, apex predators account for about 3 percent of the biomass of coral reef ecosystems, while in the more isolated Northwest Hawaiian Islands, apex predators account for 54 percent of the biomass (Craig 2012)—a significant shift in food web structure and ecosystem function. As apex predatory species become depleted, fishing effort typically expands or switches to large and then small herbivores, planktivores, and detritivores, a phenomenon known as "fishing down the food chain" (Jackson et al. 2001, 2011). Second, by targeting the largest individuals of a species, fishers can change the species' reproduction dynamics. The biggest fish produce disproportionately more offspring—that is, the relationship between body size and fecundity is strongly nonlinear (Roberts and Hawkins 2000). Moreover, in some heavily targeted species, such as parrotfishes, individual fish undergo a sex change as they mature (Roberts and Hawkins 2000). Fishing can also exert strong evolutionary pressures on fishes, selecting for individuals that reach reproductive size earlier. Thus, by targeting the largest fish, fishers may alter sex ratios, puberty size, and reproductive capacity, as well as overall abundances.

Fishers invariably catch nontarget species, or bycatch. Very few fishing methods, especially at the commercial scale, can precisely control the kinds of species caught: long lines attract seabirds, marine mammals, and sharks, as well as target fish; nets and traps capture species and juveniles that are too big to slip through the mesh. Improvements to gear, such as turtle exclusion devices on shrimp nets or escape slots on traps, can help to reduce bycatch. Banning netting in habitats where vulnerable species such as dugongs are prevalent can also reduce bycatch rates. Under most fisheries regimes, fishers throw the dead and dying bycatch back into the ocean, making accurate estimates of bycatch very difficult. Moreover, "ghost fishing" by discarded or lost fishing nets and traps can continue to catch and kill fish and other marine creatures for years (UNFAO 2005).

Finally, certain fishing methods destroy the marine habitat necessary to support healthy ocean ecosystems. Blast fishing and fishing through cyanide poisoning on coral reefs are obvious examples (McClellan 2010). More controversially, perhaps, bottom trawling is recognized as an ecosystem-destroying fishing method. As the U.S. NRC recognized in 2002, "[t]rawl gear can crush, bury, or expose marine flora and fauna and reduce structural diversity" (NRC 2002, 20). Bottom trawling and dredging have flattened the three-dimensional habitat formed by sponges, sea fans, and deepwater corals in many of the world's fishing grounds.

The cumulative impacts of fishing on marine ecosystem structure and function (and hence overall resilience) are most obvious from a historical perspective. Many of the impacts of fishing are cumulative over long periods, making them easy to miss or ignore. In 1995, Daniel Pauly coined the phrase "shifting baseline syndrome" to describe a pervasive phenomenon in fisheries management:

> Essentially, this syndrome has arisen because each generation of fisheries scientists accepts as a baseline the stock size and species composition that occurred at the beginning of their careers, and uses this to evaluate changes. When the next generation starts its career, the stocks have further declined, but it is the stocks at that time that serve as a new baseline. The result obviously is a gradual shift of the baseline, a gradual accommodation of the creeping disappearance of resource species, and inappropriate reference points for evaluating economic losses resulting from overfishing, or for identifying targets for rehabilitation measures. (Pauly 1995, 430)

A historical perspective on the status and trajectory of marine ecosystems is thus critical in assessing their resilience (Jackson et al. 2001, 2011). Moreover, historical overfishing may be the key culprit in pushing these ecosystems toward a tipping point (Hughes 1994; Steneck et al. 2011). In 2001, after examining a variety of historical records and evidence, nineteen marine scientists jointly concluded that "[e]cological extinction caused by overfishing precedes all other pervasive human disturbance to coastal ecosystems, including pollution, degradation of water quality, and anthropogenic climate change" (Jackson et al. 2001, 629). This study emphasized the significantly greater abundance of almost all fished marine species in historical times and outlined the already existing impacts of historical overfishing, theorizing that centuries of overfishing have set in motion processes that caused or contributed to current and, the authors speculated, future collapses of marine ecosystems (Jackson et al. 2001). For example, in the Pacific Northwest and Alaska, fur traders hunted sea otters "to the brink of extinction" by the 1800s in many kelp forest ecosystems, allowing sea urchins to multiply in the absence of their main predator; the sea urchins then ate the kelp, decimating the entire ecosystem (Jackson et al. 2001). This regime shift persists today, except in areas where recovery of sea otters is underway, which keeps sea urchins too scarce to overgraze the kelp.

Although numerous treaties address fishing in the high seas, under current international law (Craig 2012), regulation of fishing is primarily the responsibility of coastal nations, who can claim such governance authority out to 200 nautical miles from shore through internationally recognized exclusive economic zones (EEZs). Thus, marine biodiversity and resulting marine resilience problems arising from overfishing are appropriate immediate targets of national governance reform. At the same time, international rules regarding ocean jurisdiction also place coastal nations in the primary role of comprehensively prioritizing and regulating all direct human uses of the marine environment. An emerging tool to implement these needed governance reforms is marine spatial planning, addressed in the next section.

## Marine Protected Areas, Marine Reserves, and Marine Spatial Planning

There are many ways to influence direct human uses of the marine environment. Unfortunately, a common practice for doing so has multiplied the extent of regulatory fragmentation in marine governance—that is, the division of regulatory regimes among a variety of regulatory entities according to the particular problem being addressed (Craig 2012). At the international level, for example, pollution of the marine environment from ships is addressed through multiple treaty regimes and a variety of international organizations, while ocean dumping and land-based pollution each have their own regimes (Craig 2012). Marine species regulation is, if anything, even more fragmented, with separate treaties addressing, for example, North Pacific fur seals, Antarctic seals, seals of the Wadden Sea, Antarctic marine living resources, migratory birds, anadromous fish, southern bluefin tuna, pollock in the Bering Sea, whales, small cetaceans of the Baltic and North Seas, and sea turtles in the Americas, among many others (Craig 2012).

Regulatory fragmentation of marine governance can also be a significant problem at the national and subnational levels, as well. In the United States, for example, the U.S. Commission on Ocean Policy (USCOP) noted in 2004 that "[a]t the federal level, eleven of fifteen cabinet-level departments and four independent agencies play important roles in the development of ocean and coastal policy" and that "[t]hese agencies interact with one another and with state, territorial, tribal, and local authorities in sometimes haphazard ways" (USCOP 2004, 5). Internationally, in June 2011 the International Programme on the State of the Ocean (IPSO) recognized pervasive problems with marine governance regimes in light of the degradation of ocean ecosystems, and it recommended the rapid adoption of a more holistic approach to marine management, one that would reduce or eliminate regulatory fragmentation by addressing "all activities that impinge on marine ecosystems" (Agardy 2010).

Marine spatial planning or place-based marine management is widely hailed as a primary means of improving marine governance to promote ocean resilience (Hughes et al. 2005; McLeod and Leslie 2009; Agardy 2010; Craig 2012). Place-based marine management seeks to address all human uses of a particular marine environment in a comprehensive governance regime; in the best applications, the governance regime has as one of its top priorities the needs of the marine environment or ecosystem itself. As a result, place-based marine management is a governance measure particularly well-suited to improving marine resilience in a particular marine environment by reducing, regulating, or eliminating the direct and indirect impacts of overfishing.

Place-based marine management can occur at a variety of scales and for a variety of governance motivations, but most such management to date has started with the basic concept of a marine protected area, or MPA. Although exact definitions vary, MPAs are most essentially place-based legal protections for marine ecosystems—the ocean equivalent of terrestrial national and state parks. For example, the International Union for the Conservation of Nature (IUCN) defines an MPA to be "[a]ny area of the intertidal or subtidal terrain, together with its overlying water and associated flora, fauna, historical and cultural features, which has been reserved by law or other effective means to protect part or all of the enclosed environment" (IUCN 2010).

When it comes to classification, lumpers will almost always come up with fewer categories than splitters, and the same is true with respect to classifying protected areas (Laffoley 2008). The IUCN, for example, delineates six kinds of protected areas and intends those categories to apply on both land and sea (Dudley 2008). For purposes of this chapter, however, only three distinctions are important: the basic concept of an MPA, the more specialized concept of a marine reserve or no-take area (NTA), and the comprehensive approach of marine spatial planning or ocean zoning that envisages a network of MPAs or NTAs.

All MPAs use the law or management authority to limit ocean uses in some respect; thus, MPAs are inherently a governance tool. For example, many nations use MPAs to separate potentially conflicting uses of a marine ecosystem, such as recreational diving and commercial fishing. As Tundi Agardy has recently observed, "In planning marine protected areas, conservationists are able to root marine issues to a specific place, and in so doing, engage the public and decisionmakers in a concrete set of measures to protect that place" (Agardy 2010, x). However, when MPAs initially serve to separate uses, they can also function as part of a larger system of marine zoning, more formally known as marine spatial planning. As with conservation planning of terrestrial parks, marine spatial planning seeks to sustain ecosystems, and to protect areas essential to ecosystem processes, productivity, and function (McLeod and Leslie 2009; Agardy 2010; Craig 2012).

The most protective MPAs, generally known as marine reserves, prohibit all extractive uses of the marine ecosystem, including fishing (Craig 2005). The most protective marine reserves prohibit all access, except for scientific research—and sometimes even severely restrict research—but most tourism-related marine reserves allow nonextractive recreational uses, such as snorkeling, diving, and boating (Craig 2005).

Authorities can employ multiple types of MPAs and marine reserves in a unified ocean zoning plan, as exemplified by the Great Barrier Reef (GBR) Marine Park in Australia, established in 1976. The marine park, which is the size of Italy, was initially zoned into areas that allowed or banned trawling, commercial and/or recreational line fishing, and other activities. The most protected areas, preservation zones, prohibit access without a permit (Great Barrier Reef Marine Park Authority [GBRMPA] 2004). These no-go areas are small island rookeries for birds and turtles, accounting for less than 1 percent of the GBR Marine Park, where even a limited human presence could be damaging. Until 2004, a further 5 percent of the GBR Marine Park was zoned as no-take reserves, where commercial and recreational fishing were prohibited. However, in 2004, the GBR Marine Park was rezoned in response to growing evidence and concerns about declining abundances of corals, fishes, turtles, and dugongs. In particular, the proportion of the marine park where fishing is banned was increased from 5 percent to 33 percent, and commercial fishing was prohibited in areas with high recreational use. Although all fishing is prohibited in these marine reserve zones (NTAs), users are free to enter for recreational activities (GBRMPA 2004), and recreational fishers can fish in the remaining two-thirds of the marine park. These changes in multizone use have been very effective; the number of targeted fish has already doubled inside the new marine reserves (McCook et al. 2010). Importantly, the 2004 rezoning was accompanied by additional changes in regulations (e.g., on the number and sizes of fish that could be caught) and by a restructuring of the commercial fishing industry, interventions that are not spatially explicit but which nonetheless reinforce the effectiveness of spatial planning.

Thus, place-based marine management and especially marine reserves or NTAs have an established track record of allowing areas of depleted biodiversity to rebuild overfished stocks (Craig 2005; McCook et al. 2010). For the reasons discussed above, creating governance regimes that allow for stock rebuilding and biodiversity maintenance is an important step in protecting and enhancing marine ecosystem resilience, because such regimes allow marine ecosystems to maintain and rebuild the diversity and redundancy of ecosystem function and processes that can render such ecosystems more resilient to other stressors, such as land-based pollution and climate change. Nevertheless, as we next discuss, the relationship between place-based marine management and ocean resilience is more complex than just stock rebuilding, underscoring the importance of place-based marine management as an ocean governance improvement.

## Place-Based Marine Governance and Marine Ecosystem Resilience

In terms of improving governance to address threats to the resilience of marine ecosystems, marine reserves or NTAs most directly reduce the stresses caused by overfishing. In particular, research has demonstrated that MPAs and marine reserves that prohibit fishing and that are scientifically chosen to protect important fish habitats, such as breeding grounds or nurseries, can be quite effective in increasing both the numbers and size of targeted species of fish (Craig 2002; McCook et al. 2010; Mumby and Harbone 2010) and, more generally, in increasing biodiversity (Côté et al. 2001; Halpern 2003). In 2003, for example, Benjamin Halpern concluded that "[o]n average, creating a reserve appears to double density, nearly triple biomass, and raises organism size and diversity by 20–30% relative to the values for unprotected areas" (Halpern 2003, S125). The biggest changes, as expected, are exhibited by harvested species. As their numbers increase, other species—their prey, predators, and competitors—also change in abundance. Consequently, the recovery or preservation of targeted species has profound impacts on the structure of marine food webs and the resilience of marine ecosystems, redressing many of the biodiversity-related threats to marine resilience discussed in previous sections and making it less likely that biodiversity impacts from other stressors will push a particular marine ecosystem across a threshold and into a different and less desirable ecological state (see Jackson et al. 2001, 2011).

Networks of MPAs, especially networks of marine reserves or NTAs, seek to geographically broaden the marine environments enhanced through reduced fishing and to establish more coherent and protective human uses of marine ecosystems. A broad and international consortium of entities—for example, the World Commission for Protected Areas, the IUCN, the GBRMPA, NOAA, Natural England, the World Wildlife Fund, and the Nature Conservancy—are advocating MPA networks as a means of increasing marine ecosystem resilience (Laffoley et al. 2008). These networks link smaller individual MPAs without necessarily zoning an entire contiguous range of ocean. As the consortium explains:

> When used in isolation, small MPAs may not support fish and invertebrate populations that are large enough to sustain themselves. To ensure that young marine organisms are available to replenish and sustain populations within MPAs, the area of protection must be fairly large. However, in many regions, economic, social and political constraints make it impractical to create one single large MPA of sufficient size to support viable, self-sustaining populations of all species. Establishing networks of several to many small to moderately sized MPAs may help to reduce socioeconomic impacts without compromising conservation and fisheries benefits. Furthermore, *well-planned networks* provide important spatial links needed to maintain ecosystem processes and connectivity, as well as *improve resilience by spreading risk in the case of localized disasters, climate change, failures in management or other hazards, and thus help to ensure the long-term sustainability of populations better than single sites.* (Laffoley et al. 2008, 10, emphasis added)

In addition, the consortium identified four critical "components of a resilient MPA network": (1) "Effective management"; (2) "[r]isk spreading through inclusion of replicates of representative habitats"; (3) "[f]ull protection of critical areas that can serve as reliable sources of seed for replenishment [or] preserve ecological function"; and (4) "[m]aintenance of biological and ecological connectivity among and between habitats" (Laffoley et al. 2008, 16).

In particular, the consortium's advocacy is based on studies that indicate that marine reserves (NTAs) can increase the resilience of marine ecosystems to the impacts of climate change. In 2003, a group of seventeen coral reef researchers stated that:

> Although climate change is by definition a global issue, local conservation efforts can greatly help in maintaining and enhancing resilience and in limiting the longer-term damage from (coral) bleaching and related human impacts. Managing coral reef resilience through a network of NTAs (marine reserves), integrated with management of surrounding areas, is clearly essential to any workable solution. This requires a strong focus on reducing pollution, protecting food webs, and managing key functional groups (such as reef constructors, herbivores, and bioeroders) as insurance for sustainability. (Hughes et al. 2003)

In 2003–2005, the critical role of herbivory in the recovery of corals following large-scale bleaching was examined experimentally on the GBR, by excluding grazing parrotfishes through the use of large cages. In control areas where grazing was uninterrupted, macroalgae were scarce, and coral cover recovered quickly. In contrast, the exclusion of parrotfishes triggered a regime shift to macroalgae, and coral recruitment was suppressed (Hughes et al. 2007a). This study concluded that "management of fish stocks is a key component in preventing phase shifts and managing reef resilience. Importantly, local stewardship of fishing effort is a tractable goal for conservation of reefs, and this local action can also provide some insurance against larger-scale disturbances such as mass bleaching, which are impractical to manage directly" (Hughes et al. 2007a).

In January 2010, Peter Mumby and Alastair Harbone also observed that by "protecting large herbivorous fishes from fishing," marine reserves protecting Caribbean coral reefs would also "generate a trophic cascade that reduces the cover of macroalgae, which is a major competitor of corals," increasing the reefs' resilience to climate change (Mumby and Harbone 2010). According to their research, after the severe coral-bleaching event in 1998 in the Bahamas, coral recovery in marine reserves was significantly more extensive after two and a half years than in nonreserves, and macroalgae cover explained at least 43 percent of the variance in coral abundance (Mumby and Harbone 2010). They also concluded that "[m]arine reserves cannot protect corals from direct climate-induced disturbance, but they can increase the post-disturbance recovery rate of some corals providing that macroalgae have been depleted by more abundant communities of grazers that benefit from reduced fishing pressure" (Mumby and Harbone 2010, 4).

Of course, one of the challenges for place-based management in a climate change era is that the targeted ecosystems or marine environments are changing or moving in response to climate change impacts, making initial MPA establishment potentially less or more useful in the future. As is inherent in the research discussed, most of the research regarding place-based management, marine resilience, and climate change has occurred on coral reefs. Tropical reefs and species expand poleward into higher latitudes during warm interglacial periods, and many contemporary species are expanding their geographic ranges further into subtropical regions in response to ongoing anthropogenic warming (Figueira and Booth 2010; Yamano et al. 2011; Baird et al. 2012). Other marine ecosystems, such as kelp forests in higher latitudes, where the rate of warming is faster, are even more prone to climate change–induced geographical shifts (Craig 2012). Improving marine governance to address both these changes and resistance requires a closer examination of the law, the subject to which this chapter now turns.

## The Law of Place-Based Marine Protection and Marine Ecosystem Resilience

The oceans are a global resource, both legally and ecologically. Ocean currents connect vast regions of the seas to a much greater extent than on land, and no ocean ecosystem can ever be deemed to be completely insulated from the influence of events anywhere else on the globe. As a legal matter, this global scale of ocean dynamics suggests that governance efforts to improve the resilience of marine ecosystems must occur at multiple scales—international, regional, national, and, if national delineation of jurisdiction demands and allows, subnational (Craig and Ruhl 2010). Indeed, one of the greatest challenges for place-based approaches to marine governance is the lack of acceptance by local communities—many attempts to establish networks of marine reserves or NTAs around the world have failed because of poor compliance and public support.

This section surveys some of the important governance mechanisms and trends currently established at these various scales that might encourage (or impede) ecosystem-based, place-based governance of marine resources. This overview is necessarily brief, but it nevertheless provides a summary of existing authority (and gaps in that authority) regarding the increased and improved use of MPAs, marine reserves, and marine spatial planning aimed at improving the resilience of ocean ecosystems, especially in the face of continuing global climate change.

### *Legal Jurisdiction over the Oceans: The Basics*

To establish an MPA, marine reserve, or marine zoning plan, the relevant authority must delineate a particular area of the ocean to be protected, then specify the activities that may and may not occur within one or multiple areas. As such, the use of place-based marine governance to improve the resilience of ocean ecosystems is intimately connected to the legal rules regarding jurisdiction over the seas.

International law—specifically, the UN Convention on the Law of the Sea (UNCLOS)—allows coastal nations to exercise control over fisheries and other ecological protections in a band of waters known as an exclusive economic zone, or EEZ, which generally extends up to 200 nautical miles out from the shore. Almost all MPAs, marine reserves, and marine zoning areas are located within a specific nation's EEZ, and hence most such management areas are created through national or subnational law. For example, in 2006 Portugal finished the designation process for the Lucky Strike Hydrothermal Vents MPA in the northeast Atlantic Ocean, within Portugal's EEZ; this MPA protects twenty-one active hydrothermal "chimney" sites located at a depth of 1,700 meters. This MPA promotes the biodiversity protection goals of both the OSPAR Convention, a fifteen-party treaty to protect the environment of the North Atlantic, and Natura 2000, a network of both terrestrial and marine protected areas designed to protect Europe's natural heritage.

Indeed, most place-based marine management areas are located far closer to shore than 200 nautical miles to protect coastal habitats, such as kelp forests and coral reefs. As a result, depending on a coastal nation's internal laws and customs, MPAs and marine reserves may be created through state, territory, or municipal laws. In the United States, for example, coastal states control a band of coastal waters three nautical miles wide (Florida and Texas have broader jurisdiction in the Gulf of Mexico) along their shores. Using this authority, and acting pursuant to its Marine Life Protection Act, the state of California is developing a network of MPAs along its 1,100-mile coast, using a regional approach.

Under international law, the areas outside of any nation's EEZ are deemed the high seas, and a baseline assumption of freedom on the high seas prevails—although certain activities, such as piracy, have been illegal even on the high seas for centuries. Place-based management of the high seas is still rare and generally requires nations to act through particularized treaties. Nevertheless, such efforts are not unheard of. For example, acting pursuant to the International Convention for the Regulation of Whaling, in 1994 the International Whaling Commission and the parties to the convention established much of the Antarctic Ocean as an international whale sanctuary. Moreover, in November 2009, the Commission for the Conservation of Antarctic Marine Living Resources, which implements the 1982 Convention on the Conservation of Antarctic Marine Living Resources, established its first, 90,000-square-kilometer MPA to protect the South Orkney Islands.

Marine resources on the continental shelf can trigger another set of international law jurisdictional rules, and the establishment of an MPA to protect the Rainbow hydrothermal vent field in the North Atlantic Ocean shows how complex international law can become. The vent field was discovered in 1997, and it is located outside the 200-nautical-mile limit of Portugal's EEZ. Moreover, it is deep—the vents range from 2,270 to 2,320 meters below the surface. Hydrothermal vents are unusual marine ecosystems that are quite valuable, both biologically and, potentially, economically.

The organisms (including microorganisms) of the hydrothermal vent fields are not only adapted to complete darkness and extreme high pressure, but they also survive formidable levels of toxicity and acidity and inhabit the edges of or proximity to the vents, with water temperatures near boiling point. In these conditions, life would be impossible for most species currently living on Earth. However, hydrothermal vent fields shelter millions of animals, thus becoming what Lyle Glowka (1996) has called true "oases of life."

"Although observation and knowledge of the way hydrothermal vent fields function are still in their early stages, the scientific, ecological, and economic importance of these ecosystems is already undisputable. The wealth of the hydrothermal vent fields ranges from their singular biodiversity, to the great and growing interest that such organisms have for medical science and for industry, to the mining of economically attractive minerals—particularly polymetallic sulfides. In the near future, it is because of the unique life forms they shelter that hydrothermal vent fields may be seen to be the next great prize in the global race for natural resources. The organisms of the hydrothermal vents open up a new world of possibilities in biotechnology, as well as commercial opportunities, that cannot be ignored. The race for biological gold, that is, for genetic resources, generated in marine depths is one of the "obsessions of the beginning of the 21st century" (Ribeiro 2010, 186).

To prevent uncontrolled exploitation of this marine ecosystem, in March 2005, the World Wildlife Fund sought to establish an MPA to protect the vent field under the auspices of Annex V to the OSPAR Convention. At the same time, however, Portugal sought to establish national jurisdiction over the Rainbow hydrothermal vent field through UNCLOS's continental shelf provisions, which allow coastal nations to extend their jurisdiction over continental shelf resources beyond the 200-nautical-mile limit for ocean waters themselves. In the end, the two international law processes converged: in 2007 the parties to the OSPAR Convention recognized the designation of the Rainbow hydrothermal vent field as an MPA under Portugal's continental shelf jurisdiction and incorporated it into the OSPAR Convention MPA network for the North Atlantic (Ribeiro 2010).

International law thus provides a variety of means for nations to establish place-based marine governance regimes, and nations and international organizations are becoming more creative in using those tools to establish MPAs and other place-based governance systems, even outside any coastal nation's EEZ. However, the primary focus of place-based marine management remains the waters immediately offshore that are under the jurisdiction of particular coastal nations.

### *International Treaties*

Many international treaties protect marine living resources. Nevertheless, as is typical of ocean governance more generally, many such treaties focus on specific species or fisheries—not marine ecosystems or resilience more generally (Craig 2012). Two examples are the International Convention for the Regulation of Whaling and the various treaties seeking to protect various stocks of tuna. Nations signed the International Convention for the Regulation of Whaling beginning in December 1946, and the treaty took effect in November 1948. The convention's original purpose was to regulate quotas for harvesting whales, but in 1986, the parties imposed a worldwide moratorium on the killing of most of the large baleen whales. Although the International Whaling Commission has begun moving toward place-based protection for whales, such as through the 1994 establishment of the Southern Ocean

Whale Sanctuary, the history of the convention has been far more centered on species-specific protections (International Whaling Commission 2011).

Treaties to protect tuna stocks have been even more focused, both in terms of species and in terms of geographic coverage. In 1966, for example, several countries—including the United States, Canada, Japan, Spain, and France—signed the International Convention for the Conservation of Atlantic Tunas "to specifically address the conservation issues facing the bluefin and other highly migratory species" (Craig 2005). Australia, New Zealand, and Japan formalized the Convention for the Conservation of Southern Bluefin Tuna in 1994, focusing solely on maintaining the sustainability of the southern bluefin tuna fishery.

Nevertheless, more recent treaties do allow for and promote increased use of place-based marine governance mechanisms. For example, the Convention Concerning the Protection of the World Cultural and Natural Heritage (the World Heritage Convention) came into force in December 1975. It encourages parties to accord both emergency and long-term legal protections to places of "outstanding universal value," and while the treaty does not expressly target the ocean, a number of marine sites have been designated as World Heritage Sites, including the GBR (Australia), the Galapagos Islands (Ecuador), the Lagoons of New Caledonia (France), the Belize Barrier Reef Reserve System (Belize), Sian Ka'an (Mexico), Islands and Protected Areas of the Gulf of California (Mexico), and the Papahanaumokuakea Marine National Monument (United States).

In 1992, nations participating in the Rio Conference on Sustainable Development adopted Agenda 21, a global program for achieving sustainable development. Chapter 15 of Agenda 21 promotes conservation of biodiversity, including through "in situ conservation of ecosystems and natural habitats" (¶ 15.5). Chapter 17 more specifically addresses the "protection of the oceans, all kinds of seas, . . . and coastal areas" (¶ 17.1) and explicitly promotes integrated management of marine areas (¶ 17.5), "[c]onservation and restoration of critical marine habitats" (¶ 17.6(h)), and MPAs (¶ 17.7).

Rio conference participants also adopted the UN Convention on Biological Diversity (the Biodiversity Convention), which entered into force in December 1993. Under article 6, parties to the convention are supposed to develop national strategies to conserve and sustainably use their biodiversity, while article 8 explicitly promotes the use of protected areas. Since the second conference of the parties in 1995, these biodiversity goals have been explicitly extended to coastal and marine biodiversity through the Jakarta Mandate. In 2004, a decision of the conference of the parties implementing the Jakarta Mandate made it a high priority for party nations to establish marine and coastal management frameworks that incorporate MPAs. Moreover, at the very first conference of the parties in 1994, the International Coral Reef Initiative was announced to increase knowledge about and protections for the world's coral reefs.

The third UNCLOS came into effect in November 1994. As noted, its provisions allow coastal nations to assert jurisdiction over broad swaths of the ocean, supporting place-based marine management as a jurisdictional matter. Moreover, article 192 of the convention establishes that parties "have the obligation to protect and preserve the marine environment," although that obligation is potentially in tension with several other provisions that allow and even encourage use of the ocean's resources (Craig 2005).

At the 2002 World Summit on Sustainable Development in Johannesburg, South Africa, participating nations adopted an implementation plan for achieving sustainable development—essentially an update and refinement of Agenda 21. This implementation plan called for the worldwide establishment of MPAs to restore overfished species and protect marine biodiversity—specifically, for a worldwide network of MPAs by 2012 (Laffoley et al. 2008).

### *Regional Approaches*

The European Union (EU) is a party to the UN Convention on Biological Diversity and has committed to meeting that convention's MPA goals. One important legal vehicle for pursuing this target is the EU Habitats Directive—or, more formally, Council Directive 92/43/EEC of May 21, 1992, on the conservation of natural habitats and of wild fauna and flora. This directive created Natura 2000, a Europe-wide system of protected areas that includes many small MPAs and marine reserves. The stated goals of the marine aspects of Natura 2000 reflect international treaties, not a resilience goal: "The establishment of a marine network of conservation areas under Natura 2000 will significantly contribute, not only to the target of halting the loss of biodiversity in the EU, but also to broader marine conservation and sustainable use objectives" (Natura 2000 2007, 6). Nevertheless, in the face of climate change impacts, European efforts to create a system of MPAs and marine reserves within the twenty-two member states of the EU that border the sea are increasingly being linked to ecosystem resilience. As a 2011 news alert points out, "By protecting important habitats and ecosystem functions, such as storing carbon by the coast, MPAs can play a role in adaptation and mitigation strategies. They are also heavily influenced by climate change themselves. This could be addressed by creating 'climate-smart' MPAs. Much work has been done on ecosystem resilience and resilience toolkits in tropical regions, which could be applied to temperate and polar regions" (Environmental News Alert Service 2011).

MPAs are linked less formally in the Caribbean through the Caribbean Marine Protected Areas Management (CaMPAM) network and forum. The Caribbean Environment Programme, a subsidiary of UNEP, created CaMPAM in 1997, under the auspices of the Specially Protected Area and Wildlife Protocol of the Convention for the Protection and Development of the Marine Environment in the Wider Caribbean Region (the Cartegena Convention), an umbrella treaty that covers all aspects of marine environmental protection in that region. Since 2008, with substantial financial and technical support from the Nature Conservancy, nations operating within CaMPAM—including the Bahamas, the Dominican Republic, Jamaica, Saint Vincent and the Grenadines, Saint Lucia, Grenada, Antigua, Barbuda, Saint Kitts, and Nevis—have been pursuing the "Caribbean Challenge," aiming to protect 20 percent of the marine and coastal habitats of each nation within marine reserves (NTAs) in the Caribbean by 2020.

The Caribbean Challenge was the third such regional challenge aimed at increasing biodiversity and marine ecosystem health through the expanded use of MPAs and marine reserves. In the first, which began in 2006, the countries of Micronesia signed the Declaration of Commitment to the Micronesia Challenge. "The Micronesia Challenge is a commitment by the Federated States of Micronesia, the Republic of the

Marshall Islands, the Republic of Palau, Guam, and the Commonwealth of the Northern Marianas Islands to preserve the natural resources that are crucial to the survival of Pacific traditions, cultures and livelihood. The overall goal of the Challenge is to effectively conserve at least 30% of the near-shore marine resources (within no-take reserves) and 20% of the terrestrial resources across Micronesia by 2020" (Micronesia Challenge 2011). A year later, similar aspirations began in the Coral Triangle through the efforts of Indonesia, Timor L'Este, Papua New Guinea, the Philippines, Malaysia, and the Solomon Islands (Nature Conservancy 2011).

## *National Regimes*

The 1993 UN Convention on Biological Diversity and the 2002 World Summit on Sustainable Development have encouraged several nations to pursue national systems of MPAs and NTAs (Craig 2005), and increasingly these systems are being linked to resilience goals. Australia, for example, has been pursuing a National Representative System of Marine Protected Areas (NRSMPA) since 1991, building on its initial experimentation with the establishment of the GBR Marine Park in 1976. While the NRSMPA as a whole is a mechanism for Australia to comply with its international obligations under the Biodiversity Convention and its Jakarta Mandate (Department of Sustainability, Environment, Water, Population and Communities [DSEWPC] 2011), the system consists of MPAs established through both commonwealth (national) and state or territorial legal authority, with the commonwealth having sole responsibility for the vast majority of Australia's EEZ. Commonwealth reserves are established pursuant to the Great Barrier Reef Act of 1975 (amended in 2003) and the Environment Protection and Biodiversity Conservation Act of 1999. In 2012, the commonwealth issued a proposal to proclaim the "final" Commonwealth Marine Reserves Network, increasing the number of marine parks under its jurisdiction from twenty-seven to sixty, with a total area of 3.1 million square kilometers. Despite the "reserve" terminology, recreational fishing is permitted in two-thirds of the proposed commonwealth marine park network, and many of the parks have been zoned so that existing fishing and oil and gas industries can continue to operate largely as before. Under the proposal, 13 percent of Australia's EEZ will be closed to fishing. Nonetheless, 96 percent of the commonwealth waters within 100 km of the mainland coast will remain open to recreational fishing (DSEWPC 2012).

The 1999 Strategic Plan for Australia's NRSMPA did not explicitly mention resilience as a goal of the national system—although the stated goals are certainly compatible with promoting resilience:

> The primary goal of the NRSMPA is to establish and manage a comprehensive, adequate and representative system of MPAs to contribute to the long-term ecological viability of marine and estuarine systems, maintain ecological processes and systems, and protect Australia's biological diversity at all levels.
>
> The NRSMPA also has secondary goals: to promote integrated ecosystem management; to manage human activities; to provide for the needs of species and ecological communities; and to provide for the recreational, aesthetic, cultural and economic needs of indigenous and non-indigenous people, where these are compatible with the primary goal. (Australia and New Zealand Environment and Conservation Council 1999, 1)

However, by 2002–2003, when the GBRMPA was planning the rezoning of the GBR Marine Park, which then constituted a dominant part of the NRSMPA, it adapted a resilience framework (Olsson et al. 2008). More recently, it promulgated in 2008 the Great Barrier Reef Climate Change Action Plan, which focused on increasing the resilience of the ecosystem protected through zoning (e.g., MPAs, marine reserves), reducing commercial fishing effort, and improving water quality:

> To secure the future of the Reef it is essential for agencies responsible for managing the Marine Park and its adjacent catchment to do everything possible to restore and maintain the resilience of the ecosystem. It is critical that coordinated actions are taken to protect biodiversity, improve water quality and ensure sustainable fishing.
>
> Resilience-based management of the Reef is core business for the Great Barrier Reef Marine Park Authority. Major resilience-building

> actions already underway include the Reef Water Quality Protection Plan and the Great Barrier Reef Marine Park Zoning Plan. The emergence of climate change makes these efforts even more important, while also presenting new challenges and demanding further action. Without such action, the Reef faces a bleak future under almost all possible future climate scenarios. (GBRMPA 2007, 3)

Moreover, in 2009, the Australian Government and the Queensland Government reached a new agreement to manage the Marine Park for resilience, as a response to the threats that climate change poses for the ecosystem: "Even under the most optimistic scenarios for global reductions in greenhouse gas emissions, climate change will place substantial pressure on the Reef. What this means for the future health of the Reef will depend on its capacity to withstand and adapt to the impacts of climate change—referred to as its 'resilience'" (Australian Government and Queensland Government 2009, 4).

Australia's efforts to improve the resilience of the Great Barrier Reef in the face of climate change have also promoted a commonwealth–Queensland partnership to address some of the nonmarine stressors affecting the reef ecosystem. In particular, the commonwealth and Queensland are both financing Queensland's new regulatory program for agricultural operations in catchments that drain to the Great Barrier Reef in an attempt to reduce land-based pollution of the reef ecosystems (Craig 2012). While the agricultural program is still in its initial stages of implementation, this linking of place-based marine protection and awareness of climate change impacts to governance measures to address other, land-based stressors to marine ecosystems holds great promise as an example of how to realign regulatory fragmentation toward a common goal—and how to make governance systems more comprehensively promotive of marine resilience (Craig 2012). However, the state of Queensland has recently been criticized for poor management of coastal development and industrialization and is actively promoting unprecedented levels of coal and gas extraction that contribute to global warming and to degradation of the GBR (Douvere and Badman 2012).

In 1997, the Government of Canada enacted the federal Oceans Act, which (among other things) required the federal government to establish a Federal Marine Protected Areas Strategy (Government of Canada

2005), leading to a national system of representative MPAs. Three federal departments and agencies—Fisheries and Oceans Canada, Parks Canada Agency, and Environment Canada—implement the national system, creating three main types of MPAs (Government of Canada 2005). These include:

> Oceans Act Marine Protected Areas established to protect and conserve important fish and marine mammal habitats, endangered marine species, unique features and areas of high biological productivity or biodiversity.
>
> Marine Wildlife Areas established to protect and conserve habitat for a variety of wildlife including migratory birds and endangered species.
>
> National Marine Conservation Areas established to protect and conserve representative examples of Canada's natural and cultural marine heritage and provide opportunities for public education and enjoyment. (Government of Canada 2005, 4–5)

In 2003, a report called for the accelerated development of the federal networks of MPAs as a means to combat degradation of Canada's marine resources. In response, the Canadian federal government issued a new Federal Marine Protected Areas Strategy in 2005 (Government of Canada 2005). The 2005 strategy recognizes the more traditional biodiversity goals of MPAs. Thus:

> Around the world, marine protected areas have become increasingly regarded as a valuable conservation and protection tool. A new and innovative approach toward planning for a network of marine protected areas in Canada, including the effective use of an array of marine protected area instruments ranging from no-take marine reserves to multiple use areas, will contribute to the improved health, integrity and productivity of our ocean ecosystems. (Government of Canada 2005, 6)

Among the potential benefits of the network was the expectation that "[t]he effects of localized catastrophes, either human or naturally induced, on marine species may be reduced by establishing networks of marine protected areas over multiple ecosystems and regions, providing a buffer against localized environmental change" (Government of Canada 2005, 8). By 2010, however, the Canadian government was beginning to explicitly connect its network of MPAs to climate change and to resilience concepts:

> Marine protected areas can contribute to climate change mitigation by protecting certain marine habitats that are especially good at absorbing carbon dioxide, emitted to the atmosphere from the burning of fossil fuels, deforestation, and other human activities. For example, coastal habitats such as salt marshes, sea grasses and mangroves account for less than 0.5% of the world's seabed, but studies have shown they can store up to 71% of the total amount of carbon found in ocean sediments. Marine protected areas can also facilitate adaptation to climate change impacts, through protection of ecologically significant habitats (e.g., sources of larval supply), as well as through protection of multiple sites of similar habitat type.
>
> This increases the likelihood that at least one sample of the habitat type and its associated biodiversity will remain intact, should a catastrophic event occur in the region; thus contributing to the overall resilience (ability to adapt to change) of the marine environment. (Government of Canada 2010, 4)

Because of international law's allowance of an EEZ, "New Zealand's marine environment is more than 15 times larger than its terrestrial area" (New Zealand Department of Conservation [NZDOC] 2011). Pursuant to its 2000 biodiversity strategy, New Zealand seeks to protect 10 percent of its marine environment in a network of MPAs and marine reserves by 2020. However, as is true of many coastal nations, this goal has a biodiversity focus driven by New Zealand's international treaty commitments, and the 2005 Marine Protected Areas Policy and Implementation Plan specifically states that its goal is to "[p]rotect marine biodiversity by establishing a network of MPAs that is comprehensive and representative of New Zealand's marine habitats and ecosystems" (NZDOC and New Zealand Ministry of Fisheries [NZMOF] 2005, 6).

Nevertheless, the biodiversity strategy also recognized broader goals for New Zealand's marine ecosystems that resonate more strongly with a resilience focus. For example, the goal for 2020 is that "New Zealand's natural marine habitats and ecosystems are maintained in a healthy functioning state. Degraded marine habitats are recovering" (NZDOC and NZMOF 2005, 8). Moreover, in 2010 New Zealand released its new Coastal Policy Statement, which guides local authorities' day-to-day management of coastal resources. The first objective of this policy is "[t]o safeguard the integrity, form, functioning and resilience of the coastal environment and sustain its ecosystems, including marine and intertidal areas, estuaries, dunes, and land" by, inter alia, "protecting representative or significant natural ecosystems and sites of biological importance" (Obj. 1).

New Zealand has a number of laws that allow for the creation of MPAs and marine reserves. These include the Marine Reserves Act of 1971, which allows the NZDOC to establish marine reserves in New Zealand's territorial sea (the first 12 miles of ocean from the coast); the Marine Mammals Protection Act of 1978, which allows the NZDOC to create marine mammal sanctuaries, where fishing can be restricted; the Fisheries Act of 1983, which allows for the creation of mataitai reserves (permanent reserves important to the Maori for food gathering), taiapure (local fisheries subject to iwi or hapu [native restrictions]), and marine parks, although such use of the act was cut off through the Fisheries Act of 1996; and special MPA legislation, such as the Sugar Loaf Islands Marine Protected Area Act of 1991 and the Hauraki Gulf Marine Park Act of 2000 (Agardy 2010).

Since 2002, the New Zealand Parliament has been working intermittently on an update of the Marine Reserves Act of 1971 (Agardy 2010). The amendments, which fishers have strongly opposed, would allow for the creation of marine reserves throughout New Zealand's EEZ, and the parliament's latest report, scheduled for January 2011, has not yet materialized. The bill could also expand New Zealand's use of ocean zoning, which has recently focused mainly on nearshore marine aquaculture (Agardy 2010).

According to NOAA's National Marine Protected Area Center, "[t]he U.S. has more than 1600 MPAs. These areas cover 36% of U.S. marine waters, and vary widely in efficacy, purpose, legal authorities, managing agencies, management approaches, level of protection, and restrictions on human uses" (National Marine Protected Area Center [NMPAC]

2011). Moreover, the United States is pursuing a National System of Marine Protected Areas, which had its genesis in President Bill Clinton's May 26, 2000, Executive Order 13158. By design, this network includes MPAs and marine reserves designated at all levels of government (NMPAC 2008). To date, however, less than 1 percent of continental U.S. waters are protected through marine reserves or NTAs.

The NMPAC finally published its framework for establishing the national system in 2008, stressing that "the national system will help to increase the efficient protection of important marine resources; contribute to the nation's overall social and economic health; support government agency cooperation and integration; and improve the public's access to scientific information and decision making about the nation's marine resources" (NMPAC 2008, 8). This 2008 framework document recognized a number of benefits to the nation from the representative network of MPAs, including enhanced conservation of marine resources and better ocean planning (NMPAC 2008). Overall, "[t]he purpose of the national system is to support the effective stewardship, conservation, restoration, sustainable use, and public understanding and appreciation of the nation's significant natural and cultural marine heritage and sustainable production marine resources, with due consideration of the interests of and implications for all who use, benefit from, and care about our marine environment" (NMPAC 2008, 13).

In July 2010, President Barack Obama brought climate change, resilience, and ocean zoning into United States ocean policy through his Executive Order 13547, "Stewardship of the Ocean, Our Coasts, and the Great Lakes." In this executive order, President Obama first recognized the pervasive importance of the oceans to Americans (§ 1). The order then set out ten goals in protecting U.S. ocean ecosystems, including to: "protect, maintain, and restore the health and biological diversity of ocean, coastal, and Great Lakes ecosystems and resources; . . . improve the resiliency of ocean, coastal, and Great Lakes ecosystems, communities, and economies; . . . [and] improve our understanding and awareness of changing environmental conditions, trends, and their causes, and of human activities taking place in ocean, coastal, and Great Lakes waters" (§ 2(a)). The Ocean Stewardship Executive Order thus incorporates both climate change and resilience into its goals, a fact made clear as well in the order's purposes, which include "providing for adaptive management to enhance our understanding of and capacity to respond to climate change and ocean acidification" (§ 1). Finally, the order creates a National Ocean Council with representatives from a wide variety of federal agencies and departments (§ 4), which is charged with approving and implementing marine spatial planning in U.S. waters (§§ 4, 6). It remains to be seen how exactly the United States will incorporate its existing network of MPAs and marine reserves into this new ocean zoning plan, but it is probable that the network will be deemed to contribute to ocean resilience and to provide one basis for more comprehensive marine governance and planning.

As these examples attest, most national efforts to implement place-based marine management and to link those governance efforts to resilience goals are relatively recent and ongoing, making it difficult to evaluate their effectiveness—particularly because climate change is already changing the ecological realities of many marine ecosystems. Nevertheless, as noted, studies on coral reefs indicate that well-placed, community-supported, and enforced marine reserves and NTAs can begin to rebuild species stocks and restore ecosystem function, enhancing ecosystem resilience (Hughes et al. 2007b; McCook et al. 2010; Mumby and Harbone 2010). Australia's efforts to improve governance of agricultural runoff in order to bolster the resilience of the GBR point to the importance of reinforcing place-based marine governance by simultaneously reducing stressors from land.

Climate change, of course, remains the largely unaddressed and increasingly important threat to marine ecosystem resilience, and efforts to significantly reduce that stressor will need to be international in scope. From this perspective, place-based marine management is primarily a governance measure for climate change adaptation—that is, a potential means of improving marine ecosystems' ability to absorb climate change impacts and still remain healthy and productive ecosystems for human use (Craig 2012).

Nevertheless, to remain effective in a climate change era, place-based marine governance measures will themselves need to become responsive to climate change (Hughes et al. 2007b; Polasky et al. 2011). Robin Craig (2012) has suggested that place-based governance regimes can address climate change in three ways. First, many existing place-based governance regimes, especially for coral reefs, already reduce fishing pressures on the ecosystems and hence increase their resilience and ability to adapt to climate change. These governance regimes might thus be viewed as "accidentally" adaptive: they were established for purposes other than climate change, but the increased ecosystem resilience that they promote nevertheless increases the ecosystem's ability to cope with climate change impacts. Second, Australia's 2004 rezoning of the GBR and climate change efforts in the Coral Triangle and Micronesia reflect more purposeful attempts to incorporate place-based marine governance and the increased resilience that it can provide into climate change adaptation efforts (Craig 2012).

As noted, climate change is already changing the abundance and geographic distribution of species, and the activities of people who are reliant on marine ecosystems. Hence, place-based marine governance regimes will need to respond to these dynamics. For example, the new zoning regime for the GBR includes special management areas—zones that the Great Barrier Reef Marine Park Authority can create on a more ad hoc basis to address a variety of emerging or temporary threats to or other management problems within the reef's ecosystems (GBRMPA 2004; Craig 2012). Such provisions for increased flexibility are likely to be a crucial element of place-based marine governance in the future. Two other potential tools for making place-based marine governance climate change adaptable—in addition to normal processes for amending the relevant law—are anticipatory zoning and dynamic zoning (Craig 2012). Anticipatory zoning consists of place-based governance regimes put into place legally in anticipation of climate change impacts that will significantly change possible uses of the marine ecosystem at issue (Craig 2012). For example, in August 2009, in anticipation of melting Arctic sea ice and hence increased pressures on northern Alaska coast fisheries, NOAA in the United States created the Arctic Management Area and peremptorily prohibited all commercial fishing there (Craig 2012). Anticipatory zoning thus preserves the resilience of marine ecosystems by anticipating and reducing (or, in the case of the Arctic, eliminating) new stressors that climate change impacts are making possible.

Dynamic zoning, in turn, keys the governance regime to critical ecological factors that can shift over time, such as ocean temperature or seasonal aggregations of species or spawning events, and it shifts management requirements in accordance with that parameter. Off Hawaii, for example, TurtleWatch alerts longline fishers in real time to changes in ocean temperature bands, allowing the fishers to more effectively predict and avoid illegal encounters with endangered sea turtles, simultaneously contributing to the survival of those species and the long-term survival of the longline fishery (Craig 2012). Salmon farmers off the coast of Tasmania in Australia are calling, similarly, for seasonal (4-month) predictions of ocean temperatures so they can better locate their aquaculture pens in cooler waters (Craig 2012). Both of these efforts to establish dynamic zoning suggest it could become an important governance tool for simultaneously acknowledging climate change effects, minimizing human impacts to overstressed marine ecosystems, and cautiously promoting human economic uses of the oceans in appropriate locations. Thus, place-based marine governance will play a key role in sustaining social-ecological resilience, despite climate change's increasing impacts on marine ecosystems (Craig 2012).

## Conclusion

Fostering the resilience of the oceans and avoiding the transgression of dangerous tipping points has economic as well as ecological import. As John C. Ogden noted in 2001, "Marine ecosystems are major capital assets. In addition to providing valuable goods, such as fisheries and minerals, they provide critical life support services, such as diluting, dispersing, and metabolizing the effluents of society, thus purifying waters for recreation" (Ogden 2001, 31). Moreover, while Ogden concluded that "[t]he value of a healthy ocean is difficult to overestimate" (Ogden 2001, 31), other researchers have attempted to be more precise regarding the value of coastal and marine ecosystem services to humans. In 1997, for example, Robert Costanza and his team of ecological economists estimated that the open oceans provide ecosystem services worth US$8.4 trillion per year, while coastal oceans provide ecosystem services worth US$12.6 trillion per year (Costanza et al. 1997). Some of these services, moreover, are irreplaceably life-sustaining for humans: marine phytoplankton, algae, and other marine plants are responsible for 50–75 percent of the oxygen in the atmosphere, and fish, mollusks and crustaceans are an important source of protein for a large percentage of the world's population. For example, according to the UNFAO, "[i]n 2007, fish accounted for 15.7 percent of the global population's intake of animal protein and 6.1 percent of all protein consumed. Globally, fish provides more than 1.5 billion people with almost 20 percent of their average per capita intake of animal protein, and 3.0 billion people with at least 15 percent of such protein" (UNFAO 2011, 3).

More and more governments are beginning to embrace place-based management of marine ecosystems and to expand single MPAs and marine reserves (NTAs) into representative systems of MPAs and comprehensive ocean zoning. In Australia, Chile, New Zealand, and the Philippines, these efforts reflect legislative and political changes that began in the 1970s and 1980s (Russ and Alcala 2006; Olsson et al. 2008; Babcock et al. 2010; Gelcich et al. 2010). Moreover, many governments are also beginning to realize that these governance mechanisms can provide benefits that extend beyond "mere" restoration of fisheries and marine biodiversity, including increased resilience of ecosystems and coastal societies in the face of new climate change–caused stresses (Hughes et al. 2005).

While this evolution appears to be positive, it nevertheless remains important to remember that place-based marine management must be grounded in good science; in particular, random assortments of small MPAs may do very little to improve ecosystem health or resilience. Equally, the establishment of a large marine reserve network in far-flung corners of a nation's EEZ will have a limited benefit if most overfished coastal areas remain unprotected. Nor is place-based management a panacea for what ails marine ecosystems. Indeed, as marine governance mechanisms, MPAs, marine reserves, and ocean zoning can do little to improve land-based marine pollution or the root causes of climate change. Finally, in a climate change era, it may become increasingly important for marine management to become more flexible (Hughes et al. 2007b; Polasky et al. 2011; Craig 2012), because the impacts of climate change are likely to fundamentally alter marine ecosystems in ways that exceed "normal" variability, even for these already highly dynamic assemblages.

## References

Agardy, T. *Ocean Zoning: Making Marine Management More Effective.* London: Earthscan 2010.

Australia and New Zealand Environment and Conservation Council, Task Force on Marine Protected Areas. *Strategic Plan of Action for the National System of Marine Protected Areas: A Guide for Action by Australian Governments.* Canberra, Australia: Environment Australia, 1999.

Australian Government and Queensland Government. *Maintaining a Healthy and Resilient Great Barrier Reef: The Commonwealth and Queensland Governments' Interim Response to the Great Barrier Reef Outlook Report 2009.* Australia: Australian and Queensland Governments, 2009.

Ausubel, J. H., D. T. Crist, and P. E. Waggoner. *First Census of Marine Life 2010: Highlights of a Decade of Discovery.* Washington, D.C.: Census of Marine Life International Secretariat, 2010.

Babcock, R. C., N. T. Shears, A. C. Alcala, N. S. Barrett, G. J. Edgar, K. D. Lafferty, T. R. McClanahan, and G. R. Russ. "Decadal Trends in Marine Reserves Reveal Differential Rates of Change in Direct and Indirect Effects." *Proceedings of the National Academy of Sciences USA* 107 (2010): 18256–18261.

Baird, A. H., B. Sommer, and J. S. Maddin. "Pole-ward Range Expansion of *Acropora* spp. Along the East Coast of Australia." *Coral Reefs* 31 (2012): 1063.

Bellwood, D. R., T. P. Hughes, C. Folke, and M. Nyström. "Confronting the Coral Reef Crisis." *Nature* 429 (2004): 827–833.

Center for Marine Biodiversity, Canada. "What Is Marine Biodiversity?" (2011). http://marinebiodiversity.ca/en/what.html.

Coleman, F. C., W. F. Figueira, J. S. Ueland, and L. B. Crowder. "The Impact of United States Recreational Fisheries on Marine Fish Populations." *Science* 305 (2004): 1958–1959.

Connor, S. 1999. "Marine Life Being Massacred as the World's Oceans Are Turning Toxic." *The Independent* (London), September 3, 1999, p. 3.

Costanza, R., et al. "The Value of the World's Ecosystem Services and Natural Capital." *Nature* 387 (1997): 253–259.

Côté, I. M., I. Mosqueira, and J. D. Reynolds. "Effects of Marine Reserve Characteristics on the Protection of Fish Populations: A Meta-Analysis." *Journal of Fish Biology* 59 (2001): 178–189.

Craig, R. K. "Taking the Long View of Ocean Ecosystems: Historical Science, Marine Restoration, and the Oceans Act of 2000." *Ecology Law Quarterly* 29 (2002): 649–705.

Craig, R. K. "Protecting International Marine Biodiversity: International Treaties and National Systems of Marine Protected Areas." *Journal of Land Use and Environmental Law* 20 (2005): 333–369.

Craig, R. K. *Comparative Ocean Governance: Place-Based Protections in an Era of Climate Change.* Cheltenham, U.K.: Edward Elgar, 2012.

Craig, R. K., and J. B. Ruhl. "Governing for Sustainable Coasts: Complexity, Climate Change, and Coastal Ecosystem Protection." *Sustainability* 2 (2010): 1361–1388.

Department of Sustainability, Environment, Water, Population and Communities, Australian Government. "National Representative System of Marine Protected Areas" (2011). http://www.environment.gov.au/coasts/mpa/nrsmpa/index.html.

Department of Sustainability, Environment, Water, Population and Communities, Australian Government. "Marine Reserves and Recreational Fishing" (2012). http://www.environment.gov.au/coasts/mbp/reserves/pubs/fs-recfishing.pdf.

Douvere, F., and T. Badman. "WHC.12/36.COM/. Mission Report, Reactive Monitoring Mission to Great Barrier Reef, Australia, 6th to 14th March 2012" (2012). http://whc.unesco.org/en/documents/117104.

Dudley, N., ed. *Guidelines for Applying Protected Area Management Categories.* Gland, Switzerland: International Union for the Conservation of Nature, 2008.

Environmental News Alert Service, European Commission. "Filling in the Gaps in Marine Protected Areas" (2011). http://ec.europa.eu/environment/integration/research/newsalert/pdf/228na6.pdf.

Figueira, W. F., and D. J. Booth. "Increasing Ocean Temperatures Allow Tropical Fishes to Survive Overwinter in Temperate Waters." *Global Change Biology* 16 (2010): 506–516.

Gelcich, S, T. P., et al. "Navigating Transformations in Governance of Chilean Marine Coastal Resources." *Proceedings of the National Academy of Science USA* 107 (2010): 16751–16799.

Glowka, L. "The Deepest of Ironies: Genetic Resources, Marine Scientific Research, and the Area," *Ocean Yearbook* 12 (1996): 154–178.

Government of Canada. *Canada's Federal Marine Protected Areas Strategy.* Ottawa, Canada: Fisheries and Oceans Canada, 2005.

Government of Canada. *Spotlight on Marine Protected Areas in Canada.* Ottawa, Canada: Fisheries and Oceans Canada, 2010.

Great Barrier Reef Marine Park Authority, *Australian Government. Great Barrier Reef Climate Change Action Plan 2007–2012.* Townsville, Australia: GBRMPA, 2007.

Great Barrier Reef Marine Park Authority, Australian Government. *Great Barrier Reef Marine Park Zoning Plan 2003.* Townsville, Australia: GBRMPA, 2004.

Halpern, B. S. "The Impact of Marine Reserves: Do Reserves Work and Does Reserve Size Matter?" *Ecological Applications* 13 (2003): S117–S137.

Halpern, B. S., et al. "A Global Map of Human Impact on Marine Ecosystems." *Science* 319 (2008): 948–952.

Hughes, T. P. "Catastrophes, Phase-Shifts, and Large-Scale Degradation of a Caribbean Coral Reef." *Science* 265 (1994): 1547–1551.

Hughes, T. P. "New Marine Reserves Won't Address UNESCO's Reef Concerns." *The Conversation* June 14, 2012. http://theconversation.edu.au/new-marine-reserves-wont-address-unescos-reef-concerns-7638.

Hughes, T. P., et al. "Climate Change, Human Impacts, and the Resilience of Coral Reefs." *Science* 301 (2003): 929–933.

Hughes, T. P., D. R. Bellwood, C. Folke, R. S. Steneck, and J. Wilson. "New Paradigms for Supporting the Resilience of Marine Ecosystems." *Trends in Ecology and Evolution* 20 (2005): 380–386.

Hughes, T. P., M. J. Rodrigues, D. R. Bellwood, D. Ceccarelli, O. Hoegh-Guldberg, L. McCook, N. Moltschaniwskyj, M. S. Pratchett, R. S. Steneck, and B. Willis. "Phase Shifts, Herbivory, and the Resilience of Coral Reefs to Climate Change." *Current Biology* 17 (2007a): 360–365.

Hughes, T. P., et al. "Adaptive Management of the Great Barrier Reef and the Grand Canyon World Heritage Area." *Ambio* 36 (2007b): 586–592.

Intergovernmental Panel on Climate Change. *Climate Change 2007: Synthesis Report: Summary for Policymakers*. Geneva, Switzerland: IPCC, 2007.

International Union for the Conservation of Nature. "Fact Sheet: Marine Protected Areas—Why Do We Need Them?" (2010). http://www.iucn.org/iyb/resources/news/?4715/marine-protected-areas.

International Whaling Commission. IWC home page and associated links, visited April 2011. http://iwcoffice.org/index.htm.

Jackson, J. B. C. "What Was Natural in the Coastal Oceans?" *Proceedings of the National Academy of Sciences USA* 98 (2001): 5411–5420.

Jackson, J. B. C., et al. "Historical Overfishing and the Recent Collapse of Coastal Ecosystems." *Science* 293 (2001): 629–638.

Jackson, J. B. C., K. E. Alexander, and E. Sala. *Shifting Baselines: The Past and the Future of Ocean Fisheries*. Washington, D.C.: Island Press, 2011.

Laffoley, D., A. T. White, S. Kilarski, M. Gleason, S. Smith, G. Llewellyn, J. Day, A. Hillary, V. Wedell, and D. Pee. *Establishing Resilient Marine Protected Area Networks—Making It Happen*. Washington, D.C.: IUCN-WPCA, NOAA, and the Nature Conservancy, 2008.

Levy, J. M. "Global Oceans." In *State of the Climate in 2009*, eds. D. S. Arndt, M. O. Baringer, and M. R. Johnson, 51–78. Washington, D.C.: National Oceanic and Atmospheric Administration, 2010.

Marine Life Protection Act of 2004. California Fish and Game Code § 2850–2863. http://www.dfg.ca.gov/marine/mpa/index.asp

McClellan, K. "Coral Degradation Through Destructive Fishing Practices." *The Encyclopedia of the Earth*, 2010. http://www.eoearth.org/article/Coral_degradation_through_ destructive_fishing_practices?topic=49513.

McCook, L. J., et al. "Adaptive Management of the Great Barrier Reef: A Globally Significant Demonstration of the Benefits of Networks of Marine Reserves." *Proceedings of the National Academy of Science USA* 107 (2010): 18278–18285.

McLeod, K., and H. Leslie. *Ecosystem-Based Management for the Oceans*. Washington, D.C.: Island Press, 2009.

Micronesia Challenge. "The Micronesia Challenge: About Us" ( 2011). http://www.micronesiachallenge.org.

Millennium Ecosystem Assessment. *Ecosystems and Human Well Being: Current State and Trends*, vol. 1. Washington, D.C.: Island Press, 2005.

Mumby, P. J., and A. R. Harbone. "Marine Reserves Enhance the Recovery of Corals on Caribbean Reefs." *Public Library of Science (PLoS) One* 5 (2010): 1–7.

National Marine Protected Area Center, National Oceanic and Atmospheric Administration. *Framework for the National System of Marine Protected Areas of the United States of America.* Silver Spring, MD: NMPAC, 2008.

National Marine Protected Area Center, National Oceanic and Atmospheric Administration. National Marine Protected Areas Center home page, viewed April 2011. http://www.mpa.gov.

National Research Council. *Effects of Trawling and Dredging on Seafloor Habitat.* Washington, D.C.: National Academies Press, 2002.

National Research Council. *Mississippi River Water Quality and the Clean Water Act: Progress, Challenges, and Opportunities.* Washington, D.C.: National Academies Press, 2008.

Natura 2000. *Guidelines for the Establishment of the Natura 2000 Network in the Marine Environment. Application of the Habitats and Birds Directives.* Brussels, Belgium: European Union, 2007.

Nature Conservancy. "The Coral Triangle: The Coral Triangle Initiative: Reefs, Fisheries, and Food Security" (2011). http://www.nature.org/ourinitiatives/regions/asiaandthepacific/ coraltriangle/overview/coral-triangle-initiative.xml.

New Zealand Department of Conservation. "Marine reserves & other protected areas" (2011). http://www.doc.govt.nz/conservation/marine-and-coastal/marine-protected-areas.

New Zealand Department of Conservation. "Coastal Policy Statement" (2010). http://www.lincoln.ac.nz/Documents/LEaP/nz-coastal-policy-statement-2010.pdf

New Zealand Department of Conservation and New Zealand Ministry of Fisheries. *Marine Protected Areas Policy and Implementation Plan.* Wellington, New Zealand: NZDOC and NZMOF, 2005.

Obama, Barack. "Stewardship of the Ocean, Our Coasts, and the Great Lakes" Executive Order 13547 (2010). http://www.whitehouse.gov/the-press-office/executive-order-stewardship-ocean-our-coasts-and-great-lakes

Ogden, J. C. "Maintaining Biodiversity in the Oceans." *Environment* 43 (2001): 28–32.

Olsson, P., T. P. Hughes, and C. Folke. "Navigating the Transition to Ecosystem-Based Management of the Great Barrier Reef, Australia." *Proceedings of the National Academy of Science USA* 105 (2008): 9489–9494.

Pauly, D. "Anecdotes and the Shifting Baseline of Fisheries." *Trends in Ecology and Evolution* 10 (1995): 430.

Polasky, S., S. R. Carpenter, C. Folke, and B. Keeler. "Decision-Making Under Great Uncertainty: Environmental Management in an Era of Global Change." *Trends in Ecology and Evolution* 26 (2011): 398–404.

Ray, G. C., and J. F. Grassle. "Marine Biological Diversity: A Scientific Program to Help Conserve Marine Biological Diversity Is Urgently Required." *BioScience* 41 (1991): 453–457.

Ribeiro, M. C. The "'Rainbow': The First National Marine Protected Area Proposed Under the High Seas." *International Journal of Marine and Coastal Law* 25 (2010): 183–207.

Roberts, C. M., and J. P. Hawkins. *Fully-Protected Marine Reserves: A Guide.* Washington, D.C.: World Wildlife Fund Endangered Seas Campaign, 2000.

Royal Society. "Ocean Acidification Due to Increasing Atmospheric Carbon Dioxide." Cardiff, U.K.: Clyvedon. 2005.

Russ, G. R., and A. C. Alcala. "No-Take Marine Reserves and Reef Fisheries Management in the Philippines: A New People Power Revolution." *Ambio* 35 (2006): 245–254.

Spotts, P. N. "Climate Change Brews Ocean Trouble: Scientists Tie Global Warming to Increased Upwelling of Deep Ocean Water, which Can Create Crippling Aquatic Dead Zones." *Christian Science Monitor* March 8, 2007. http://www.csmonitor.com/2007/0308/p13s01-sten.html.

Steneck, R. S., et al. "Creation of a Gilded Trap by the High Economic Value of the Maine Lobster Fishery." *Conservation Biology* 25 (2011): 904–912.

United Nations Conference on Environment and Development. *Agenda 21.* Rio de Janeiro, Brazil. (1992). http://sustainabledevelopment.un.org/content/documents/Agenda21.pdf.

United Nations Education, Scientific, and Cultural Organization. "Marine Biodiversity." *CONNECT: International Science, Technology & Environmental Education Newsletter* 21 (1996): 1–2.

United Nations Environment Programme. "Marine Pollution and Coastal Development" (2011). http://www.grida.no/publications/rr/in-dead-water/page/1250.aspx.

United Nations Food and Agriculture Organization. "Issues Fact Sheets: Ghost Fishing" (2005). http://www.fao.org/fishery/topic/14798/en.

United Nations Food and Agriculture Organization. *The State of World Fisheries and Aquaculture 2010.* Rome, Italy: UNFAO, 2011.

U.S. Commission on Ocean Policy. *An Ocean Blueprint for the 21st Century. Final Report.* (2004). http://govinfo.library.unt.edu/oceancommission/documents/full_color_rpt/000_ocean_full_report.pdf.

U.S. Environmental Protection Agency. "Coastal Zones and Sea Level Rise" (2011). http://www.epa.gov/climatechange/effects/coastal/index.html.

Yamano, H., K. Sugihara, and K. Nomura. "Rapid Poleward Range Expansion of Tropical Reef Corals in Response to Rising Sea Surface Temperatures." *Geophysical Research Letters* 38 (2011): L04601.

# Resilience and Water Governance

## *Addressing Fragmentation and Uncertainty in Water Allocation and Water Quality Law*

BARBARA A. COSENS AND CRAIG A. STOW

The U.S. EPA reports that almost half of the nation's rivers and two-thirds of its lakes are use-impaired due to poor water quality (U.S. EPA 1998, 2002, 2010; Houck 2002). The Western Water Policy Review Advisory Commission identified both poor water quality and unhealthy aquatic systems among the water challenges facing the West (Western Water Policy Review Advisory Commission 1998). The water quality impairment is caused both by chemical pollution and physical alteration of streams. Nutrients and excess sediment impair water quality in 30 percent of the nation's streams (U.S. EPA 2011). In the Great Basin nearly two-thirds of the native fish are either listed under the Endangered Species Act (ESA) or considered of concern by the U.S. Fish and Wildlife Service (USFWS). Water development is considered second only to the introduction of nonnative fish in causing these problems (Doremus 2001).

In 1906, Justice Holmes, writing for a majority of the U.S. Supreme Court in an opinion that denied an injunction against the disposal of Chicago's sewage into the Missouri River, stated, "It is a question of the first magnitude whether the destiny of the great rivers is to be the sewers of the cities along their banks or to be protected against everything which threatens their purity," thus requiring clear proof of the source of serious harm if the court is to act in the absence of Congressional legislation (*Missouri v. Illinois* 1906). It would be sixty-six years before Congress would act.

By many accounts, the efforts under the 1972 Federal Water Pollution Control Act, commonly known as the Clean Water Act (CWA), to clean up point source discharges from municipal sewage and industrial activities under the National Pollutant Discharge Elimination System program have been successful (e.g., Houck 2002; Adler et al. 1993). Similarly, by all accounts and relying on the statistics above, it is also clear that the cleanup of non–point source pollution is insufficient to achieve the goal of the CWA to "restore and maintain the chemical, physical and biological integrity of the Nation's waters" (CWA § 1251(a)). Diffuse sources of pollution resulting from land use activities, including agriculture, timber harvest, and development in the floodplain; dewatering of streams for beneficial use of water; and loss of the ability of streams to handle sediment input due to modification of floodplains and adjacent wetlands, continue to present challenges to efforts to achieve water quality standards. It is not that the CWA does not address non–point source issues. Whether due to the inadequacy of the approach or of its implementation, it has not yielded the results sought of fishable, swimmable rivers.

A report of the National Research Council (NRC) to Congress in 2001 identified uncertainty as a key barrier to management decision making and improved water quality under the CWA and recommended that the states and EPA move forward with management actions "while making substantial efforts to reduce uncertainty" (NRC 2001). Further, the NRC specifically recommended against the practice of measuring the success of the program to address non–point source pollution (also referred to as the total maximum daily load [TMDL] program) by the attainment of administrative goals, but rather by measurable improvements in water quality. In addition, the NRC report and a report of an expert panel, including coauthor Stow, assembled by the Center for the Analysis and Prediction of River Basin Environmental Systems at Duke University, recommended the use of adaptive implementation to address uncertainty (NRC 2001; Shabman et al. 2007).

In this chapter, we identify two areas that must be addressed in reforming water law and water quality law if the nation's stated goals are to be achieved:

(1) Fragmentation. Lack of integration between regulation of point source and non–point source pollution, non–point source pollution and land use regulation, and water quality and water allocation regulation not only places an absolute limit on the ability of current regulation to achieve its goals, but imposes high costs and diminishing returns on any efforts toward those goals under existing regulation.

(2) Uncertainty. The diversity of both the physical and biological features of water bodies and of the sources of non–point source pollution require site-specific responses. Yet the cost of characterizing water quality on a water body by water body basis and the failure to follow up on efforts to achieve water quality standards with monitoring of results inhibits improvements in water quality characterization and identification of land use changes that are effective in moving toward water quality goals. In contrast to the early expectations of a fairly straightforward calculation of pollutant loads (Houck 2002), studies have shown that even modeling of relatively data-rich systems shows variance in excess of 50 percent (Stow et al. 2003).

Integration of water resource management and adaptation in the face of uncertainty require new approaches to natural resource management and regulation. Resilience theory provides a foundation for relating these changes to the complexity of the social-ecological system and for the necessary communication across disciplines to define those changes. Resilience theory as applied to ecological systems addresses a system's ability to respond to change while continuing to provide, or to shift to a state in which it will provide, a full range of ecosystem services or function (Walker and Salt 2006). On paper, the traditional approach to natural resource management involving single-variable optimization can be implemented by individual resource management or regulatory agencies acting at a single scale. But the complexity of social-ecological systems will lead to unexpected and unwanted results if management and regulation continue down that path.

The concept of resilience made its appearance in the study of ecological systems in the work of C. S. Holling (1973) and his recognition that ecosystems are not static. When applied to ecological systems without a human component, resilience theory focuses on both the capacity of the system to return to its prior level of self-organization following a disturbance and the degree to which that capacity is influenced by or sensitive to changes at smaller and larger scales (Gunderson and Holling 2002; Walker et al. 2004). Resilience theorists call on adaptive management, in which the natural adaptive abilities of an ecosystem are emphasized and promoted over the active management, control, and resource exploitation of the system to foster ecological resilience (Folke et al. 2005). The term "adaptive implementation" used by Shabman et al. (2007) incorporates the concepts of adaptive management but applies them in a setting in which the agency must implement a specific regulatory statute rather than acting under its general management authority over a particular resource or area of land. Both concepts allow for incremental action in the face of uncertainty, along with constant monitoring and feedback to improve management or regulatory decisions.

We use resilience theory to explore the legal changes and implementation approaches necessary to achieve adaptive implementation of the CWA and integrated water resource management within the current U.S. framework, which places water quality oversight at the federal level, water allocation at the state level, and most land use decisions at the local level. We begin by exploring fragmentation and uncertainty issues in current water management and regulation.

## Water Management: Sources of Fragmentation and Uncertainty

Discussion of means to address fragmentation and uncertainty in water management must begin with an explanation of how the current system arose in the United States and how that leads to fragmentation and lack of adaptability in the face of uncertainty. History matters, because it informs the choice of approaches that may be politically acceptable. In addition, the greater the understanding of how the current system arose, the greater the chance of avoiding unintended consequences in changing it. We begin with the law applicable to water quality, follow this with discussions of management of aquatic species and then water allocation, and conclude with land use regulation.

### *Water Quality*

Historically water quality issues were considered a matter of state law. Dilution was the primary focus of water quality control, and a public nuisance suit was the primary legal remedy to the impact on downstream states when Justice Holmes wrote the words quoted above. Several events and reports spurred federal action. In January 1969 a Union Oil well off the coast of Santa Barbara, California, blew out. In June 1969 the Cuyahoga River in Cleveland, Ohio, burned (Adler et al. 1993). A 1971 report titled *Water Wasteland*, compiled by researchers for Ralph Nader, detailed fish kills, water courses and beaches unfit for swimming, contaminated drinking water, and loss of bird species due to toxic chemicals in water (Adler et al. 1993). That same year, similar findings were presented in the Second Annual Report of the President's Council on Environmental Quality (Adler et al. 1993). Leaving water quality to the states was apparently not working. Federal action up until this time had focused primarily on financial assistance for sewage treatment plants.

Water quality standards under the CWA are set on a water body basis and are tailored to the particular use of that water body, including recreation and habitat for endangered species (33 USC § 1313). Because water quality standards are based on use, it is theoretically those who control its use who should set those standards; at least this was the argument states made to Congress as the CWA was being crafted (Houck 2002). Governors seeking to retain state control provided testimony to Congress that ranged from the 1970s equivalent of the need for local knowledge and flexibility; to the constitutionality of federal interference with state control of any but interstate waters, an argument that persists today (e.g., see *Rapanos v. United States*, 2006); to claims that states were already doing a great job (Houck 2002), an argument made despite the fact that it was inconsistent with the scientific data before Congress. An additional argument made was that a standards-based program that can take into account the individual assimilative capacity of specific water bodies was the appropriate approach (Houck 2002). As will be discussed below, despite this argument, states have failed to consider the assimilative capacity of a water body when allocating water for use. Rarely is pollutant load a consideration when approving a new water diversion right, despite the fact that increased diversion reduces the ability of the water body to dilute pollutants. This might make sense in the water abundance of the eastern United States, but its proponents were not limited to the nonarid states.

Despite the tremendous success of the National Pollutant Discharge Elimination System program, the nation's waters remain polluted. Houck provides a list of agricultural and silviculture impacts in various states, then concludes that these sources:

> have several characteristics in common. Individually small, it is their cumulative impacts that are the problem, and we have not yet in any medium found an easy way to persuade people to fix problems for which they are only a small contributing factor. Furthermore, although individually small, these sources are supported by industries with a political lock on state legislatures and, in some cases, on Congress as well. (Houck 2002)

Most effort to abate non–point source pollution has proceeded under the entirely voluntary provisions of section 319, which provides both technical assistance and grants for state management programs (33 U.S.C. § 1329). Yet another provision holds more potential for clout—section 303(d).

States setting standards had been the approach of the Water Quality Act of 1965, which was applicable to interstate waters, but the arguments in favor of keeping it prevailed only for what was viewed as a backup provision—section 303(d), applicable only if a water body did not meet water quality standards with regulation of point sources. The TMDL program under section 303(d) embodies an approach "that both the states and pollution dischargers insisted on" (Houck 2002). Section 303(d) was a provision that appeared and was treated as an afterthought in passage of the 1972 CWA (Houck 2002), yet it contains more clout than originally thought. Under section 303(d), once standards are set, TMDLs are set on a water body by water body basis for each pollutant found in excess of the applicable standard (33 U.S.C. § 1313(d); Houck 2002). Loads are then allocated between point and non–point sources as appropriate. Despite the existence of the provision in the CWA since 1972, nothing happened for a decade. Because the provision was assumed not to trigger EPA action until a state submitted TMDLs, states could sit back and do nothing (Houck 2002). A series of citizen suits in the late 1980s challenged this assumption and prevailed on the issue of the EPA's need to act if the state failed to submit TMDLs (e.g., *Alaska Center of the Environment v. Reilly,* 1991; *Alaska Center for the Environment v. Reilly,* 1992); this was followed by a successful challenge to the assumption that a state could submit anything and get off the hook (*Idaho Sportsmen's Coalition v. Browner* 1996). Thus, the arduous process of characterizing the water quality of the nation's waters began, only to find that this process of characterizing the water quality of every water body within a state involves an enormous undertaking and is fraught with uncertainty, as discussed below (Shabman et al. 2007).

### *Aquatic Species*

Similar to water quality regulation, take of fish and wildlife has traditionally been a matter of state concern (Goble et al. 2006), whereas the preservation of the habitats various species rely upon has often occurred at the federal level (e.g., Migratory Bird Treaty Act of 1918; Land and Water Conservation Fund Act of 1965). With much of the funding for state fish and wildlife agencies coming from hunting and fishing licenses, state management focuses on game species and sport and commercial fisheries. With a few exceptions, federal land preservation rarely focused on biodiversity and generally preserved land already in federal ownership. The primary tool for restricting development of private land, land use zoning, is rarely used for purposes of biodiversity protection.

For most aquatic species and nongame species, the approach of states and federal land preservation once again fell short. Increasing human population, an even greater rate of increasing consumption, and the resulting loss of habitat caused the loss of "over five hundred species formerly found in the United States . . . [with] an additional 47 percent of the species unique to this country" at risk (Goble et al. 2006). The Biological Resources Division of the U.S. Geological Survey (USGS) identifies freshwater fish as the single most endangered vertebrate group in the United States (Doremus 2001).

Similar to the history of federal regulation of water quality, Congress stepped in to fill the gap with passage of the ESA in 1973. The ESA combines both time-honored approaches: restriction on take (sections 7 and 9) and preservation of habitat (section 4) (Goble et al. 2006). For listings of aquatic species and designation of their critical habitat, the USFWS lists resident fish species and NOAA Fisheries lists ocean and anadromous fish species.

The regulation of water quality and listing of aquatic species, both at the federal level, has led to some degree of coordination. Listed species must be included as one of the "uses" for which water quality standards are designated (CWA §§ 1251(a)(2), 1313(c)(2)(A)). However, as described above, achieving these standards remains elusive and, as described below, efforts are hampered by uncertainty.

### *Water Allocation*

Although the system of water allocation differs between the eastern and western United States, the fact that water allocation remains at the state level, water quality regulation at the federal level, and land use decisions at the local level is equally applicable. For purposes of this chapter, western water allocation can simply be considered the worst case scenario in terms of separation of water allocation law from other management and regulatory efforts aimed at the water resource.

Western water allocation arose both as a means to manage water in an arid region (e.g., Stegner 1953) and to provide certainty to investment for extraction and development of natural resources (e.g., *Irwin v. Phillips* 1855). In the late 1800s, John Wesley Powell recognized that rivers would control western development and recommended to Congress that the federal government draw jurisdictional boundaries along topographic divides (Stegner 1953; Reisner 1987). He also noted that the cost of water development would require formation of collectives or irrigation districts for the control of land and water (Stegner 1953). Congress did not take his recommendations in defining political boundaries. States ignored much of his advice concerning shared development and shared water use.

Under U.S. Supreme Court interpretation of federal law, water allocation and management is a matter of state law (*California Oregon Power*

*Co. v. Beaver Portland Cement Co.* 1935; *United States v. Rio Grande Irrigation Co.* 1899). Without the foresight that this land would come to be home to a population that must share its water resources for economic pursuits as diverse as irrigation and fishing or the havoc that dewatering of streams would play on habitat, state courts sought a means of allocation that protected investment (Wilkinson 1985; Dellapenna 2000). The result: the doctrine of prior appropriation, followed in some form by most western states (Hutchins 1971).

Under prior appropriation, the right of the earliest appropriator on a stream is satisfied first; junior appropriators take the remaining water. Shortages are not shared among appropriators. Due to the high variability in annual water supply, the fact that the largest use, irrigation, does not require a constant diversion of water, and the fact that state water allocation criteria generally do not require consideration of instream flow, state agencies generally approve water allocation permits to the point that all water is allocated in the wettest year. As drought becomes an increasingly frequent occurrence, this practice leads to dewatering of many streams at certain times of the year in areas of irrigated agriculture.

The allocation of appropriative water rights to individuals rather than geographically related communities, as recommended by Powell, also significantly reduced the possibility that allocation and management decisions would be made for the good of the community as a whole or for the long-term health of the riparian habitat (Tarlock 2000). Additionally, the practice inhibits any effort to adapt as new knowledge on the effects of diversion is gained. The relative control over water diversion of the individual, as opposed to the state, is implemented for the most part as though diversion decisions are unrelated to water quality. States that have requirements for consideration of impact on water quality (e.g., Montana Code § 85-2-311(f)), generally limit it to the water quality required by another appropriator. Some states include criteria that could be interpreted more broadly to address the affect on water quality in general, such as Idaho's public interest criteria (Idaho Code § 42-203A(5)(e)). In either case, it is an inquiry that takes place prior to issuing a permit to divert water and is a one-time consideration, after which the permit is viewed as a property right. In the two states used as examples, water allocation and water quality decisions are housed in separate agencies; thus, imposing an additional barrier to thorough review of allocation decisions in light of water quality impacts.

In addition, growing urban demand for water in the western United States and increasing call for dedication of water to instream flow for ecological purposes has run headlong into this rigid system of property rights. States authorize water transfers and increasingly have established water banks to facilitate the movement of water from one use to another (e.g., Dellapenna 2000). Because the voluntary nature of water transfers may prevent water availability for a new use, some states allow development of the new use but require mitigation for effects on existing rights. Mitigation measures can be voluntarily agreed to among affected parties; a party could consider water quality in the mitigation process. However, mitigation or augmentation may also be approved by the state over the protest of an existing right holder. In such a case, water quality is not a consideration (e.g., Idaho Administrative Code § 37.03.11.043; Colorado Revised Statutes § 37–92–305).

State courts considering whether the reliance of a pollutant discharge permit on mixing to achieve water quality standards can be grounds for denial of an allocation permit have characterized the issue as whether the discharge permit gives rise to a water right (e.g., *City of Thornton v. Bijou Irrigation Co.* 1996). This leads to the logical answer—no. In contrast, if water quality and water quantity are integrated, it might be more appropriate to ask whether the public interest requires that water for dilution remain in the stream, thus limiting stream allocation with water quality discharge in mind.

Application of the take provisions of the ESA to reduce water diversion has run headlong into the use of the system of property rights to allocate water in the western United States. The property right to water recognized in the western United States is a mere use right and thus differs from a property right to land, which is the right to possess and exclude others from the actual land. Idaho law describes the nature of the property right as it was understood at common law:

> All the waters of the state, when flowing in their natural channels, including the waters of all natural springs and lakes within the boundaries of the state are declared to be the property of the state . . . and the right to the use of any of the public waters which have heretofore

> been or may hereafter be allotted or beneficially applied, shall not be considered as being a property right in itself, but such right shall become the complement of, or one of the appurtenances of, the land or other thing to which, through necessity, said water is being applied. (Idaho Code § 42–101)

Yet despite this very clear difference between a use right and a right to the actual water, recent court decisions have interpreted *any* diminishment in use due to application of the ESA to be a Fifth Amendment Taking because the water itself is taken, rather than looking to the change in value of the "land or other thing to which . . . [it] is being applied" (see *Casitas Municipal Water District v. United States*, 2008). This view prevents any coordination between water allocation under existing rights and determination of flow needs under the ESA or water quality standards set for listed species without payment to the water right holder.

### *Land Use*

During colonial times, land use regulation occurred at multiple levels of government (e.g., Hart 1995–1996). However, since at least the recognition of the constitutionality of local land use zoning (*Village of Euclid v. Ambler Realty Co.* 1926), land use decisions have been considered a matter of state law and are generally delegated to local subdivisions of the state through state constitutions or statutes (e.g., Heim 2005; Baker and Rodriguez 2009). In addition, a majority of states recognize municipal home rule by either statutes or constitutions, which effectively allows a municipality to act for the health and welfare of its community even in the absence of state delegation, as long as the action does not conflict with state or federal law. Some states have interpreted home rule authority to allow land use zoning that is not consistent with state-mandated comprehensive plans (e.g., Heim 2005; Baker and Rodriguez 2009). Although a viable means to ensure tailoring to local needs, economics, and values, lack of integration with other levels of government decision making impedes efforts to address non–point source pollution.

As noted above, the primary sources of non–point source pollution are related to land use activities. EPA's *Watershed Approach Framework* recognizes the relation between non–point source pollution and land use and calls for a local, watershed-based approach to assure tailoring to both the characteristics of the local pollution problem and local social and economic needs (U.S. EPA 1996). However, the section on implementation includes nothing about the use of local land use planning and regulation as a tool, leaving implementation to voluntary measures (U.S. EPA 1996).

Surprisingly, interpretation by courts of Fifth Amendment takings by restriction on land use to prevent take of critical habitat is not as limited as the interpretation applied to water rights, with the analyses appropriately looking at the degree in reduction in value of the particular property in question (Meltz 2009). However, it is the absence in coordination between ESA designations of critical habitat and local land use planning that is a greater hindrance. Development in the floodplain has substantial impact on temperature, sediment load, and other factors that affect aquatic species (*NWF v. FEMA* 2004; National Marine Fisheries Service Northwest Region 2008; Williams 2011). Yet there appears to be no requirement for coordination between local floodplain ordinances and efforts to meet water quality standards.

The relation between land use zoning and water allocation is also problematic. Many states have comprehensive approaches for water allocation and management. The lower courts in at least one state have struck down local attempts to use land use zoning to protect water quality and supply as preempted by state law (e.g., *Ralph Naylor Farms, LLC v. Latah County* 2006; *Eagle Creek Partners, LLC v. Blaine County* 2008).

This description of the fragmentation of water management illustrates the difficulty of implementing a flexible approach. Authority divided among multiple federal, state, and local entities to use greater flexibility without robust coordination could create havoc. Yet our political system, the recognition of property rights to water, and investment dependent on the current regulatory approach prevent wholesale implementation of a new approach. In this chapter, we will look at changes to improve flexibility and integration within the framework of the existing system. The next major section turns to resilience theory and network analysis to inform these changes, but first we consider the added complexity of uncertainty.

### *Uncertainty*

The problems raised by fragmentation of water quality regulation, species management, water allocation, and land use planning are compounded by uncertainty in both the characterization of aquatic health and the ecological and social consequences of particular actions. The problem of uncertainty extends to each of the areas of regulation addressed, but will be discussed here in the context of water quality.

The process of characterizing the water quality of every water body within a state involves an enormous undertaking and is fraught with uncertainty (Shabman et al. 2007). Added to the uncertainty of characterizing the water quality of a water body is the uncertainty associated with characterizing the sources of non–point source pollutants. As noted by Houck (2002):

> The Achilles' heel of water quality standards-based regulation has always been the difficulty of ascribing and quantifying environmental effects for particular discharge sources. There is always another possible source, or another possible reason . . . [a]nd when we come to more complex biological impacts such as the fate and effects of nutrients, particularly those effects hundreds of miles downstream, we are beyond any pretense of precise mathematics for cause and effect decisions. The question is whether we are also, for these same reasons, beyond the reach of law. (58)

Houck concludes that with the CWA requirement for a margin of safety in any TMDL order, we should not be "beyond the reach of the law," because the margin can account for the degree of uncertainty. The question we seek to explore in this chapter is not the legal requirement but the practical question of how best to proceed in the face of uncertainty. Recognizing that it is not enough simply to have the authority to act, we seek positive results.

Two related theories inform the discussion of how to address fragmentation and uncertainty: network and resilience theory. The following sections describe how these theories inform solutions to the problem followed by specific application to water management and regulation.

## Network Theory: Integration across Multiple Entities and Jurisdictions

The study of resilience in social-ecological systems has led to the development of the concept of the adaptive cycle to describe the state and evolution of a self-organizing system and panarchy theory to describe the overlapping and multiscalar structure of adaptive cycles linked across scales (Holling 2001; Gunderson and Holling 2002). The adaptive cycle consists of: (1) a growth phase leading to; (2) increasing conservation, which at some point may become sufficiently rigid that a small perturbation leads to; (3) collapse or "release," a chaotic phase resulting in; (4) reorganization and innovation in a context in which the cost of failure is low. Some innovation will succeed, leading back to (1) growth (Holling 2001). The idea that management for resilience could mean allowing collapse of a system does not instill faith in the approach by those who recognize stability as one of the key factors in an economic system. But panarchy recognizes that adaptive cycles occur at many scales and feedback occurs across scales. Higher, slower cycles may provide stability for smaller scales to engage in innovation and adaptation while minimizing the risk of collapse (Holling 2001; Gunderson and Holling 2002). Innovation and adaptation at smaller scales can feed back to the maintenance of stability at larger scales. Viewed from the perspective of the U.S. system, stable federal and state law can provide room for local innovation. Experimentation at the scale of a watershed does not involve the degree of risk or political aversion present if experimentation was undertaken at the scale of a river basin. For example, local changes in tillage practice followed by monitoring of both pollutant results and economic viability can, if successful, provide information to facilitate adoption elsewhere. Moving from structural to nonstructural flood control (i.e. reconnecting the river to the floodplain) can have both water quality and habitat benefits, yet could be highly risky if undertaken on a large scale without substantial data on the resulting flood control benefits. Nested scales of management authority thus allow for local adaptation while providing large scale stability.

Furthermore, the concept of panarchy cautions that there will be linkage among the results of actions at different scales whether or not strict legal lines are drawn between authorities and jurisdictions. Thus, a local land use decision to build in a river floodplain may lead to increased need for investment in storage across an international boundary. If the ecological system will not allow us to ignore the linkage among scales, then it is advisable to approach the linkages among agencies and jurisdictions deliberately and build them to facilitate adaptation.

Analogous to the concept of nested scales in ecological systems, integration of water resource management requires moving from a focus on efficiency and clean lines between jurisdictional authorities, to a focus on diversity, redundancy, and multiple levels of management, including a role for local knowledge and local action. The adaptive state of systems at scales above and below the scale of a system of interest may enhance or detract from the resilience of the system of interest (Walker et al. 2004; Walker and Salt 2006).

> In applying this concept to a river basin, while coordination may be needed across the entire basin, issues that arise do not always require action at such a large scale. Designation of an entity with authority at the scale of the particular ecological system can serve as the mechanism for coordination at that scale, but it is not a replacement for diversity in governance at multiple scales (Huitema et al. 2009). Scholars recognize that a scale appropriate for one problem may not be relevant for another (Ruhl and Salzman 2010). In addition, complex systems do not always have clearly identifiable scales for governance (Ruhl and Salzman 2010). In contrast to the difficult search for the appropriate match in scale, resilience thinking rejects the call for a single, efficient level of management and instead calls for multiple overlapping authorities. At the same time, coordination across scales is necessary. As noted by Lee (1993) in reference to the Columbia River Basin: Each of the major uses of the basin's resources is managed by a different constellation of human institutions, each set of

managers guards its rights and prerogatives, and none is sufficiently powerful to bring the others to heel. Multiple management of multiple uses produces a tragedy of the commons. (28)

Instead, a networked approach informed by resilience allows response at different scales across different entities depending on the source and impacts of the problem. Importantly, it leads to consideration of minor changes to the existing diverse system, rather than an entirely new approach.

Complicating the scale issue even more are situations in which the source of the problem and the negative impact occur at different scales, thus removing any incentive for action at the scale of the problem source (Long 2010). This situation highlights the need for response capacity at multiple, linked scales. In the United States, we have seen the results of this mismatch in the form of backlash to some of the environmental laws passed in the 1970s. For example, the failure of states to take action led to federal regulation to achieve clean water, (Federal Water Pollution Control Act (CWA §§ 1251–1387), clean air (Clean Air Act of 1963 [CAA]), and species protection (ESA). Yet although the scale of the problem is federal, the source may be local land use planning in the case of non–point source pollution (Adler et al. 1993), or local development of wetlands important to filtration of polluted water and flood mitigation on a larger scale (*Rapanos v. United States* 2006; Cosens 2008), or local development that will endanger an obscure species important to biodiversity in general (*National Association of Home Builders v. Babbitt* 1997). Problems such as non–point source pollution that result from the cumulative effect of many small activities, some located far from the point at which the cumulative effect begins to violate water quality standards, require both local capacity to act and enhanced monitoring to tie source to impact.

All too frequently, those of us in the academy propose governance changes based on the ideal end point we seek to achieve without regard to where we are now. The resistance to change and slow incremental process with which it occurs in a democratic system, coupled with the fact that the slow pace ensures deliberation and consideration of equity and justice, is possibly what led to Winston Churchill's famous quote that "democracy is the worst form of government, except for all the rest that have been tried." Rather than propose an entirely new form of nested levels of governance, this article seeks to make use of the existing diversity and multiple scales of water governance while developing means to integrate across and within various scales. In doing so, we propose process changes coupled with more minor changes to substantive law.

Substantive laws govern what is managed, who is regulated, and the goal of that management or regulation. Administrative law governs how these functions are implemented (Stewart 2003). The term "administrative law" is used in this article to refer to any law governing the process of agency or governing body decision making. For example, administrative law is used here in reference to U.S. domestic law to include not only the Administrative Procedure Act, but procedural requirements in substantive law, such as the requirement for development of an Environmental Impact Statement in the National Environmental Policy Act of 1970 (NEPA §§ 4321 et seq.) or the requirement of consultation in the ESA (§§ 1531 et seq.).

Three changes that will not disrupt the existing legal system that establishes and allocates authority among institutions are needed to facilitate adaptable response at multiple scales: (1) enhanced monitoring; (2) increased local capacity; and (3) vertical and horizontal networks across jurisdictions and scales to allow coordinated response among overlapping authorities without high transaction costs. Because enhanced monitoring is also a major factor in addressing uncertainty, it will be discussed below. The following paragraphs explore the use of networks to allow flexibility in the scale of response, multiscalar response, and enhanced flow of resources to the local level.

The need for local capacity with robust networks to multiple levels can be seen more clearly if the example used is a sudden, high-risk situation. It is on the local level, not the level of an entity like the U.S. Department of Homeland Security, that a major portion of the resources are needed for response to events like Hurricane Katrina or 9/11. Yet without the link to assistance both from other communities and from state, national, and international levels, a disaster of such magnitude will be beyond the capacity of any local government.

In addition, studies of postdisaster short and long-term relief indicate that networks for coordination and clear definition of roles must be addressed prior to a disaster if local organizations are to be effectively used in providing relief (Stys 2011). Taking the example of emergency response further, a proven and highly robust system for multiagency/jurisdiction networking is presented to the general public on national news on an annual, if not more frequent, basis. The example is the incident command system for multijurisdictional response to a large-scale, often mass casualty, emergency. The incident command system is a highly robust process for multientity response to an emergency, in which the scale and timing was highly uncertain prior to its occurrence (e.g., see U.S. Department of Homeland Security; U.S. Forest Service 2000). Rather than create a new agency at the scale of every conceivable emergency, the incident command system provides a means for rapid crisis response across multiple agencies at the same level and through multiple levels of agencies.

The incident command system works on the theory that rapid coordination requires that no more than seven people report to any one person. The incident commander is at the top level of response and is selected based on the nature and location of the emergency. No more than seven people or entities report to the incident commander, no more than seven to each of those seven, and so on, until the on-the-ground response to, for example, a wildfire, a flood, or an earthquake may involve hundreds or even thousands of people. In one author's (Cosens) experience as a search and rescue volunteer, the initial hours or even days of response to a large-scale emergency are often chaotic, as response personnel move into position, assess the scope of the problem, and identify the chain of command. However, within a remarkably short period of time, given the level of uncertainty involved, a relatively smooth operation emerges in which information and coordination of decisions in response to changes in the problem flow rapidly within and between levels. Although referred to by some as a "command and control" approach (Stys 2011), the incident command system is highly adaptive to meet the type, location, and scale of a disaster. The result is a clear line of hierarchical authority leading to the "command and control" description. This hierarchical approach fits the emergency nature of the situation, but it is the ease of formation of networks for flow of information and resources made possible by the conscious focus on flexible cross-entity and cross-scale coordination that serves as a lesson for water management.

Other than a flood, earthquake, landslide, wildfire, or volcanic eruption, the types of change and uncertainty in a river basin do not occur on the timescale of an emergency. Fortunately, examples of establishment of networks for management of slower processes also exist. Efforts to coordinate across different levels and authorities are reflected in the move toward integrated water resource management (Global Water Partnership 2000, 2002; European Union Council Directive 2000). A formal structure for integrated water resource management does not currently exist in the United States. Ruhl and Salzman describe a process in the Mississippi River Basin and Gulf of Mexico in which "weak ties" are formed among individuals working at various levels of government and nongovernmental organizations as a solution to the scale of response needed for a water quality issue in which the pollution sources are diffuse and basin-wide (Ruhl and Salzman 2010).

Effective networks rely on two primary components: (1) the existence of certain social skills among people within the network; and (2) facilitation of, and removal of barriers to, cross-agency and cross-scale interaction. In their work on adaptive governance, Folke et al. (2005) note that success in managing ecological systems for resilience often depends on the involvement of key personality types such as *mavens* ("altruistic individuals, with social skills, who serve as information brokers, sharing and trading what they know"), *connectors* ("individuals who know lots of people not only by numbers but the kind of people they know and in particular the diversity of acquaintances"), and *entrepreneurial leaders* (creative decision makers willing to risk being the first to try something). This is consistent with author Cosens's observation of multijurisdictional water negotiations, in which success is often determined by key personalities involved (Cosens 1998).

Administrative law and institutional structure cannot mandate the individuals involved, but could be designed to maximize diversity, thus increasing the likelihood that these personality types are represented. In addition, organizational structures may be set up to provide positions and incentives for people who play the roles of maven, connector, and entrepreneurial leader within governmental entities. Clearly, there is also a role for our institutions of higher education in educating students with the skills necessary to bring people together and to communicate across boundaries.

That certain types of people are needed to form transjurisdictional networks is not a satisfying answer in itself. Any solution formalized in an administrative process must strike a balance between facilitation of network formation and avoidance of a rigid structure that cannot adapt to changing scale and type of problem. We can take lessons from experience with systems like the incident command system for emergency response to improve the capacity of networks to form and act adaptively and to facilitate that improvement by incorporating the following elements into administrative law and resource allocation.

First, coordination and communication among different entities works better if its requirement is explicit (Bingham 2009). Thus, it must be a written legal requirement that is assigned to a specific position created within each entity without imposing rigid requirements on the position. Establishment of a network framework upfront can also avoid transaction costs encountered with the ad hoc development of a network after a problem is identified (Huitema et al. 2009).

Second, practice improves response. Under the incident command system, the operation appears to be much smoother in response to incidents, such as wildfire, that occur somewhat predictably on an annual timescale, than it is in response to rare events, such as a hurricane or earthquake. In the context of river management, this could translate into frequent information sharing among entities as a building block in the relationships necessary for multijurisdictional decision making.

Third, substantial resources must be devoted to the local level, since local actors are likely to be a component of response to any scale of problem. The current structure of resource availability, both with respect to funding and people, for entities that manage natural resources may need to be inverted, with greater resources made available at the local rather than at the national level. Once again, the flow of resources in response to an emergency serves as an illustration of the need. Few question the need for the largest number of people responding to a wildfire to be those on the ground. This same thinking should be imported to resource management in general.

Fourth, harmonization of methods and regulations in the area of overlap will result in more effective networks (Zaring 2009). While noting the important role of social networks in environmental compliance, Bodin and Crona (2009) note that not all networks are equal in their effectiveness. By studying the topology of networks, their work has begun to identify key characteristics. Although intuitively we might consider a cohesive group to be more likely to achieve self-regulation, complex problems require a balance between cohesion and diversity of network membership to foster creative solutions, use of local knowledge, and adaptive capacity. In addition, ties to external sources for knowledge and resources and to assure legitimacy are necessary. Thus, the current diversity in entities with water management authority on any particular river basin is actually an asset. Rather than remove the diversity already present, simply altering process through administrative law to harmonize methods and regulations will enhance network response.

Finally, attention must be paid to the difference between formal and informal networks. For effective management across multiple entities, networks with low transaction costs are needed (Huitema et al. 2009). One way to achieve this goal is to build network formation into the administrative process rather than leaving it to be formed on an ad hoc basis when a problem arises. Models for coordination across entities and scales imbedded in administrative law could greatly reduce these transaction costs. At the same time, caution is warranted in any attempt to formalize the interaction across entities and scales before understanding existing informal networks for communication and action. Research by Bodin and Crona (2009) suggests that informal networks appear to be more successful than an imposed structure. Informal network formation can be facilitated through capacity building, identification of influential actors through use of social network analysis prior to establishing lines of coordination, encouraging broad participation, and providing a forum for communication (Bodin and Crona 2009). It may be more important to remove legal barriers to network formation than to mandate a particular structure.

## Resilience Theory: Adaptation in the Face of Uncertainty

As noted earlier, resilience as applied to ecological systems addresses a system's ability to respond to change while continuing to provide, or returning to a state in which it will provide, a full range of ecosystem services or functions (Walker and Salt 2006). Resilience theory provides a framework for moving from management through optimization to a more adaptive form of management based on recognition of the complexity of an ecological system and the associated uncertainty.

To apply resilience theory to governance, it is also necessary to define the use of terms related to resource management. Much of the literature calls on adaptive management to achieve resilience (Lee 1999; Folke et al. 2005; Huitema et al. 2009). The term adaptive management has been used to describe a process of learning through monitoring ecosystem response to a particular action followed by incremental change in the action based on what is learned (Lee 1999; Folke et al. 2005; Huitema et al. 2009), and generally applies to management action by a single entity. Under adaptive management, the natural adaptive abilities of an ecosystem are emphasized and promoted over the active management, control, optimization, and resource exploitation of the system (Folke et al. 2005). Continual and artificial maintenance of an ecosystem within human-defined parameters is less desirable. Instead, natural disruptions to the ecosystem are allowed to take place. The ecosystem becomes more resilient through natural disruptions (Ruhl 2011).

Adaptive implementation is a similar concept, but applies to the implementation of a regulatory statute such as the CWA rather than general management authority over a particular ecosystem (Shabman et al. 2007). Fostering the natural adaptive abilities of an ecosystem and providing a system of management or regulatory implementation that adapts to the results lessens the need to predict the full implications of a management or regulatory activity prior to taking action. Uncertainty is accounted for in the incremental adjustment to change. Adaptive implementation will require three key components: (1) authority to make incremental adjustments to decisions based on monitoring, (2) monitoring, and (3) implementation at the smallest relevant scale with accompanying capacity building. Of course, overarching all of this is the need for "buy-in" from stakeholders, especially the regulated community. One component of achieving this may be to allow implementation measures to go either way as the result of monitoring—that is, to become more or less stringent depending on results.

The current gap between the recognition by resilience scientists that adaptive management is an appropriate management tool in the face of uncertainty and the achievement of practical adaptive management implementation arises from a failure to recognize and integrate the social component. It is not enough that monitoring and incremental adjustment will provide the best results in the face of uncertainty. The decisions on whether to use adaptive management, what to monitor, and how to make incremental adjustment must be done in a manner that will foster political acceptance. Thus, to achieve implementation of adaptive management requires attention to the process of governance used to carry it out. In this context, "governance [or adaptive governance in the context of implementation of adaptive management] is the process of resolving trade-offs and of providing a vision and direction for sustainability, management is the operationalization of this vision . . ." (Boyle and Pond 2001, 122). By its definition, governance will involve trade-offs and thus may not lead to the perfect scientific result for the ecological system. But the flaw in implementing adaptive management without the integration of the social system is that it makes the same mistake as traditional management, by optimizing for a subset of the system—the ecosystem. Coupled with adaptive management, an appropriate form of governance addresses the entire social-ecological system. This raises the question: What are the components of governance necessary to gain political acceptance for implementation of adaptive management? Cosens (2010, 2013) has developed a framework for legitimacy as a term to describe the necessary components in adaptive governance elsewhere and that work will be summarized and applied in the following paragraphs.

It is a basic tenet of political theory that people seek legitimacy in the actions of those who govern them (Bodansky 1999). Legitimacy can be thought of in simple terms as a measure of how persuasive the basis for a particular action is (Bodansky 1999). Legitimacy includes the justification for assertion of authority and has both popular and normative aspects (Bodansky 1999). Thus, an assertion of authority must be both justified (normative) and perceived to be justified (popular) to be legitimate (Bodansky 1999).

Democracy therefore emerges as a system with a high level of legitimacy. Through the process of electing those who govern, people consent to their leadership. In casting a vote, a person is reflecting his or her perception of who can best govern. But when democratic nations move from implementation of the law by elected officials to delegation of authority to administrative agencies or appointed governing bodies, they dilute the direct connection between the elected official and the voters affected by regulation. As the administrative state has grown, administrative law governing the process by which agencies or appointed bodies take action has developed to fill this gap in direct accountability. The direct accountability gap increases with the scale of governance (Esty 2006). Thus, local agencies may have a higher perception of legitimacy in a river basin than federal or international entities. As the scale is reduced to the local level, fewer formal protections are needed to assure accountability to the regulated public. The introduction of flexibility to water resource management to allow adaptive management can challenge traditional sources of legitimacy, thus presenting a barrier to adoption of new approaches. To achieve acceptance of governance that implements adaptive management, administrative law becomes a tool for enhancing legitimacy.

Daniel Esty (2006) outlines five sources of legitimacy applicable to administrative entities in addition to the democratic process, three of which will be addressed here: (1) results-based: legitimacy derived from the fact that decisions are based on objective expertise and the results can be determined to be good; (2) order-based: legitimacy based on the fact that rules are clear, stable, and publicly available; and (3) deliberative: legitimacy based on the inclusion of a public dialogue in the process of decision making. The following discussion will use these sources of legitimacy to discuss the components necessary for adaptive implementation. A more thorough discussion of legitimacy and its relation to management for resilience is the topic of articles by Cosens (2010, 2013) and will be relied upon here.

(1) *Results-based legitimacy* reliant on an agency's scientific expertise began in the United States with Gifford Pinchot's call for scientific, federal management of the forests and establishment of the National Forest Organic Act in 1897 (§§ 473–478, 479–482, 551, as amended in 1905, 1911, 1925, 1962, 1964, 1968 and 1976). The infusion of science into decision making has at its very core the belief that the process will be more objective and the results better. However, reliance on scientific expertise is increasingly questioned as a source of legitimacy, as agency

science shows vulnerability to politicization (Wagner 1995; Doremus and Tarlock 2005; Ruhl and Salzman 2006; O'Reilly 2007; Cosens 2008). Recent environmental and natural resources laws impose requirements for use of "best available science" (e.g., Safe Drinking Water Act of 1974), as trust for objective science-based decision making without legally imposed criteria has eroded (Wagner 1995; Doremus and Tarlock 2005; Ruhl and Salzman 2006; O'Reilly 2007; Cosens 2008). However, it is not clear that such requirements are sufficient to address the growing politicization of agency science in the United States.

Adaptive implementation, if properly formulated, can actually provide a new approach to results-based legitimacy and stem the tide of politicized science (Camacho 2009). Adaptive implementation requires that the results of an agency action be monitored and that the action be adjusted based on the monitoring. Although substantial resources may be allocated to studying a problem in the process of developing a solution, data is rarely gathered to verify the results of a particular action, such as implementation of certain tillage practices to reduce sedimentation, once it is taken (Shabman et al. 2007; Camacho 2009). Furthermore, even if data were collected following agency action, agencies rarely have the authority to modify the action without going back through the rulemaking process or, in some cases, without new legislation.

Reluctance to provide authority for flexible implementation lies, in part, in failure to characterize monitoring as an essential component of program implementation and failure to tie adjustment based on monitoring to agency accountability. By imposing a requirement that progress toward a particular goal, such as meeting water quality standards, must be accounted for and adjustment made in the face of new data, the use of science to achieve the goals of an interest group rather than the goal of a statute can be reduced. In this way, legitimacy is served by adaptive implementation. By placing funding for monitoring at the center of measures to assure agency accountability rather than as an add-on study, legitimacy is served by adaptive implementation.

(2) *Order-based legitimacy* captures the concepts of stability and finality, or at least predictive capability, regarding application of the law. It can be found in the move from administrative action on a case-by-case basis to rule making. By promulgating rules to govern implementation of laws passed by Congress, agencies are more likely to apply

the law equally and uniformly to the regulated community. In addition, procedures governing rule making require notice and allow comment on proposed rules prior to finalization (Administrative Procedure Act of 1946), thus assuring public input and public notice on how a law will be implemented. In recent years, the move to negotiated rule making combines aspects of order-based legitimacy and deliberative legitimacy; it will be discussed below as a recommended aspect of adaptive implementation. Increasing adaptive capacity while retaining the benefits of the rule-making process requires inclusion of monitoring and adjustment within the rule itself. By maintaining a process of notice and comment, or negotiated rule making in the process of adjustment, legitimacy may still be served. However, a larger problem is the possibility of instability resulting from more flexible management of resources relied on for economic pursuits, such as hydropower, commercial fishing, and irrigation.

The expectations that rules will be stable and that finality can be achieved may present the most significant barriers to authorization of the flexibility needed to implement adaptive implementation. The desire for finality lies at the core of the desire to use implementation of "best management practices" as a proxy for achieving water quality standards. The communication gap between those advocating adaptive implementation/management, and those seeking finality is fundamental to many legal battles concerning natural resources and the environment and can be characterized as a basic conflict between science and law. In simple terms, science is a search for the truth (a process), whereas litigation is a search for finality (resolution) (Cosens 2008). Scientific inquiry has no statute of limitations, no concept of res judicata. Scientific methodology is a process of disproving what was formerly thought to be true, of re-investigating questions thought solved, of re-interpreting information in light of new discoveries. In contrast, civil litigation is designed to finalize a dispute, to provide a forum where, no matter how flawed the inquiry, a peaceful final resolution of a dispute can be achieved. Similarly, the current process of rule making is designed to visit the issue once and then proceed under a final rule rather than to re-examine and adjust to the results of the rule's implementation.

The finality of civil litigation serves those with economic interests in the resource by providing stability for investment, whereas science serves those concerned with sustaining the resource itself by continuing the search for the true impacts of human action on the ecological system. But a re-examination of what is actually happening in natural resources and environmental law suggests that, despite the goals of civil litigation, finality is not being achieved. The fact that one side (the environmental side) of these issues seeks equal treatment for goals not served by the civil litigation (a true understanding of human impacts on the environment) destabilizes the system. Once a court provides a final answer, the issues will be revisited with another legal theory. Once the judicial system is exhausted, the issues will be revisited in the legislature. Once the political system is exhausted, the issues will be addressed through civil disobedience. Thus, while the process of civil litigation works well to help people resolve issues of liability for personal injury or for interpretation of a contract, it may not be the best approach when the viability of an ecosystem and the economies dependent on it are at issue. It is this reality, that the current system is not providing what people seek, that encourages a more incremental and collaborative approach.

Legitimacy in the incremental approach of adaptive implementation will require affording equitable treatment to both the economic need for finality and the progress toward the true system understanding needed to address environmental concerns. Some current examples suggest that management relying on monitoring for adjustment uses a biological time frame when the management action is developed by a science-based agency (e.g., USGS Patuxent Wildlife Research Center, basing monitoring on biological goals) and social time frames when the action is negotiated (e.g., Snake River Water Rights Settlement 2004, placing a thirty-year time frame on a biological opinion to provide stability for water users in the region). Yet to foster ecological resilience while maintaining legitimacy, both ecological and social time frames must be considered when setting the pace of incremental change. In addition, short-term human interests that tend to coincide with the elections cycle must not control the pace, yet must be factored in when seeking new approaches to management.

(3) *Deliberative legitimacy* is reflected in the growing expectation and provision for public comment in numerous aspects of agency decision making, from the requirement of notice and comment in rule making to the increasing use of public meetings to gain support for a decision.

In the United States, the passage of the National Environmental Policy Act (NEPA) of 1969, can be considered the major turning point in public involvement in agency decision making (Hirt and Sowards 2011). Unlike the requirements of the Administrative Procedure Act that meetings and records be open to the public (which is certainly an important aspect of legitimacy), NEPA imposes the affirmative duty on agencies to develop, analyze, and provide to the public for comment, information on the environmental impact of major federal actions. (NEPA § 4332). Although NEPA does not impose any substantive requirement to choose the most environmentally sustainable alternative (*Vermont Nuclear Power Corp. v. NRDC* 1978), it arms the public with the information necessary to participate in shaping the decision through the political process. In that context, it is one factor in local capacity building.

Adaptive implementation provides an excellent opportunity to employ some of the newer methods of public involvement (Bingham 2009). It lends itself to use of a procedure similar to negotiated rule making, in which the agency collaborates with the regulated community and interested parties (e.g., a watershed committee) to develop a rule for reaching decisions on incremental changes in management. Small-scale (spatial and temporal) impacts may be reflected best in local knowledge, thus a more collaborative process may improve the knowledge base for the decision. A more collaborative approach to management could also make use of the interagency networks discussed above in the context of response scale to allow coordination of adaptation and avoid unintended consequences of agency action. In addition, the network approach may be necessary to allow for manageable public input on jurisdictionally complex water basins.

While adaptive implementation in the face of uncertainty presents a theoretical approach to managing the uncertainty associated with understanding the water quality and sources of pollution to our water resources, without careful attention to the legitimacy of the decision making that imposes it (i.e., adaptive governance), adaptive implementation will remain a theoretical approach. By assuring that decisions involving the goals of water management and choices in making incremental change are based on and incorporate (1) adequate monitoring and scientific analysis (results-based legitimacy); (2) notice and opportunity for input to decision making and management changes based on a time frame that takes both the biological and social needs into account (order-based legitimacy); and (3) capacity building and development of collaborative processes at the scale of the water system to be managed (deliberative legitimacy), adaptive implementation can move from theory to impact.

## Conclusion

Despite almost forty years of management under the federal CWA, the goal of full-use attainment in the nation's aquatic systems remains elusive. Fragmentation of authority among federal, tribal, state, and local jurisdictions, and uncertainty in both the characterization of water bodies and responses to management actions to improve water quality are primary barriers to achieving and maintaining the full array of ecosystem services possible from the nation's water bodies. Efforts to integrate water management and regulation must proceed against the very real backdrop of the existing fragmented system. Not only is it a purely academic exercise to contemplate design of governance and implementation of regulation starting from a clean slate, in seeking resilience in the social-ecological systems involved, modularity and functional redundancy in authority is beneficial. Rather than propose a new system, this chapter recommends integration through network formation. This will require (1) evaluation of existing informal networks among authorities, (2) removal of barriers to the smooth functioning of networks, (3) provision of authority where existing networks are absent, and (4) assistance where necessary to build local capacity.

Progress to reduce uncertainty and improve return on investment in efforts to restore healthy aquatic systems requires a reframing of the dialogue surrounding regulatory implementation and agency accountability. Resources spent on monitoring results must be viewed as a long-term investment by attaching a dollar value to information and providing the authority for adjustment of regulation as new information becomes available that changes our understanding of system behavior and as societal needs evolve. The short-term expenditure on collection of data will lead to efficiency in identification of appropriate measures and will reduce the current politicization of agency science by assuring that constant progress toward statutory goals is achieved. It is a necessary investment not only in our own future but as a legacy for our children.

## References

Adler, R. W., J. C. Landman, and D. M. Cameron. *The Clean Water Act: 20 Years Later*. Washington D.C.: Island Press, 1993.

Alaska Center of the Environment v. Reilly, 762 F.Supp. 1422 (W.D. 1991).

Alaska Center for the Environment v. Reilly, 796 F.Supp. 1374 (W.D. 1992).

Baker, L. A., and D. B. Rodriguez. "Constitutional Home Rule and Judicial Scrutiny." *Denver University Law Review* 86 (2009): 1337–1424.

Bingham, L. B. "Collaborative Governance: Emerging Practices and the Incomplete Legal Framework for Public and Stakeholder voice." *Journal of Dispute Resolution* 2 (2009): 269–325.

Bodansky, D. "The Legitimacy of International Governance: A Coming Challenge for International Environmental Law?" *American Journal of International Law* 93 (1999): 596–624.

Bodin, O., and B. I. Crona. "The Role of Social Networks in Natural Resource Governance: What Relational Patterns Make a Difference?" *Global Environmental Change* 19 (2009): 366–374.

Boyle, M., J. Kay, and B. Pond. "Monitoring in Support of Policy: An Adaptive Ecosystem Approach." In *Encyclopedia of Global Environmental Change*, ed. T. Munn, 116–137. New York: Wiley, 2001.

California Oregon Power Co. v. Beaver Portland Cement Co., 295 U.S. 142 (1935).

Camacho, A. E. "Adapting Governance to Climate Change: Managing Uncertainty Through a Learning Infrastructure." *Emory Law Journal* 59 (2009): 1–77.

Casitas Municipal Water District v. United States, 543 F.3d 1276 (D.C. Cir. 2008).

City of Thornton v. Bijou Irrigation Co., 926 P.2d 1 (Colorado 1996).

Clean Air Act of 1970 (amended 1990), 42 U.S.C. § 7401 et seq.

Colorado Revised Statutes, § 37.

Cosens, B. "The 1997 Water Rights Settlement Between the State of Montana and the Chippewa Cree Tribe of the Rocky Boy's Reservation—The Role of Community and of the Trustee." *Journal of Environmental Law and Policy* 16 (1998): 255–295.

Cosens, B. "Resolving Conflict in Non-Ideal, Complex Systems: Solutions for the Law-Science Breakdown in Environmental and Natural Resource Law." *Natural Resources Journal* 48 (2008): 257–301.

Cosens, B. "Transboundary River Governance in the Face of Uncertainty: Resilience Theory and the Columbia River Treaty." *University of Utah Journal of Land, Resources, and Environmental Law* 30 (2010): 229–265.

Cosens, B. "Legitimacy, Adaptation and Resilience in Ecosystem Management." *Ecology and Society* 18, no. 1 (2013): 3. http://dx.doi.org/10.5751/ES-05093-180103

Dellapenna, J. W. "The Importance of Getting Names Right: The Myth of Markets for Water." *William and Mary Environmental Law and Policy Review* 25 (2000): 317–377.

Doremus, H. "Water, Population Growth, and Endangered Species in the West." *University of Colorado Law Review* 72 (2001): 361–414.

Doremus, H., and A. D. Tarlock. "Science, Judgment, and Controversy in Natural Resource Regulation." *Public Land & Resources Law Review* 26 (2005): 1–37.

Eagle Creek Partners, LLC v. Blaine County, No. CR 2007–670 (Fifth Judicial District of Idaho May 6, 2008).

Endangered Species Act of 1973, 16 U.S.C. §§ 1531–1544.

Esty, D. C. "Good Governance at the Supranational Scale: Globalizing Administrative Law." *Yale Law Journal* 115 (2006): 1490–1562.

European Union Council Directive 2000/60/EC, 2000 O.J. (L. 327) 1–73 (EC). http://ec.europa.eu/environment/water/water-framework/index_en.html.

Federal Water Pollution Control Act [commonly known as the Clean Water Act of 1972] 33 U.S.C. §§ 1251–1387.

Folke, C., T. Hahn, P. Olsson, and J. Norberg. "Adaptive Governance of Social-Ecological Systems." *Annual Review of Environmental Resources* 30 (2005): 441–473.

Global Water Partnership. *Integrated Water Resources Management.* TAC Background Papers, no. 4, 2000. http://www.gwp.org/Global/GWP-CACENA_Files/en/pdf/tec04.pdf.

Global Water Partnership. "ToolBox: Integrated Water Resources Management" (2002). http://www.gwptoolbox.org/index.php?option=com_content&view=article&id=8&Itemid.

Goble, D. D., J. M. Scott, and F. W. Davis. *The Endangered Species Act at Thirty,* vol. 1: *Renewing the Conservation Promise.* Washington, D.C.: Island Press, 2006.

Gunderson, L. H., and C. S. Holling. *Panarchy: Understanding Transformations in Human and Natural Systems.* Washington, D.C.: Island Press, 2002.

Hart, J. F. "Colonial Land Use Law and Its Significance for Modern Takings Doctrine." *Harvard Law Review* 109 (1995–1996): 1252–1300.

Heim, M. R. "Home Rule: A Primer." *Journal of the Kansas Bar Association* 74 (2005): 26–38.

Hirt, P. W., and A. M. Sowards. "The Past and Future of the Columbia River." In *The Columbia River Treaty Revisited: Transboundary River Governance in the Face of Uncertainty,* ed. B. Cosens, 115–136. Corvallis: Oregon State University Press, 2011.

Holling, C. S. "Resilience and Stability of Ecological Systems." *Annual Review of Ecology and Systematics* 4 (1973): 1–24.

Holling, C. S. "Understanding the Complexity of Economic, Ecological and Social Systems." *Ecosystems* 4 (2001): 390–405.

Houck, O. A. *The Clean Water Act TMDL Program: Law, Policy, and Implementation,* 2d ed. Washington, D.C.: Environmental Law Institute, 2002.

Huitema, D., E. Mostert, W. Egas, S. Moellenkamp, C. Pahl-Wostl, and R. Yalcin. Adaptive water governance: Assessing the institutional prescriptions of adaptive (co-) management from a governance perspective and defining a research agenda, *Ecology and Society* 14, no. 1 (2009): 26. http://www.ecologyandsociety.org/vol14/iss1/art26.

Hutchins, W. A. *Water Rights Laws in the Nineteen Western States*, vols. I and II. Misc. Pub. No. 1206, Natural Resource Economics Division, Economic Research Service, U.S. Department of Agriculture, 1971.

Idaho Administrative Code § 37.

Idaho Code § 42.

Idaho Sportsmen's Coalition v. Browner, 951 F.Supp. 962 (W.D. 1996).

Irwin v. Phillips, 5 Cal. 140. (1855).

Land and Water Conservation Fund Act of 1965, 16 U.S.C. §§ 4601-4-4601-11.

Lee, K. N. *Compass and Gyroscope: Integrating Science and Politics for the Environment.* Washington, D.C.: Island Press, 1993.

Lee, K. N. "Appraising Adaptive Management." *Conservation Ecology* 3, no. 2 (1999): 3. http://www.consecol.org/vol3/iss2/art3.

Long, J. "From Warranted to Valuable Belief: Local Government, Climate Change, and Giving Up the Pickup to save Bangladesh." *Natural Resources Journal* 49 (2010): 743–800.

Meltz, R. *The Endangered Species Act (ESA) and Claims of Property Rights "Takings."* Congressional Research Service Report 7–5700. Washington, D.C.: CRS, 2009. http://ncseonline.org/NLE/CRSreports/09Mar/RL31796.pdf.

Migratory Bird Treaty Act of 1918, 16 U.S.C. §§ 703–712.

Missouri v. Illinois, 200 U.S. 496 (1906).

Montana Code § 85.

National Association of Home Builders v. Babbitt, 130 F.3d 1041, 1059 (D.C. Cir. 1997), *cert. denied* 524 U.S. 927 (1998).

National Environmental Policy Act of 1969, 42 U.S.C. §§ 4321–4375.

National Forest Organic Act of 1897, 16 U.S.C. §§ 473–478, 479–482, 551, June 4, 1897, as amended 1905, 1911, 1925, 1962, 1964, 1968, and 1976.

National Marine Fisheries Service Northwest Region. "Endangered Species Act—Section 7 Consultation Final Biological Opinion and Magnuson-Stevens Fisher Conservation and Management Act Essential Fish Habitat Consultation: Implementation of the National Flood Insurance Program in the State of Washington Phase One Document—Puget Sound Region" (2008). http://online.nwf.org/site/DocServer/NMFS_Puget_Sound_nfip-final-bo.pdf?docID=10561.

National Research Council. *Assessing the TMDL Approach to Water Quality Management.* Washington, D.C.: National Academies Press, 2001.

NWF v. FEMA, 345 F.Supp. 2d 1151 (W.D. Wash. 2004).

O'Reilly, K. "Science, Policy, and Politics: The Impact of the Information Quality Act on Risk-based Regulatory Activity at the EPA." *Buffalo Environmental Law Journal* 14 (2007): 249–287.

Ralph Naylor Farms, LLC v. Latah County, No. CV 2005–670 (Second Judicial District of Idaho May 9, 2006).

Rapanos v. United States, 547 U.S. 715 (2006).

Reisner, M. *Cadillac Desert: The American West and Its Disappearing Water.* New York: Viking Penguin, 1987.

Ruhl, J. B. "General Design Principles for Resilience and Adaptive Capacity in Legal Systems." *North Carolina Law Review* 89 (2011): 1373–1401.

Ruhl, J. B., and J. Salzman. "In Defense of Regulatory Peer Review." *Washington University Law Review* 84 (2006): 1–61.

Ruhl, J. B., and J. Salzman. "Climate Change, Dead Zones, and Massive Problems in the Administrative State: Guidelines for Whittling Away." *California Law Review* 98 (2010): 59–120.

Safe Drinking Water Act of 1974, 42 USC § 300g–1.

Shabman, L., D. Reckhow, M. B. Beck, J. Benaman, S. Chapra, P. Freedman, M. Nellor, J. Rudek, D. Schwer, T. Stiles, and C. Stow. *Adaptive Implementation of Water Quality Improvement Plans: Opportunities and Challenges.* Durham, N.C.: Nicholas Institute, Duke University, 2007.

Snake River Water Rights Settlement of 2004. Mediator's Term Sheet of April 20, 2004. https://www.idwr.idaho.gov/waterboard/WaterPlanning/nezperce/pdf_files/complete-agreement.pdf.

Stegner, W. *Beyond the Hundredth Meridian: John Wesley Powell and the Second Opening of the West.* Lincoln: University of Nebraska Press, 1953.

Stewart, R. "Administrative Law in the Twenty-First Century." *New York University Law Review* 78 (2003): 437–460.

Stow, C. A., C. Roessler, M. E. Borsuk, J. D. Bowen, and K. H. Reckhow. "A Comparison of Estuarine Water Quality Models for TMDL Development in the Neuse River Estuary." *Journal of Water Resources Planning and Management* 129 (2003): 307–314.

Stys, J. J. *Non-Profit Involvement in Disaster Response and Recovery.* Report prepared for the Center for Law, Environment, Adaptation and Resources (CLEAR) at the University of North Carolina School of Law (2011). http://www.law.unc.edu/documents/clear/nonprofit.pdf.

Tarlock, A. D. "Reconnecting Property Rights to Watersheds." *William and Mary Environmental Law and Policy Review* 25 (2000): 69–112.

United States v. Rio Grande Irrigation Co., 174 U.S. 690, 706 (1899).

U.S. Administrative Procedures Act of 1946, 5 U.S.C. §§ 551–559, 701–706, 1305, 3105, 3344, 5372, 7521.

U.S. Department of Homeland Security. "National Incident Management System" http://www.fema.gov/national-incident-management-system.

U.S. Environmental Protection Agency. "Watershed Approach Framework" (1996). http://water.epa.gov/type/watersheds/framework.cfm.

U.S. Environmental Protection Agency. "National Aquatic Resource Surveys: An Update" (2011). http://water.epa.gov/type/watersheds/monitoring/upload/nars-progress.pdf.

U.S. Environmental Protection Agency, "Water Quality Assessment and Total Maximum Daily Loads Information" (1998, 2002, and 2010). http://www.epa.gov/waters/ir/index.html.

U.S. Forest Service. "International Programs: Disaster Mitigation Program" (2000). http://www.fs.fed.us/global/aboutus/dmp/welcome.htm.

U.S. Geological Survey. "Patuxent Wildlife Research Center. Managers' Monitoring Manual." http://www.pwrc.usgs.gov/monmanual.

Vermont Nuclear Power Corp. v. NRDC, 435 U.S. 519 (1978).

Village of Euclid v. Ambler Realty Co. 272 U.S. 365, 397 (1926).

Wagner, W. E. "The Science Charade in Toxic Risk Regulation." *Columbia Law Review* 95 (1995): 1613–1723.

Walker, B., and D. Salt. *Resilience Thinking: Sustaining Ecosystems and People in A Changing World*. Washington, D.C.: Island Press, 2006.

Walker, B., C. S. Holling, S. R. Carpenter, and A. Kinzig. "Resilience, Adaptability and Transformability in Social-Ecological Systems." *Ecology and Society* 9, no. 2 (2004):5. http://www.ecologyandsociety.org/vol9/iss2/art5.

Western Water Policy Review Advisory Commission. *Water in the West: Challenge for the Next Century.* Report pursuant to: Western Water Policy Review Act of 1992. P.L. 102–575, Title XXX. Albuquerque: University of New Mexico, 1998.

Wilkinson, C. F. "Western Water Law in Transition." *University of Colorado Law Review* 56 (1985): 317–345.

Williams, A. "Floodplain Delineation Methodology Utilizing LIDAR Data with Attention to Urban Effects, Climate Change, and Habitat Connectivity in Lapwai Creek, Idaho." Master's thesis, University of Idaho, 2011.

Zaring, D. "Three Challenges for Regulatory Networks." *International Lawyer* 43 (2009): 211–217.

# Institutionalized Cooperation and Resilience in Transboundary Freshwater Allocation

OLIVIA ODOM GREEN AND CHARLES PERRINGS

Of all the expected impacts of climate change, the most significant is likely to be the change in the availability of freshwater associated with changing precipitation. The most recent Intergovernmental Panel on Climate Change (IPCC) report notes that all regions of the world are expected to experience a net negative impact on water resources and freshwater ecosystems. The impact is likely to be highly variable, with some regions experiencing declining runoff and others experiencing increasing runoff. Many areas are projected to experience increased variability in precipitation and hence in water supply, water quality, and flood risks (IPCC 2007). Individual countries are already developing strategies to adapt to or to mitigate the consequences of changes in water availability, including the adoption of water conservation measures and water conservation and adaptive management practices, together with the development of novel institutions—including water markets. Internationally, however, there are fewer options. Transboundary water resources are typically governed by international treaties and by the mechanisms established under those treaties (e.g., transboundary water commissions).

In this chapter, we consider the role of transboundary water agreements in the resilience of social-ecological systems to water shocks and stresses, where *resilience* is taken to mean the capacity of the system to maintain functionality over a particular disturbance regime (Holling 1973; Perrings 1998). Transboundary agreements take many forms, some being more effective than others at adapting to environmental and social change. Two attributes of a social-ecological system determine its resilience. One is the adaptive capacity of both components of the system (Carpenter et al. 1999; Elmqvist et al. 2003; Scheffer and Carpenter 2003; Kinzig et al. 2006). For freshwater ecosystems, this depends on the flow regimes, sediment and organic inputs, nutrient flows, and biotic assemblages (Baron et al. 2002). For the institutions governing transboundary freshwater systems, resilience depends on treaty mechanisms that account for fluctuations in the water cycle. A second important attribute of coupled systems is their robustness: the properties of the system that allow it to accommodate perturbations without additional adaptation (Webb and Levin 2005). Using the results of a content analysis, we evaluate these attributes in a sample of seventy-three international water treaties.

Formal management regimes governing shared river basins, as designed in international water treaties, are instrumental in mitigating potential conflicts and disputes between riparian states and in guiding societal interaction with water (Yoffe et al. 2003). Institutions complement ecosystem services by restricting consumptive use and contamination and through smoothing out variability via infrastructure at their disposal. However, analysis of resilience of institutions tends to focus on levels of conflict and cooperation over the shared resource, not the ecological impact of the management approach. Here, we attempt to bridge the divide between what treaty elements promote resilience for the institution, as measured by conflict, while also promoting resilience in the environment, as measured by the ecosystem's capacity to cope with disturbance while maintaining stability.

First, we give a brief summary of the principles guiding international water allocation. Second, three allocation frameworks are discussed and evaluated for their capacity to contain risk using the results of a content analysis of seventy-three water allocation treaties and how they contribute to international water-related conflict. Third, other mechanisms for treaty design are defined and their application assessed using the results of the same survey. Fourth, we discuss how treaties can integrate adaptive management principles in order to promote resilience to climatic shocks. We conclude that, to be resilient to environmental change, treaties must incorporate principles that promote institutional capacity to manage conflict and iterative governance strategies, most likely through establishment of joint water commissions bound to implementing adaptive management processes.

## Principles of International Water Law

Numerous principles guide the development of international water law. Each treaty is unique in its application of principles, as each state has its own unique set of claims, rights, and needs, and each water body displays a unique set of characteristics. The particular mix of treaty provisions has profound implications for its capacity to enhance resilience of social-ecological systems (SES). Certain principles favor states based on hydrography (upstream versus downstream position), and others tend to favor states based on chronology (first to make claim or establish use). The constellation of riparians influences levels of cooperation and environmental impact. For example, an upstream riparian has little incentive to cooperate, because they can externalize the negative impacts of resource exploitation by flushing it all downstream. Thus, in a vacuum, upstream states prefer a unilateral order and are reluctant to engage in institutionalized cooperation, especially if they derive no benefits from downstream states (e.g., transportation, anadromous fish migration). Turkey exemplifies this sort of behavior along the Tigris–Euphrates basin. Conversely, downstream states tend to be more willing to institutionalize cooperation, because they experience the flushed externalities of upstream neighbors and therefore want some level of control over the resource upstream. For example, countries that have ratified the 1997 United Nations (UN) Convention on Non-Navigational Use of International Watercourses (UN Convention, described later in this chapter) are largely downstream states, such as Iraq. However, this trend may not hold true dependent on where power is concentrated in the basin. A hegemonic downstream state, such as Egypt on the Nile, is less likely to cooperate with its less powerful upstream neighbors, because those states may not make enough use of the resource to impact the downstream state, and because the more powerful state may be able to limit that use by other means (intimidation; Zeitoun and Warner 2006). However, (benign) hegemons may also use their power to create incentives to cooperate, such as South Africa in the Southern African Development Community Protocol on Shared Water Resources (Turton and Funke 2008).

The end product usually reflects a compromise between extreme positions. Because negotiations are a give-and-take process, extreme claims rarely, if ever, take their pure form. Even more, some arguments have never been applied and only serve as extreme negotiation positions (LeMarquand 1976; Matthews 1984). The environmental impact of each claim is an important consideration, as some principles facilitate resource stewardship, while some arguments encourage unbounded exploitation. Below is a brief summary of these concepts and their potential environmental consequences. For a more detailed description of international water law concepts and their origins, see Dellapenna (1995, 2001) and Barberis (1991).

### *Absolute Territorial Sovereignty*

Absolute territorial sovereignty, commonly referred to as the Harmon Doctrine after its founder, U.S. Attorney General Judson Harmon, claims that a state has an absolute right to all water flowing within its borders. This extreme position largely favors upstream claimants and is hydrographical in nature. Because of its grave consequences to downstream states and the environment, application of this claim is extremely rare and has only been applied in treaties relating to upstream tributaries (McCaffrey 1996). Harmon's own successor rejected the theory, and the United States eventually repudiated this position in its negotiations with Mexico over the Rio Grande (McCaffrey 1996). Still, it lives on as an initial position for upstream users and as a rationale for the three nonsignatories of the UN Convention—Turkey, China, and Burundi—that are all upstream states. Interestingly, the concept is still argued domestically in the United States. The principle was recently invoked by the governor of Georgia in a statement claiming that Georgia was entitled to use all the water that falls within its borders, despite the U.S. Supreme Court's rejection of the principle when Colorado invoked it for greater rights to the Arkansas River against Kansas in 1902 (*Kansas v. Colorado* 1902; Gleick 2009).

With respect to the environment, pure application of the Harmon Doctrine could be devastating. Granting a state the right to divert all water within its territory creates exploitation incentives void of any incentive to conserve and could dry up the entire resource. For example, the U.S. state of Georgia, the upstream state on the Apalachicola–Chattahoochee–Flint river system has been embroiled in controversial litigation recently with its downstream states, Alabama and Florida. Georgia claims an absolute right to use all the water that flows within its borders. Alabama and Florida claim that Georgia's overuse has reduced instream flows and contaminated the system to a point that has decimated endangered downstream estuarine fisheries, especially shellfish beds (*Georgia v. U.S. Army Corps of Engineers* 2002; *In re Tri-State Water Rights Litigation* 2011).

### *Absolute Riverain Integrity*

Absolute riverain integrity, also referred to as absolute territorial integrity, is the contrasting and equally extreme argument of international water law. States invoking this argument claim that each riparian has a right to the natural unobstructed flow of a river as it runs through its territory. This argument largely favors downstream states in humid climates and is hydrographical. Upstream states are prevented from diverting and consuming any flow, because the farthest downstream state is entitled to enjoy the entire flow. However, because the farthest downstream state has no obligation to another downstream state, they may be entitled to dewater. Much like the Harmon Doctrine, this argument is not integrated into treaties in practice and serves mostly as a strong initial position for negotiations (Odom and Wolf 2011).

The environmental impact of applying this principle may seem minimal, but restrictions against diverting flow may not curb instream development or restrict any consumption or development of the farthest downstream state. In addition, an upstream state may be entitled to install run-of-the-river hydropower dams that do not divert water but impede aquatic migration and alter sedimentation patterns, among other effects. Likewise, upstream states may not be required to prevent contamination as long as the volume of flow is unaltered.

### *Limited Territorial Sovereignty*

Limited territorial sovereignty, a hybrid of the extreme claims, provides for balanced use of a water body restricted by principles of equity (i.e., one state's use does not cause significant harm to another state's reasonable use). This doctrine is guided by vague principles catalogued in many international conventions, namely the UN Convention and the Helsinki Rules on the Uses of the Waters of International Rivers, and many of the factors are discussed below.

The UN Convention, adopted in 1997 by the UN General Assembly, specifically focuses on international transboundary water resources. The UN Convention codifies many of the principles deemed essential by the international community for the management of shared water resources, such as equitable and reasonable utilization of waters with specific attention to vital human needs, protection of the aquatic environment, and the promotion of cooperative management mechanisms. The convention also incorporates provisions concerning data and information exchange and mechanisms for conflict resolution. Once ratified, the UN Convention will oblige a legally binding framework, at least upon its signatories, for managing international watercourses. Even without ratification, its guidelines are being increasingly invoked in transboundary negotiations in which its general concepts are bolstered by basin-specific details to govern a particular watercourse.

Adopted by the International Law Association in 1966, the Helsinki Rules describe guidelines on how international surface waters and groundwater should be used. The Helsinki Rules led to the Berlin Rules on Water Resources of 2004, which address all freshwater resources (not limited to international waters). Both sets of rules emphasize sustainable management practices that minimize environmental harm. Interpretation of factors and the relative weight assigned to each varies dependent on the parties. Hydrohegemony scholars argue that the vagueness of these factors gives too much leeway for the more powerful states to grant disproportionate weight to those factors that favor themselves

(Zeitoun and Warner 2006). However, an equitable distribution of water requires that institutions not operate in favor of any particular state.

Guiding principles include equity in interpretation of law, restrictions on use to benefit other states, reciprocity in adherence to agreements, and peaceful dispute resolution. These principles guide water allocations based on equitable and reasonable use, the obligation to not cause harm, and past use (historic rights), but vague and occasionally contradictory language results in varied and conflicting interpretations of the principles (Biswas 1999). Each principle and its environmental impact is described below.

### EQUITABLE AND REASONABLE USE

Equitable and reasonable use is a principle of customary international law accompanied by a long list of factors to determine what is equitable or reasonable. Its general acceptability stems from its vagueness. The UN Convention and the Helsinki Rules provide guidance on how the principle is to be implemented. Relevant factors, such as level of harm and demand, are identified and calculated, and the conclusion is reached on the basis of the sum or a cost–benefit analysis. Environmentally, this approach depends on the threshold of allowable harm or what level of harm is deemed "reasonable." Typically, use is limited by an obligation to not prevent another state from enjoying its equitable use of the resource, discussed below.

### OBLIGATION TO NOT CAUSE HARM

The obligation to not cause harm is normally qualified as the obligation to avoid *significant* harm and holds that riparian users must take all appropriate action to prevent causing significant harm to other riparian users, not necessarily the environment. Legitimate water projects that might negatively impact other riparian states and their development potential may be enjoined by this principle. Some scholars attribute its controversial implications as the reason the UN Convention has not been ratified (McCaffrey 2001). Typically, upstream states oppose this provision;

however, harm does not only flow downstream. A downstream state may cause significant harm to an upstream state by development projects that restrict upstream migration of anadromous species, impound flow, impact navigation, or cut off future plans for upstream development because that future development may harm the downstream project.

Environmentally, restriction on harming the environment seems to promote environmental integrity. However, what counts as harm turns on the interpretation of significant harm and on the parties' values. Environmental harm may not be of great concern, and thus, environmental degradation would not constitute a significant harm to be avoided.

### HISTORIC RIGHTS

Historic rights grant priority to existing uses over proposed uses. Chronology is key. Prior uses are usually deemed inherently superior to proposed uses and granted the volume necessary to continue. This likely stems from the defensibility of prior uses as an expression of need, especially when protecting native rights. Downstream states in arid or exotic watersheds often invoke this principle in order to defend infrastructure. For example, Egypt has a long history of development on the Nile compared with the relatively recent projects begun upstream.

Environmentally, this approach can be harmful, because no regard is paid to the type of use, impacts, or benefits. While upstream development may be curbed in order to protect pre-existing downstream projects or uses, this doctrine also creates a strong incentive to develop first and fast. There is no incentive to cooperate on joint projects, which tend to take longer to build up. Instead, each state may develop similar projects, whereas a joint project with benefit sharing would have had less impact on the environment. In addition, strict adherence to priority by date leaves no room for changed values over time (Wilkinson 1989).

## *Prioritization of Uses*

Prioritization of uses sets up a hierarchy of uses with those serving vital human needs above all others. Agriculture, energy production, and industrial uses usually follow domestic use, depending on the weight given to certain hydrographic and sociopolitical factors and the values of the region. For example, the two sets of boundary waters agreements between the United States and Canada, and the United States and Mexico prioritize differently, due to the amount of water available along each border region and the dominant uses in the region: the former prioritizes first domestic and sanitary, then navigation, electricity generation, and irrigation; the latter gives descending weight to domestic, agriculture, electric power, other industry, navigation, fishing, and other beneficial uses.

Environmental concerns and instream flows may be included in the priority list, on par with environmentally harmful uses, but priority to instream use is rare. Generally, uses with the worst environmental record come after those with less harmful effects (domestic). However, relatively intensive uses, such as industrial and agricultural, are usually granted medium to high priority. The relatively low priority granted to environmental use is a reflection of the relatively recent change in societal values regarding instream flow.

## Allocation Frameworks

Water allocation frameworks generally reflect (1) the rights of each party; (2) the wants and needs of each party; and (3) in some cases, an evaluation of benefits to be shared. Some treaties consist of a mix of the first two approaches, with some allocations based on territorial entitlements and others granted based on existing uses or other measurements of need. The third category may also influence the allocation scheme but requires a paradigm shift from the rights-based mentality, as political boundaries are lifted and parties become citizens of the watershed. Depending on how rights, needs, or benefits are defined, and how nature is valued and assessed, each allocation approach may fall anywhere along the resilience spectrum.

A content analysis was conducted using the Transboundary Freshwater Dispute Database (TFDD) of water-related treaties and events. The database is housed at Oregon State University under the direction of Dr. Aaron T. Wolf and is widely considered to be the most comprehensive collection of water treaties, water-related events, and geospatial data of transboundary water resources. The full results of this survey, including detailed statistical analysis, are published in Dinar et al. (2010).

### *Entitlement-Focused Negotiation*

Sovereigns argue over entitlements based on relative riparian position or past use. Claims based on absolute principles may seem easy to calculate, as they would constitute 100 percent of flow, either to be used as water flowing within borders as absolute territorial integrity dictates or to be kept instream as absolute riverain integrity requires. However, claims based on extreme principles are politically impossible to measure. As discussed above, the absolute principles rarely form the basis of an allocation, so the relative ease of calculation is negated by the difficulty in persuading the other side to grant full allocation of a stream. Thus, any allocation based on rights is necessarily tempered from the absolute.

Institutions that allocate based on property rights tend to generate less conflict, but also are less adaptive to changing conditions. India and Pakistan, for example, divide the tributaries of the Indus River according to territory, as agreed in the Indus Waters Treaty of 1960 between Pakistan and India, brokered by International Bank for Reconstruction and Development (World Bank). India is entitled to 100 percent of the eastern tributaries; Pakistan enjoys unrestricted use of 100 percent of the western tributaries, and India agrees to allow those western tributaries to flow unimpeded through Indian territory (with some minor exceptions). Scholars note the resilience of this institution by pointing out that the terms were honored in the midst of armed conflict (Kirmani 1990; Alam 2002). However, as the international community trends toward integrated watershed management and catastrophic flooding in Pakistan in 2010 spurred calls for increased storage upstream (Kiani 2011), this rigid allocation scheme hinders the institution's ability to adjust to changed values and demand. See, for example, the difficulty in agreeing on development of the Kishanganga Dam. The states could not agree on the construction of a hydropower facility on a western tributary in India, invoking the intervention of the International Court of Arbitration, which recently enjoined the project (*The Nation* 2008, 2011). Indian development plans also commonly prompt accusations of violative flow impoundment, especially during drought (Jamil 2008).

Ecologically, decades of relative peace over the Indus River led to many large-scale developments in a river valley already known for its vast network of canals. Without any environmental protection provisions, both states have been free to utilize their full entitlements, which are 100 percent of flow. Dewatering, contamination, and instream infrastructure has led to ecological collapse, including endangerment of the charismatic Indus River dolphin (Reeves et al. 1991; Braulik 2006). In this instance, an institution capable of containing the level of conflict/number of grievances expressed has nevertheless led to a loss of ecological resilience.

### *Need (Current Use)-Based Allocation*

Need-based or current use–based allocation is the most common approach applied in international negotiations. Of the water allocation agreements that explain why water is to be allocated a certain way (fifty treaties of a sample of seventy-three treaties), most base allocations on current use (mostly for irrigation, domestic uses, and hydropower generation) and project that use based on growth predictions. Current use is the more sensible unit of measurement once negotiators have moved past abstract absolute rights, because it can be concretely measured based on irrigable acreage, population, use, environmental requirement, or any other mutually agreeable parameter. Historic uses are typically granted the necessary water to continue, because they are simple to quantify. This approach may include a discussion of priorities, beneficial and reasonable use, environmental harm, and equity.

The resilience of need-based allocation mechanisms depends on the characterization and prioritization of need and the inclusion of adaptability mechanisms (discussed below). Calculations limited to irrigable acreage or population are vulnerable against even foreseeable changes in population and demand, and exhibit much less resilience with respect to less predictable disturbances, such as climate change, natural disaster, epidemic, or regulation (e.g., species protection that requires increased instream flow à la the U.S. Endangered Species Act of 1973). Content analysis of the water allocation data set reveals that most need-based treaties are not so rigid as to require an allocation made based on a certain need to be used strictly for that purpose. Need is typically used to quantify the distribution, and once the distribution is agreed upon, the states have flexibility in utilization, within reason. For example, the treaty governing the Incomati/Inkomati River between Swaziland and South Africa allocates a specific amount for afforestation in both states based on the amount of land to be planted, a rare nod to environmental need in a water treaty. Using that water for any purpose other than afforestation would violate the spirit and terms of the allocation provision. On the other hand, an allocation based on population and per capita consumption would not necessarily be violated if the allocation is distributed among other uses.

### *Benefit Sharing*

Benefit sharing requires that the basin be managed as a single unit, the aggregate benefits then being distributed to each country according to some negotiated principle of proportionality. Such an approach shifts the analysis from the resource itself to the benefits of utilization. Thus, the product to be shared or allocated expands beyond water to include goods and commodities, such as agricultural products, hydroelectricity, and, conceivably, benefits from nonuse, such as conservation, instream flows, and improved water quality (Sadoff and Grey 2002).

Consider the following example. By lifting the border between the upstream and downstream states, water managers can collaborate on the optimal location for a reservoir. Consider a downstream state that experiences frequent flooding and seeks to construct a reservoir, but the state's territory is mostly floodplain, so any reservoir would be shallow, have a large footprint, and store less water than a reservoir with a smaller footprint in a deep valley in the headwaters. In addition, downstream development would impede the migration of anadromous fish. At the same time, a canyon in an upstream state may be a more suitable location for a storage facility that would benefit the downstream state in the form of flood protection without imposing downstream physical barriers to fish migration. Upstream development would also cause significant environmental harm, harm which must be addressed if the SES is to be resilient. The upstream facility may be able to provide greater benefit (i.e., store more water) with potentially less cost (i.e., flooded land and evaporation). Then, whatever benefits the downstream state reaps can be paid forward to the upstream state, perhaps in collaborative measures to protect upstream migration or as financial side payments. For instance, as part of the Columbia River Treaty, the United States pays Canada for the benefits of flood control and Canada diverts water between the Columbia and Kootenai for hydropower purposes as part of a larger scheme of sharing hydropower benefits, whereby Canada releases water for hydropower production in the United States. Half of the electricity is either delivered to Canada, or the United States pays Canada for its value (Wolf 1999).

The assumption that the benefits of river basin management are shared between countries implies a cooperative approach on the part of participating countries. The conditions in which countries will cooperate are quite limited, as are conditions in which international agreements are self-enforcing (Barrett 2003; Wu and Whittington 2006; Touza and Perrings 2011). Where they exist, however, self-enforcement mechanisms have been shown to correlate with a reduction in conflict over the resource (Dinar et al. 2010). Integration of various elements of water resource planning across multiple scales, known as *integrated water resource management*, tends to be decentralized and more legitimate, as it incorporates an array of stakeholder interests (Blomquist et al. 2005). However, recent research has shown that integrated water resource management may be at odds with the flexibility required of adaptive management practices (Engle et al. 2011) (but see Chapters 4 and 5, this volume). Of water quantity–related treaties (a broader group than those that outline allocations), ten move beyond needs and discuss sharing benefits from and beyond the river in an equitable allocation of water, hydraulic development, reforestation, flood control, capital, and other commodities—their "basket of benefits" or "benefit-shed." Governance of the Aral Sea is a prime example.

The 1998 treaty over the Naryn Syr Darya Cascade Reservoirs in the Aral Sea basin reflects the heterogeneous distribution of natural resources in the region and attempts to share them equitably through integrating regional fuel, energy, agriculture, and water economies. The Naryn River originates in the Tian Shan Mountains of Kyrgysztan, where opportunities for storage and hydropower generation are greater than in the broader, downstream valleys of Uzbekistan and Kazakhstan, where cotton production is vital to the regional economy. Kyrgysztan and Tajikistan store water over the winter and forgo potential hydropower generation from winter releases in order to provide Uzbekistan and Kazakhstan with a dependable source of irrigation in the growing season. In exchange for the agricultural and flood control benefits provided from the storage and release of water from Kyrgyzstan's Toktogul Reservoir (the largest storage facility on the Naryn River) and electricity provided from hydropower generation in the growing season, Uzbekistan provides electric power, natural gas, and fuel oil, while Kazakhstan provides electric power and thousands of tons of coal to Kyrgyzstan during the season of storage. Additionally, Kyrgyzstan agrees to reduce its national energy consumption by 10 percent in order to reduce demand on the energy sector. Tajikistan was later added in an effort to create a single economic zone addressed to water and energy issues in 1999.

Although many may still question the resilience of the current resource allocation in the area, the institution is a marked improvement from the nonintegrated, independent exploitation and historic Soviet misuse. The collapse of the Aral Sea, one of the world's most shocking man-made environmental disasters, decimated the region in both ecological and economic terms. Integrated development has tempered the dewatering, and the self-enforcing aspects of the arrangement constitute institutionalized cooperation between states. States continue to disagree over development projects and impacts (*The Telegraph* 2010), but the presence of the treaty has increased the institutional capacity of the states to cope with change and conflict (Yoffe et al. 2003). In addition, the physical Aral may be better off, because increased storage in winter may keep more water instream during irrigation season. Increased fossil fuel consumption in lieu of hydropower production may tip the environmental footprint of this arrangement into a less stable state (albeit hydropower has its own set of impacts that can erode resilience), but the agreed 10 percent consumption reduction signals some degree of conservation awareness.

## Resilient Treaty Design

The unique hydrology, climate, economics, and politics of each international basin preclude blueprints for building resilient institutions, and each water treaty comprises a unique set of provisions to govern the resource. Water can be distributed in many different ways, some simple, some complex, some resilient, some not. Management regimes can be iterative, information shared, conflicts foreseen, extreme events planned for, systems monitored, and joint commissions formed. Or not. Treaty design, not the mere presence of a treaty, is paramount to the stability of the institution (Dinar et al. 2010).

While blueprints may not exist, patterns emerge indicating that assemblage of certain provisions correlate with institutional resilience and adaptive capacity (Dinar et al. 2010). An institution's adaptive capacity can be forecast by the presence of certain provisions and, equally important, the institution's ability to effectively mobilize these mechanisms in response to disturbance (Adger et al. 2011). Adaptive capacity provisions require resources such as information, learning capacity, and human and social capital, all of which may be accounted for in treaty design (Tol and Yohe 2007).

Since 1820, more than 450 water treaties and other water-related agreements have been signed, more than half of which were concluded after 1950. Despite growth in treaty formation and increased calls for resilience, a review of treaties from the last sixty years reveals an overall lack of mechanisms designed to enhance resilience to water shocks. For example, states rarely explicitly delineate water allocations, the most conflictive and volatile issue area between co-riparian states (Giordano and Wolf 2003). Further, the treaties that do specify quantities often do by fixed volumes, ignoring dynamic hydrology, values, and needs. Likewise, water quality provisions play a minor role in co-riparian agreements historically (Giordano 2003). Enforcement mechanisms are also absent in a large percentage of the treaties. Thus, even if a treaty acknowledges the need to be resilient facing variable water availability, there may not be any means of ensuring each state will respond to disturbance according to treaty principles.

### *Water Allocation Method*

The method of water allocation embodied in a treaty has profound implications for the resilience of the social-ecological system served by the treaty to both climatic and other shocks (Dinar et al. 2010; Odom and Wolf 2011). States share water by many means, often as a reflection of the negotiation process and historic relations, and the degree to which the environment is considered varies greatly. Methods of allocation vary from simple volumes to a complex calculus of flows, availability, and time of year. Disturbingly, the treaty record is replete with allocations that do not allow for the vagaries of nature and the scientific unknown and that often lead to tense political standoffs and environmental degradation.

Most commonly, states divide water based on fixed volumes, the least adaptive form of allocation. Rigid entitlements leave no flexibility to account for hydrologic variability, and the IPCC predicts such rigidity will lead to increased international tension:

> One major implication of climate change for agreements between competing users (within a region or upstream versus downstream) is that allocating rights in absolute terms may lead to further disputes in years to come when the total absolute amount of water available may be different. (McCarthy et al. 2001, 225)

In low-water years, if states divert their full entitlement, water levels fall, perhaps to the point of dewatering, and strain the aquatic ecosystem. Another likely drought scenario challenges institutional resilience when an upstream state diverts its full allotment, leaving insufficient flow for the downstream state to satisfy its entitlement. Without flexibility in allotments or flow requirements at some downstream point (e.g., the upper basin states of the Colorado River must release enough water from Glen Canyon Dam so that 7.5 MAF/year flows to the lower basin at Lees Ferry), allocation by strict volume leads to vulnerability in institutions and social-ecological systems.

Of the sample of water allocation treaties ($n = 73$), twenty-eight treaties allocated water based on fixed quantities, with sixteen of those not providing flexibility in the allocation mechanism. Of the twelve remaining treaties, nine acknowledge variability and base the fixed quantity on availability. For example, the 1994 treaty between Lebanon and Syria over the Al-Asi River (Orontes) provides 80 MCM to Lebanon when the amount of water in Lebanon is 400 MCM or greater. When the annual quantity falls below 400 MCM, the year is considered "rainless," and the Lebanese share is reduced by 20 percent. The remaining three treaties in the data set allocate fixed quantities that may be recouped over a particular period of time. The adaptability of this method depends in part on the period of time. In low-water periods, recoupment over a course of days does not account for seasonal variation. In prolonged drought, more generous recoupment periods of several years may still result in failure to meet treaty obligations.

Agreements that reduce diversions during low-water years provide for more adaptable institutions and are less likely to dewater a stream. However, states must agree on what constitutes a low-water year and utilize accurate assessment methods (Kilgour and Dinár 2001). For example, the dry season immediately after the 1996 treaty between Bangladesh and India over the Ganges River, a treaty that allocates quantities and percentages according to the water level at Farakka Barrage and took 35 years to negotiate, the water level fell below the minimum accounted for in the treaty, partly due to the data used to establish minimum flows, and the institution was strained. The parties used historical hydrological data that did not account for increased upstream diversions or changing climate patterns. As a result, the institution faltered, and Bangladesh, the downstream state, faced water scarcity. By relying on historical data and past trends to predict future patterns, the institution failed to appreciate the complexity of the social-ecological system along with its tipping points (Liu et al. 2007; Lenton 2011).

Most treaties (55 percent or forty out of seventy-three) allocate based on fixed quantities, percentages, or both. Percentages account for changes in availability due to physical or social disturbance, and unless the entire flow is allocated to states (e.g., 50/50), grant proportional protection to instream flow. Whether that protection is adequate to support a resilient social-ecological system depends on hydrology and the reserved amount. If 30 percent of flow is reserved instream, that may not be adequate for habitat, even in a wet year. On the other hand, leaving 100 percent of flow instream may not be sufficient in a severe drought. Institutionally, allocation by percentage is less likely to lead to state grievances (Dinar et al. 2010).

### Extreme Events Provision

While the resilience of a system is measured by its ability to cope with change (Yoffe et al. 2003), water allocation treaties regularly ignore the inevitability of extreme events, especially drought, despite research demonstrating that the intensity of state grievances is robustly correlated with the degree of flow variability in the basin (Dinar et al. 2010). Extreme fluctuations in flow variability can be built into the allocation mechanism or dealt with in separate provisions or through separate treaties. Only twelve treaties in the TFDD collection of water quantity, flood protection, and hydropower treaties ($n$ = 102) contain drought adaptability mechanisms. These mechanisms range from vague provisions to consult and cooperate in the event of drought (e.g., Agreement between the Governments of the Republic of Portugal, the People's Republic of Mozambique and the Republic of South Africa Relative to the Cahora Bassa Project, signed May 2, 1984) to more specific pledges to cease all irrigation pumping if flow falls below certain levels (e.g., Complementary Settlement to the Agreement of Cooperation between the Government of the Eastern Republic of Uruguay and the Government of the Federal Republic of Brazil for the Use of Natural Resources and the Development of the Cuareim River Basin, signed May 6, 1997). Presence of an adaptability mechanism to drought has been shown to affect the intensity of state grievances (Dinar et al. 2010). Notably absent are provisions to protect the environment in times of drought.

Floods that cross boundaries tend to account for a disproportionate share of casualties and are more severe in magnitude of loss (Bakker 2009a). Many treaties in the TFDD collection concern flood control, mostly in the form of joint infrastructure projects (e.g., reservoirs, levees, river straightening), and these treaties tend to be single-issue agreements that outline specifics of flood control mechanisms, not vague references. However, agreement over flood control mechanisms does not robustly correlate with fewer state grievances (Dinar et al.

2010). In addition, Bakker (2009a) found that international river basins with water institutions experienced higher flood magnitudes than those basins without institutions.

### *Data Sharing, Joint Monitoring, and Information Exchange*

Monitoring and data sharing is crucial for iterative governance of a dynamic resource. In contentious basins, agreements to exchange data can be the first step in building a cooperative governance structure (Sadoff et al. 2008). When ecosystem knowledge is shared among a variety of stakeholders, institutions enable responsive and flexible solutions at the appropriate scale (Brown 2002). To ward off the tendency to "hunker down," whereby states are less willing to share information when facing disturbance, the agreement to share must be institutionalized (Anderies et al. 2004; Putnam 2007).

Unfortunately, though many treaties include monitoring and data-sharing provisions, information sharing among states is relatively low (Grossman 2006). Whether caused by a lack of monitoring or technology, incompatibility between databases, or a lack of willingness to share, restricted data sharing has exacerbated the ecosystem harm of water system infrastructure, such as the effect of dams on the Senegal River (Grossman 2006). The Senegal River is managed by a joint commission that is responsible for operating a database of flow information. One contributing factor to the data-related problems of the Senegal is that an upstream riparian, Guinea, is not a member of the joint commission and does not contribute inflow data, resulting in data gaps. Data gaps weakened the capacity of dam operation models to optimize both water allocations and environmental health as designed. As a result, additional agreements are necessary to define the roles of all contributing organizations and to create a central river basin information system (Grossman 2006).

### *Joint Commission*

Recognizing the restraints of the negotiation process, treaties often establish joint commissions to manage shared resources. Of the world's

276 transboundary river basins, 78 are represented by an international commission and even within these, few include all nations riparian to the affected basins, precluding the integrated basin-wide management advocated by the international community (Bakker 2009b). The scale of operations ranges from day-to-day decisions concerning operation of infrastructure to occasional discussion of new development projects. In some cases, joint commissions enjoy some degree of independence from their respective national governments and may be less bound by political relations and realities. For example, the Joint Water Committee, established by the Israel–Jordan Treaty of Peace to resolve conflict without amending the treaty, to monitor water quality, and to share data, mitigated a rhetorically heated conflict between the riparian states during the severe drought of 1998–1999 by modifying allocations (Odom and Wolf 2011). The treaty provides for Jordanian water storage (winter) and release (summer) in the Sea of Galilee/Kinneret/Lake Tiberias/Lake of Gennesaret system, which lies completely within Israel. In return, Israel secured fifty-year leases for wells and agricultural land in Jordan. The treaty also allocates fixed volumes of the Jordan and Yarmouk Rivers. When Israel threatened to renege on its delivery schedule, protests erupted in Amman and the King of Jordan issued harsh words, challenging the stability of peace between the nations. The treaty was silent with respect to flow variability, especially such extreme fluctuations as those experienced during a drought. Likewise, the Joint Water Committee did not have expressed authority to modify allocations. Nonetheless, the committee intervened to mitigate the conflict, exhibiting institutional capacity to manage the disturbance (Odom and Wolf 2011).

International river management organizations that attempt to tackle multiple collective-action problems, such as environmental conservation and water allocation, face increased challenges to their effectiveness, but many argue that integration of all related problems is necessary for effective governance (Kliot et al. 2001; Dombrowsky 2007; Sadoff et al. 2008). Thus, an organization that focuses solely on how to allocate water and has no jurisdiction to mitigate or even monitor environmental impacts of withdrawals can only be effective to a limited extent. Conceivably (and perhaps even likely), withdrawals made without concern for environmental harm will decimate the source and render the organization ineffective. If addressing multiple issues reduces the effectiveness of an organization but a singular focus leads to exacerbation of related problems, there must be a trade-off made whereby the most important related problems, such as sustainability and ecosystem resilience, are addressed along with resource distribution (Schmeier 2010).

### *Dispute Resolution Mechanism*

If all other treaty mechanisms fail to avoid conflict, a previously agreed upon method of dispute resolution will, at the very least, provide some guarantee of cooperation. How dispute resolution influences the effectiveness of an institution varies by basin and method. Disputes may be solved by consultation, arbitration through a neutral third party (e.g., UN, World Bank, International Court of Arbitration/Justice), diplomatic channels, or a joint commission, among other means. This mechanism may act as a fail-safe against defection and may provide states with at least a minimal incentive to cooperate, depending on where a dispute may ultimately be settled. For example, if a state is only bound to vague commitments to resolve conflict diplomatically, they may stall or act uncooperatively with little penalty. On the other hand, if disputes are to be settled on the world stage by a neutral arbitrator, the stakes for acting in bad faith may be higher.

## Treaties to Promote Resilience in Social-Ecological Systems

Since the 1992 UN Rio Declaration on Environment and Development proclaimed twenty-seven principles to guide sustainable development, states have increasingly incorporated environmental protection into treaties, but many of the recent environmentally conscious treaties merely establish a spirit of cooperation over ecosystem protection without obligating riparians to fulfill any specific ecological threshold or dictating any particular process for protection or monitoring.

Of the water quantity–related treaties in the TFDD collection, eleven include explicit provisions for environmental protection, and eight of these were signed in the past twenty years. These treaties have the primary purpose of either allocating water or establishing a framework to allocate water while protecting ecosystems. In our survey of water allocation mechanisms, only three treaties justified allocations for environmental reasons, compared with twenty-three (32 percent) that gave no justification, and thirteen (18 percent) and eleven (15 percent) that justified allocations for irrigation and hydropower, respectively. For example, the 1992 agreement between South Africa and Swaziland over the Incomati/Inkomati allocates fixed quantities of water that vary according to availability, with additional allocations reserved for afforestation based on the area of land to be planted. It should be noted that the environmentally related justification is for only one aspect of the allocation.

It follows that adaptive management principles in international agreements are required if they are to accommodate the effects of changing climate on variability in water availability. However, few treaties actually spell out iterative processes (Wolf 2007; Drieschova et al. 2008, 2011; Eckstein 2010). Political difficulties prevent treaties themselves from being iterative. While open, negotiations may go back and forth, and there may be room for iterative give and take as information is exchanged. However, once closed, treaties may be effectively locked until they expire, often a period of decades. Another challenge for integrating adaptive management practices would be convincing state water ministers to leave room for flexibility and error in the management of a vital resource. Given the scrutiny under which negotiations proceed, negotiators would likely prefer to leave experimentation in the hands of the trusted technocrats they appoint (Arvai et al. 2006; Gregory et al. 2006). Thus, for iterative processes to influence governance of international watercourses, the process must be built into the organization established to jointly manage the resource. Mechanisms such as monitoring and information sharing must be institutionalized to bind the river basin organization to these procedures.

Unlike U.S. administrative law, which requires detailed front-end procedures, treaties are typically brief and lack long descriptions of process. Instead, they often establish a joint management commission that could (at least in theory) be mandated to operate according to resilience principles. This would require the regulatory structure to be adaptive—to be an experimentation-based learning process consisting of continuous monitoring of the transboundary water body (i.e., the social-ecological system), data review, and the regulatory adjustments needed to promote the desired goal, all at the appropriate scale (Ruhl 2005; Bruch 2009). Many of these principles, such as data accumulation, information sharing, and cooperative management are not foreign to the international water community. By mandating an adaptive management approach through monitoring, assessment, and adjustment provisions, treaties may increase their institutional capacity to cope with societal and ecological change.

Modern technology, especially information-gathering technologies such as those prescribed in iterative approaches, can change the law. International water law scholar Joseph Dellapenna contends that technology changes the problems international law must address, the range of solutions, and the intellectual structures of international legal institutions (Dellapenna 2000). Institutionalizing adaptive management techniques may change the law by identifying new problems posed by international water development, as data accumulation may reveal a problem in water development that was previously unknown, present new solutions as continuous experimentation results in surprising results, and change the structure of treaties as joint commissions are structured to operate based on iterative processes.

## Conclusion

To manage a transboundary water body for resilience in a social-ecological system, an institution must be capable of coping with change. There is no ideal model for this. However, certain principles have been shown to reduce conflict over shared natural resources—flexible allocations that reflect the variability of availability, joint commissions that cooperatively manage the resource, dispute resolution mechanisms that give certainty to the process of conflict, and provisions that proactively determine operational adjustments in times of extreme disturbance and stress. These provisions and their application vary in how they protect the environment, and environmental protection should not be traded away for institutional stability.

As such, institutions must be designed to cope with or reduce conflict, while also maintaining ecological integrity. After all, what good is a peaceful water organization that has no water to manage? To this end, collaborative management should integrate iterative processes so that as social and ecological systems alter one another, they also coevolve into more resilient regimes.

## Acknowledgments

This research was conducted with the support of an appointment to the Research Participation Program at the National Risk Management Research Laboratory administered by the Oak Ridge Institute for Science and Education. The views expressed in this paper are those of the authors and do not necessarily represent the views or policies of the U.S. Environmental Protection Agency.

## References

Adger, W. N., et al. "Resilience Implications of Policy Responses to Climate Change." *Wiley Interdisciplinary Reviews: Climate Change* 2 (2011): 757–766.

Alam, U. Z. "Questioning the Water Wars Rationale: A Case Study of the Indus Waters Treaty." *Geographical Journal* 168 (2002): 341–353.

Anderies, J. M., M. A. Janssen, and E. Ostrom. "A Framework to Analyze the Robustness of Social-Ecological Systems from an Institutional Perspective." *Ecology and Society* 9, no. 1 (2004): 18. http://www.ecologyandsociety.org/vol9/iss1/art18.

Arvai, J., et al. "Adaptive Management of the Global Climate Problem: Bridging the Gap Between Climate Research and Climate Policy." *Climatic Change* 78 (2006): 217–225.

Bakker, M. "Transboundary River Floods: Examining Countries, International River Basins and Continents." *Water Policy* 11 (2009a): 269–288.

Bakker, M. "Transboundary River Floods and Institutional Capacity." *Journal of the American Water Resources Association* 45 (2009b): 553–566.

Barberis, J. "The Development of International Law of Transboundary Groundwater." *Natural Resources Journal* 31 (1991): 167–186.

Baron, J. S., N. L. Poff, P. L. Angermeier, C. N. Dahm, P. H. Gleick, N. G. J. Hairston, R. B. Jackson, C. A. Johnston, B. G. Richter, and A. D. Steinman. "Balancing Human and Ecological Needs for Freshwater: The Case for Equity." *Ecological Applications* 12 (2002): 1247–1260.

Barrett, S. *Environment and Statecraft: The Strategy of Environmental Treaty-Making*. Oxford: Oxford University Press, 2003.

Biswas, A. K. "Management of International Waters: Opportunities and Constraints." *International Journal of Water Resources Development* 15 (1999): 429–441.

Blomquist, W., A. Dinar, and K. Kemper. "Comparison of Institutional Arrangements for River Basin Management in Eight Basins." World Bank Research Working Papers, Washington D.C., 2005.

Braulik, G. T. "Status Assessment of the Indus River Dolphin, *Platanista gangetica minor*, March–April 2001." *Biological Conservation* 129 (2006): 579–590.

Brown, K. "Innovations for Conservation and Development." *Geographical Journal* 168 (2002): 6–17.

Bruch, C. "Adaptive Water Management: Strengthening Laws and Institutions to Cope with Uncertainty." In: *Water Management in 2020 and Beyond*, eds. A. K. Biswas, C. Tortajada, and R. Izquierdo, 89–113. Berlin: Springer, 2009.

Carpenter, S., D. Ludwig, and W. Brock. "Management of Eutrophication for Lakes Subject to Potentially Irreversible Change." *Ecological Applications* 9 (1999): 751–771.

Dellapenna, J. "Building International Water Management Institutions: The Role of Treaties and Other Legal Arrangements." In: *Water in the Middle East: Legal, Political, and Commercial Implications*, eds. J. A. Allan and C. Mallat, 55–89. New York: Tauris Academic Studies, 1995.

Dellapenna, J. "Law in a Shrinking World: The Interaction of Science and Technology with International Law." *Kentucky Law Journal* 88 (2000): 809–883.

Dellapenna, J. "The Customary International Law of Transboundary Fresh Waters." *International Journal of Global Environmental Issues* 1 (2001): 264–305.

Dinar, S., O. Odom, A. McNally, B. Blankespoor, and P. Kurukulasuriya. "Climate Change and State Grievances: The Resiliency of International River Treaties to Increased Water Variability." Institute of Advanced Study Insights Working Paper Series, Durham University, Durham, N.C., 2010.

Dombrowsky, I. *Conflict, Cooperation and Institutions in International Water Management: An Economic Analysis*. Cheltenham, U.K.: Edward Elgar, 2007.

Drieschova, A., I. Fischhendler, and M. Giordano. "The Role of Uncertainties in the Design of International Water Treaties: An Historical Perspective." *Climatic Change* 105 (2011): 387–408.

Drieschova, A., M. Giordano, and I. Fischhendler. "Governance Mechanisms to Address Flow Variability in Water Treaties." *Global Environmental Change: Human and Policy Dimensions* 18 (2008): 285–295.

Eckstein, G. "Water Scarcity, Conflict, and Security in a Climate Change World: Challenges and Opportunities for International Law and Policy." *Wisconsin International Law Journal* 27 (2010): 409–461.

Elmqvist, T., C. Folke, M. Nystrom, G. Peterson, J. Bengtsson, B. Walker, and J. Norberg. "Response Diversity, Ecosystem Change, and Resilience." *Frontiers in Ecology and the Environment* 1 (2003): 488–494.

Endangered Species Act of 1973, 16 U.S.C. §§1531–1544.

Engle, N. L., O. R. Johns, M. C. Lemos, and D. R. Nelson. "Integrated and Adaptive Management of Water Resources: Tensions, Legacies, and the Next Best Thing." *Ecology and Society* 16, no. 1 (2011): 19. http://www.ecologyandsociety.org/vol16/iss1/art19.

Georgia v. U.S. Army Corps of Engineers, 302 F.3d 1242 (11th Cir. 2002).
Giordano, M. "Managing the Quality of International Rivers: Global Principles and Basin Practice." *Natural Resources Journal* 43 (2003): 111–136.
Giordano, M. A., and A. T. Wolf. "Sharing Waters: Post-Rio International Water Management." *Natural Resources Forum* 27 (2003): 163–171.
Gleick, P. "Whose Water Is It? Water Rights in the Age of Scarcity." *San Francisco Chronicle* August 2, 2009.
Gregory, R., D. Ohlson, and J. Arvai. "Deconstructing Adaptive Management: Criteria for Applications to Environmental Management." *Ecological Applications* 16 (2006): 2411–2425.
Grossman, M. "Cooperation on Africa's International Waterbodies: Information Needs and the Role of Information-Sharing." In: *Transboundary Water Management in Africa— Challenges for Development Cooperation*, eds. W. Scheumann and S. Neubert, 173–235. Bonn, Germany: German Development Institute, 2006.
Holling, C. S. "Resilience and Stability of Ecological Systems." *Annual Review Ecology and Systematics* 4 (1973): 1–23.
In re Tri-State Water Rights Litigation, 644 F.3d 1160 (11th Cir. 2011).
Intergovernmental Panel on Climate Change. *Climate Change 2007: Synthesis Report*. Geneva: IPCC, 2007.
Jamil, M. "Violation of the Indus Waters Treaty." *The Nation* October 28, 2008. http://www.nation.com.pk/pakistan-news-newspaper-daily-english-online/columns/28-Oct-2008/Violation-of-Indus-Waters-Treaty.
Kansas v. Colorado, 185 U.S. 125 (1902).
Kiani, K. "IRSA Chief Calls for More Reservoirs." *Dawn News* [Islamabad] October 1, 2011. http://dawn.com/2011/10/01/irsa-chief-calls-for-more-reservoirs.
Kilgour, D. M., and A. Dinar. "Flexible Water Sharing Within an International River Basin." *Environmental and Resource Economics* 18 (2001): 43–60.
Kinzig, A. P., P. Ryan, M. Etienne, T. Elmqvist, H. Allison, and B. H. Walker. "Resilience and Regime Shifts: Assessing Cascading Effects." *Ecology and Society* 11, no. 1 (2006): 20. http://www.ecologyandsociety.org/vol11/iss1/art20.
Kirmani, S. "Water, Peace and Conflict Management: The Experience of the Indus and Mekong River Basins." *Water International* 15 (1990): 200–205.
Kliot, N., D. Shmueli, and U. Shamir. "Institutions for Management of Transboundary Water Resources: Their Nature, Characteristics and Shortcomings." *Water Policy* 3 (2001): 229–255.
LeMarquand, D. "Politics of International River Basin Cooperation and Management." *Natural Resources Journal* 16 (1976): 883–901.
Lenton, T. M. "Early Warning of Climate Tipping Points." *Nature Climate Change* 1 (2011): 201–209.
Liu, J., et al. "Complexity of Coupled Human and Natural Systems." *Science* 317 (2007): 1513–1516.
Matthews, O. P. *Water Resources: Geography and Law*. Washington D.C.: American Association of Geographers, 1984.

McCaffrey, S. "The Harmon Doctrine One Hundred Years Later: Buried, not Praised." *Natural Resouces Journal* 36 (1996): 549–90.

McCaffrey, S. *The Law of International Watercourses: Non-Navigational Uses.* Oxford: Oxford University Press, 2001.

McCarthy, J., O. F. Canziani, N. A. Leary, D. J. Dokken, and K. S. White, eds. *Climate Change 2011: Impacts, Adaptation, and Vulnerability.* Cambridge: Cambridge University Press, 2001.

Odom, O., and A. T. Wolf. "Institutional Resilience and Climate Variability in International Water Treaties: The Jordan River Basin as "Proof-of-Concept." *Hydrological Sciences Journal* 56 (2011): 703–710.

Perrings, C. "Resilience in the Dynamics of Economy-Environment Systems." *Environmental and Resource Economics* 11 (1998): 503–520.

Putnam, R. D. "*E pluribus unum*: Diversity and Community in the Twenty-first Century, The 2006 Johan Skytte Prize Lecture." *Scandinavian Political Studies* 30 (2007): 137–174.

Reeves, R. R., A. A. Chaudhry, and U. Khalid. "Competing for Water on the Indus Plain: Is There a Future for Pakistan's River Dolphins?" *Environmental Conservation* 18 (1991): 341–350.

Ruhl, J. B. "Regulation by Adaptive Management—Is It Possible?" *Minnesota Journal of Law, Science and Technology* 7 (2005): 21–57.

Sadoff, C., T. Greiber, M. Smith, and G. Bergkamp. *Share: Managing Water Across Boundaries.* Gland, Switzerland: International Union for Conservation of Nature and Natural Resources, 2008.

Sadoff, C. W., and D. Grey. "Beyond the River: The benefits of Cooperation on International Rivers." *Water Policy* 15 (2002): 119–131.

Scheffer, M., and S. R. Carpenter. "Catastrophic Regime Shifts in Ecosystems: Linking Theory to Observation." *Trends in Ecology & Evolution* 18 (2003): 648–656.

Schmeier, S. "Navigating Cooperation Beyond the Absence of Conflict: Mapping Determinants for the Effectiveness of River Basin Organizations." *International Journal for Sustainable Societies* 4 (2010): 11–27.

*The Nation.* "No Consensus on Kishan Ganga Dam. Lahore, Pakistan." June 3, 2008. http://www.nation.com.pk/pakistan-news-newspaper-daily-english-online/politics/04-Jun-2008/No-consensus-on-Kishan-Ganga-Dam.

*The Nation.* "India Stopped from Building Kishaganga Dam. Islamabad." September 25, 2011. http://www.nation.com.pk/pakistan-news-newspaper-daily-english-online/politics/25-Sep-2011/India-stopped-from-building-Kishanganga-Dam.

*The Telegraph.* "Aral Sea 'one of the planet's worst environmental disasters.'" April 5, 2010. http://www.telegraph.co.uk/earth/earthnews/7554679/Aral-Sea-one-of-the-planets-worst-environmental-disasters.html.

Tol, R., and G. Yohe. "The Weakest Link Hypothesis for Apative Capacity: An Empirical Test." *Global Environmental Change* 17 (2007): 218–227.

Touza, J., and C. Perrings. "Strategic Behavior and the Scope for Unilateral Provision of Transboundary Ecosystem Services That Are International Environmental Public Goods." *Strategic Behavior and the Environment* 1 (2011): 89–117.

Turton, A., and N. Funke. "Hydro-Hegemony in the Context of the Orange River Basin." *Water Policy* 10 (2008): 51–70.

Webb, C. T., and S. A. Levin. "Cross-System Perspectives on the Ecology and Evolution of Resilience." In: *Robust Design: A Repertoire of Biological, Ecological, and Engineering Case Studies*, ed. E. Jen, 151–172. Oxford: Oxford University Press, 2005.

Wilkinson, C. "Aldo Leopold and Western Water Law: Thinking Perpendicular to the Prior Appropriation Doctrine." *Land and Water Law Review* 24 (1989): 36–74.

Wolf, A. T. "The Transboundary Freshwater Dispute Database Project." *Water International* 24 (1999): 160–163.

Wolf, A. T. "Shared Waters: Conflict and Cooperation." *Annual Review of Environment and Resources* 32 (2007): 241–269.

Wu, X., and D. Whittington. "Incentive Compatibility and Conflict Resolution in International River Basins: A Case Study of the Nile Basin." *Water Resources Research* 42 (2006): W02417.

Yoffe, S., A. T. Wolf, and M. Giordano. "Conflict and Cooperation over International Freshwater Resources: Indicators of Basins at Risk." *Journal of the American Water Resources Association* 39 (2003): 1109–1126.

Zeitoun, M., and J. Warner. "Hydro-Hegemony—A Framework for Analysis of Trans-Boundary Water Conflicts." *Water Policy* 8 (2006): 435–460.

# Ecosystem Services, Ecosystem Resilience, and Resilience of Ecosystem Management Policy

J. B. RUHL AND F. STUART CHAPIN III

Two emerging theoretical models have captivated ecological science and policy over the past decade. One is the concept of ecosystem services, which focuses on the benefits that people derive from ecosystems, including the flows of economically valuable services to human populations (Costanza et al. 1997; Daily 1997; Ruhl et al. 2007; Ruhl and Salzman 2007). The other is resilience theory, which explores how natural and social systems withstand disturbances over time (Gunderson and Holling 2002; Gunderson et al. 2010). In this chapter, we examine the connections between the two.

Ecosystem services theory and resilience theory have both gained tremendous stock in ecosystem management policy over the past decade (Ruhl and Salzman 2007; Benson and Garmestani 2011). The reason is simple—each resonates firmly in the modern conception of ecosystems as complex adaptive systems (Levin 1999). Ecosystem services theory merges the disciplines of ecology, geography, and economics to gain a better understanding of how complex ecological landscapes produce a natural economy that sustains human and social capital (Ruhl et al. 2007). Resilience theory studies the social-ecological interface to gain a better understanding of how dynamic forces in nature affect social systems, and vice versa (Benson and Garmestani 2011). Standing alone, each of these theoretical models has established substantial independent credibility throughout academic, government, and private research bodies. Less attention has been paid, however, to the relationship between ecosystem services theory and resilience theory. Are the two mutually antagonistic or will one support application of the other? How will knowing about one model influence how the other is developed and used? Given how important ecosystem services theory and resilience theory have become to environmental policy, it is fitting to devote attention to these questions.

We approach the topic from two perspectives. The first is to ask whether using the concept of ecosystem services in policy can support the resilience of ecological systems. Is there anything to be gained in sustainable ecosystem management by applying ecosystem services theory, and are there any risks from doing so? The second perspective is to ask whether using the concept of ecosystem services in ecosystem management policy can support the resilience of policy itself. Will using ecosystem services concepts help ecosystem management policy deliver on its goals over sustained time frames in the face of massive perturbations, such as global climate and economic changes?

The reason for dividing the analysis into these two perspectives is that promoting resilience of ecosystems, if that is our policy goal, does not necessarily demand that we achieve resilience in ecosystem management policy, and likewise, achieving resilience in ecosystem management policy does not necessarily confer ecosystem resilience. The two, quite simply, are not the same or even necessarily compatible, and this is often overlooked in resilience theory literature. It may be desirable to support resilience in both ecosystems and ecosystem management policy, but there may be complex trade-offs between the two that make mutual optimization unattainable. How and where we apply ecosystem services theory, therefore, may depend on our overall objectives for how and where we desire resilience in nature and in policy.

Our discussion opens with a background on the features of ecosystem services theory relevant to the resilience context. The two perspectives on how applying ecosystem services in policy could support or undermine resilience are then explored, resilience of ecosystems first, followed by resilience in ecosystem management policy. Given what is learned through those two interrogations, the chapter concludes with some thoughts about how and where ecosystem services theory can most usefully be employed to support the policy goals of resilience of ecosystems and resilience in ecosystem management policy.

## Major Themes and Implications of Ecosystem Services Theory

The concept of ecosystem services is a classic old/new idea. Plato expounded on the values humans derived from healthy ecosystems, as did the influential geographer George Perkins Marsh centuries later (Mooney and Ehrlich 1997). It is obvious that healthy, sustainable ecosystems are good not just for bugs and bunnies, but for humans as well, so what is new and important about making that observation for purposes of environmental policy in the twenty-first century?

The answer can be traced to several trends that have unfolded since the dawn of the modern era of American environmental law and policy in the early 1970s. Before 1970, environmental law consisted primarily of the common law of nuisance and smatterings of ad hoc state and federal legislative measures having to do mostly with public lands and natural resource extraction (Lazarus 2004). There was no broad, coherent environmental policy at national or state levels. With public attention focused on environmental disasters such as Love Canal, however, beginning in 1970 Congress and President Nixon in short order hammered out a series of landmark environmental protection statutes covering, among other things, water pollution, air pollution, coastal zone management, environmental impact assessment, federal public land management, and endangered species protection (Lazarus 2004). This new statutory regime set in motion several related policy debates that did not crescendo until the mid-1990s and which bore directly on the emergence of ecosystem services theory.

The first policy debate concerned the balance between what is often described (usually derisively) as command-and-control regulation of land use and pollution behavior versus market-based instruments, such as pollution trading and wetlands banking (Freeman and Kolstad 2007). Fueled by the apparent success of the sulfur-dioxide emissions trading program designed to control acid rain, which was the result of 1990 amendments to the Clean Air Act, sentiment in favor of relying more on the market-based approaches to achieve environmental policy goals had grown strong by the mid-1990s. Much of this debate centered on the nation's primary ecological regulation statute, the Endangered Species Act of 1973, which was under intense fire in Congress during the early 1990s (Ruhl 1998).

Another policy tension building during this period was between the rise of economic cost–benefit analysis as a method of developing and assessing policy choices and the growing discontent with its incomplete (some would say cavalier) treatment of environmental values (Ackerman and Heinzerling 2004). The premise of cost–benefit analysis as practiced at the time was that overall efficiency should be an important driver in how policy decisions are made and implemented and that careful comparison of the costs and benefits of alternative policy choices could usefully inform how best to achieve the efficiency goal. The problem was that the science of cost–benefit analysis was not very advanced when it came to describing the benefits of maintaining ecological integrity. The result was that many cost–benefit analyses simply described benefits such as flood protection and groundwater recharge from forests and other ecosystem resources as "unquantified benefits" that had little influence in the analysis, given the hard numbers that could be assigned to costs such as lost jobs and other economic impacts (Ackerman and Heinzerling 2004).

The third policy debate centered around the theories of ecosystem management and adaptive management, which were gaining force in the early 1990s, particularly as they were proposed for application to federal public lands (Grumbine 1994). There was growing concern that the "multiple-use" mandate of public land units such as national forests and grazing lands had been implemented by the land management agencies with too strong a bias toward resource use and extraction at the expense of sustainable ecological integrity. By the mid-1990s, this increasingly vocal movement for a "nature first" ecosystem-scale approach to public land management had grown in strength (Christensen et al. 1996; Brooks et al. 2002).

Perhaps one reason the concept of ecosystem services captured attention during this time frame was its ability to inform and possibly help resolve each of these policy tensions. Unlike the musings of Plato and Marsh, modern ecosystem services theory is a rigorous scientific discipline merging ecology, economics, and geography to go beyond simply observing the importance to humans of healthy ecosystems, to actually describing what those values are, their ecological sources, and their human beneficiaries. By focusing on the economic value supplied to human populations from intact, healthy, sustainable ecological resources, ecosystem services theory matched up well with the trend toward market-based thinking in environmental policy. Also, by driving research toward improving the capacity to describe those economic values, ecosystem services theory could provide a much stronger voice for ecological values in cost–benefit analysis. And the idea that intact ecosystems provide economic values to potentially distant populations across many scales reframes the debate over multiple use of public lands by clarifying a broader set of trade-offs between resource depletion and resource protection.

The stage was therefore well set for ecosystem services concepts to hit the scene with a splash in 1997, first in an article in *Nature* (Costanza et al. 1997) and later that same year in a more comprehensive book titled *Nature's Services* (Daily 1997). In the *Nature* article, lead author Robert Costanza and his research team estimated the global economic value of ecosystem services at over US $30 trillion annually, and *Nature's Services,* edited by Gretchen Daily, surveyed an array of ecological systems such as wetlands and forests as sources of numerous ecosystem services streams. These groundbreaking works were instrumental in placing the concept of ecosystem services firmly on the map several years later when the United Nations Millennium Ecosystem Assessment (MEA) organized its reports around the theme (MEA 2003). The volume of literature developing and applying the theory of ecosystem services has exploded since then (Ruhl et al. 2007; Gomez-Baggethun et al. 2010).

The MEA reports were influential in developing the now widely accepted typology of ecosystem services, under which benefits flow to human populations in four streams: (1) *provisioning services* are commodities, such as food, wood, fiber, and water; (2) *regulating services* moderate or control environmental conditions, such as flood control by wetlands, water purification by aquifers, and carbon sequestration by forests; (3) *cultural services* include recreation, education, and aesthetics; and (4) *supporting services,* such as nutrient cycling, soil formation, and primary production, make the previous three service streams possible (MEA 2003). To put these in a concrete context, think of the public lands management debate mentioned above, in which advocates of ecosystem management theory accused the traditional agency management regimes of overly favoring resource use and extraction. Translated into the ecosystem services lexicon, that debate focuses on the balance between provisioning services (e.g., timber harvests and water diversions) and readily monetized cultural services (e.g., hunting and off-road vehicles) on the one side, and regulating services (e.g., carbon sequestration and groundwater recharge), supporting services (e.g., nutrient cycling and soil formation), and those cultural services not readily monetized (e.g., sense of place, aesthetic value) on the other. What this framework adds to that debate is the clear sense that public lands not only provide value through on-site use and extraction, but also supply value to potentially large and distant human populations through off-site flows of regulating and supporting services (Ruhl 2010b). (For the sake of simplicity, reference to cultural services in the remainder of this chapter applies to those that are readily monetized rather than the full suite of cultural services.)

The capacity of ecosystem services theory to reframe such debates is a potentially powerful force in environmental policy, but it is for that very reason that the theory has attracted critics as well as advocates. Advocates of using ecosystem services concepts in environmental decisions make the case that to the extent money talks, as it increasingly has through the trends toward market-based approaches, cost–benefit analysis, and multiple-use land management, failing to frame environmental protection in economic terms is folly—the environment will not get a seat at the table. But critics argue that a larger seat may come at too great a cost. One objection is that going down the ecosystem services road will serve only to commodify nature and crowd out other bases for resource protection, such as intrinsic values, ethical concerns, and social equity (McCauley 2006; Kosoy and Corbera 2010). Another is that managing resources to produce ecosystem services may lead to optimization strategies favoring particular services that have a high current value, such as carbon sequestration, and thereby possibly diminish other services, such as groundwater recharge (Dasgupta 2001; Bennett et al. 2009).

There is something of merit behind both views. Ecosystem services theory truly can make a difference in how resource allocation decisions are viewed and made. Unlike provisioning and cultural services, the market value of which is embedded in commodity prices and entry fees and thus easily measured and monitored, regulating and supporting services tend to behave more like nonmarket public goods (Lant et al. 2008). It is difficult for the owner of honeybee habitat to charge for pollination services enjoyed on distant farms or for the farmers to charge for groundwater recharge enjoyed by the honeybee habitat owner. Likewise, landowners and urban dwellers enjoying services flowing from other parcels see them as "free" and thus have little incentive to invest in the resources providing them.

Landowners thus have incentives to optimize provisioning and cultural services available to them, but little incentive to provide regulating and supporting services that benefit people in other places. In this respect, for example, a farm is like any other ecological unit—changes in one ecosystem usually affect other ecosystems, however we draw the boundaries. But as highly managed ecological units, farms significantly tilt the production frontier for ecosystem services toward provisioning and cultural services and away from regulating and supporting services (Ryszkowski 1995; Bjorklund et al. 1999). Ecological practices at a cornfield are designed to produce corn efficiently within the relevant regulatory and economic environment. Putting aside the question of whether regulation of farms has established appropriate environmental performance baselines, unless farmers are paid to provide off-site regulating and supporting services such as carbon sequestration and groundwater recharge, one would not expect to find significant flows of off–site regulating and supporting services from farms, except as they might be incidental to optimizing the on-site provisioning and cultural services upon which the farmer makes a living. Seen from this perspective, ecosystem services theory focuses environmental policy on salient questions (such as how to shift the farm services production frontier to a more multifunctional mix) and toward effective approaches (probably by using incentives and other payment mechanisms) (Ruhl 2008; Yang et al. 2010). Careful analyses of the identity and location of beneficiaries and providers of different types of ecosystem services thus set the stage for considering a range of policies and payment schemes to address different types of trade-offs among ecosystem services (Bennett et al. 2009; Kinzig et al. 2011; Seppelt et al. 2011).

But there may be danger in overplaying the ecosystem services card. There is an old adage in resource management that "if you manage for ducks, you get ducks," and the same could be the result of making ecosystem services production the driving goal of policy decisions. For example, managing for particular ecosystem services outputs might or might not produce the biodiversity characteristics we wish to maintain for other purposes (Duarte 2000; Balvanera et al. 2001; Bhaskar and Adams 2009). Similarly, the drive to commodify services, such as through the practice of allowing landowners to sell "credits" for a broad array of "stacked" services generated through resource conservation, has become the subject of intense controversy over whether it truly will lead to sound ecosystem management decisions (Kosoy and Corbera 2010; Kinzig et al. 2011). The fear is that landowners will chase dollars, which may lead to more conservation, but will it be the desired conservation?

Another critique of ecosystem services theory is that despite garnering much attention among academics and policy makers, it has yet to gain much traction in hard law (Salzman 1998; Fischman 2001; Ruhl and Gregg 2001; Ruhl et al. 2007). Even when ecosystem services have been recognized in statutes, agency regulations, or judicial opinions as legally relevant, very little detail has been offered regarding how to implement the concept through legal rules and standards. As an example, although the U.S. Army Corps of Engineers and Environmental Protection Agency recently promulgated regulations requiring that provision of ecosystem services be maintained or replaced at wetland impact sites, the regulations contain absolutely no details on which services matter; how, where, and when they must be measured; or how the agencies will ensure that land developers meet the requirement (Ruhl et al. 2009). Although the lack of progress on integrating ecosystem services concepts into environmental law can be attributed at least in part to the need for more focused, policy-relevant research, the enthusiasm that greeted the emergence of ecosystem services theory in the 1990s has dampened as it becomes increasingly clear that the knowledge base is still limited and translating the theory into enforceable law may be more difficult than thought.

We do not, however, intend to resolve the debate over ecosystem services theory or propose specific legal text. Rather, the focus here is to ask how using ecosystem services concepts to shape environmental decisions could promote or undermine resilience in the environment and in environmental policy. For that purpose, the foregoing discussion illuminates five key features of ecosystem services theory warranting particular emphasis.

First, ecosystem services theory is fundamentally anthropocentric in focus—it is about how ecological systems deliver economic value *to humans*. Two ecologically indistinguishable wetlands, for example, may deliver vastly different ecosystem services profiles based on proximity to human populations. While this does not necessarily mean one wetland is "better" than the other, it does mean they are different at least in their proximity to people, and this difference may influence ecosystem management policy.

Second, notwithstanding this anthropocentric focus, ecosystem services theory is not intended to displace other reasons for managing and protecting ecological integrity, such as the intrinsic value of species or ethical beliefs about human responsibilities toward nature. Knowing the ecosystem services profile of a wetland parcel adds to the information available for making ecosystem management decisions without detracting from any other scientific, economic, or moral basis for decision making (Costanza et al. 1989).

Third, where ecosystem services theory pushes hardest on conventional economic thinking about the environment is with respect to regulating and supporting services. Indeed, it is unlikely that ecosystem services theory offers much besides semantics when it comes to managing provisioning and cultural services already deeply embedded in economic markets and resource management policy. Calling corn a "provisioning service" or hunting a "cultural service" hardly changes the policy dialogue. Regulating and supporting services, by contrast, have been largely ignored as explicit benefits in markets and in policy, and thus present the greatest opportunity for scientific and policy advancement through attention to their service values (Lant et al. 2008).

Fourth, just as ecosystem services theory does not displace other bases for environmental protection decisions, neither does it displace other bases for economic development decisions. Even if all regulating and supporting ecosystem services values were somehow integrated into market prices, the fact that maintaining ecological integrity of a wetland parcel provides economic value to some humans does not mean the owner will favor that use over, say, a housing subdivision. In the absence of policies imposing regulatory constraints on land use decisions, therefore, ecosystem services will have to compete in the market along with alternative uses of land and resources.

Fifth, ecosystem services theory exposes and recognizes inherent trade-offs. It exposes trade-offs environmental policy has been making without the benefit of complete information about consequences, such as through the cost–benefit analysis practice of dumping ecosystem services into the "unquantified benefits" category. But it also recognizes trade-offs between ecosystem services, such as the tension between provisioning/cultural services and regulating/supporting services, as well as between beneficiary populations, such as the trade-off that could occur between favoring the global benefit of carbon sequestration at the expense of the local benefit of groundwater recharge (Jackson et al. 2005; Kinzig et al. 2011).

The overarching principle to draw from these features is that ecosystem services theory is not a panacea. It does not offer a complete theory of ecological or economic decision making, or even one with clear and unequivocal outcomes. Rather, ecosystem services theory expands the analytical devices available for decision making, which in some contexts may clarify decisions and in other contexts may cloud them with yet more uncertainty and controversy. This chapter isolates just one of the many parameters over which these effects could play out—resilience as a policy goal for the environment and for environmental policy.

## Ecosystem Services and Ecological Resilience

Ecosystem services theory has only recently been put into practice and is not yet a dominant force in the hard law of environmental decision making. Asking what ecosystem services theory offers the policy goal of resilience thus remains itself largely a theoretical question, particularly given that the exact dimensions and properties of resilience are themselves not very far developed beyond theory. We can adopt the conventional general definition of resilience as the flexibility of an ecological or social-ecological system to adjust to unforeseen shocks and stresses and to sustain its fundamental structure, function, and identity (Holling

1973; Folke 2006; Chapin et al. 2011), but there is no definitive set of principles for measuring and characterizing ecosystem resilience in the field, much less for establishing policy goals about ecosystem resilience. Even assuming we could measure, monitor, and manipulate ecosystem resilience with precision, that would not settle the policy question of what quality and quantity of resilience to achieve for any given ecosystem. Resilience is just one among many possible policy goals for ecosystem management and in many cases is unlikely to be the goal that drives decision making compared with, say, food production or endangered species protection. In some cases, as in degraded mine tailings or overgrazed pastures, resilience of the current ecosystem state may not even be a desirable policy goal (Benson and Garmestani 2011; Ruhl 2011). Indeed, the challenges of ecological restoration are to overcome the resilience of degraded systems (Hobbs and Cramer 2008).

We do not comprehensively define the dimensions of ecosystem resilience or chart the way for ecosystem resilience policy. Nevertheless, asking how ecosystem services theory relates to ecosystem resilience theory requires some basic sense of what ecosystem resilience is about. To lend some framework to this analysis, therefore, we employ a model of system resilience developed by Alderson and Doyle (2010) that unpacks the well-known engineering/ecological resilience distinction (Holling and Gunderson 2002) into a more refined typology of attributes. Under their model, five key properties of a system contribute to the capacity to endure through surrounding change:

*Reliability* involves robustness to component failures. *Efficiency* is robustness to resource scarcity. *Scalability* is robustness to changes to the size and complexity of the system as a whole. *Modularity* is robustness to structured component rearrangements. *Evolvability* is robustness of lineages to changes on long timescales (Alderson and Doyle 2010).

Ecosystems exhibit all of these properties in one degree or another (Pimm 1984; Milne 1998; Levin 1999). A forest, for example, may experience isolated fires that damage portions of the landscape that are repaired over time through reliability resilience. During drought, efficiency resilience directs how the forest ecosystem responds to diminished water resources, with some species succumbing to the stress before others. Forests have scalability resilience, as in a nested system of watersheds within larger watersheds, through which drought might devastate some smaller watersheds while only moderately impairing the larger watershed system as a whole. Forests also can respond to stress with modularity resilience by transitioning components on micro-levels, as may happen when a wetland area transforms to upland in sustained drought, while not losing overall forest system integrity at larger scales. Finally, forests evolve as even natural disturbances and long-term changes in exogenous conditions such as climate introduce new species, change watercourses, and alter temperature regimes.

The overall resilience of an ecosystem to stress and perturbation at different spatial and temporal scales is a function of how these five resilience qualities are packaged over any given space and time span. Although it would be ideal to maximize all five properties of resilience in a natural or social system, it is far more likely that trade-offs will limit that possibility, forcing difficult decisions about system design. For example, reliability and efficiency appear most in keeping with the concept of engineering resilience, whereas scalability, modularity, and evolvability match up with ecological resilience (Holling and Gunderson [2002, 28] describe engineering resilience as focusing on "maintaining *efficiency* of function," whereas ecological resilience "focuses on maintaining *existence* of function"). A system that is highly efficient in using scarce resources might as a consequence have less modularity due to lack of redundancy in important system functions. A recurrent system design question, therefore, is which of these resilience properties to favor.

That is, of course, an ever-present and complex policy decision in environmental management. Simply put, what kind of an ecosystem are we trying to attain and maintain through our management strategies? The answer, naturally, will differ based on circumstances and often will be highly contested for a given ecological unit. This chapter, therefore, assumes no fixed set of principles or goals. Rather, our objective is to ask a focused question: Assuming that achieving and maintaining a prescribed quality of ecological resilience is one of the policy goals of a particular environmental management program, how might using the theory and practice of ecosystem services contribute to or detract from attaining that goal? The following discussion examines that question through the model of the five resilience properties.

### *Reliability*

Over the past few decades, ecologists have increasingly moved toward the model of ecosystems as complex adaptive systems in which disturbance regimes play a pivotal role in sustaining system dynamics (Levin 1999). Flood, fire, drought, and storm shake up ecosystems by introducing diversity, putting pressure on weak links, and triggering feedback processes. The capacity to "bounce back" from such disturbances is the hallmark of resilience in a complex system, and that is no less the case for ecosystems. Far from existing in fragile states of perfect equilibrium, as anyone who has witnessed the carpet of green emerging only weeks after a forest fire will attest, ecosystems exhibit tremendous capacity to rebuild after even severe disturbances. This capacity is tied in part to the diversity and redundancy of ecosystem components and processes, which are in turn promoted (and tested) by disturbance regimes.

Disturbance is thus, ironically, a source of sustainability for ecosystems by maintaining a diversity of ecosystem components from which ecosystems can rebuild. Of course, as the geologic record all too clearly reveals, ecosystems are not impervious to disturbance; nonlinear tipping points may be encountered, such as after periods of protracted drought or cold, that overwhelm redundancies and feedback processes critical to system reliability and "flip" an ecosystem into an entirely new and starkly different set of system dynamics.

A central question around which much of environmental policy revolves is how humans contribute to maintaining or degrading ecosystem reliability from small to global scales. How will a land development project interfere with wetland processes and make it less likely the larger wetland system will withstand the next severe drought? How can conservation measures be taken to restore reliability to a degraded wetland system? What are the consequences to reliability mechanisms in marine and other ecosystems of continued rise in anthropogenic releases of carbon dioxide into the atmosphere? Humans can hardly escape the reality that we are a disturbance regime unlike any other for ecosystems (Vitousek et al. 1997), but unlike fire and flood, we can ask what kind of disturbance regime we want to be—the kind that contributes to ecosystem dynamics or the kind that pushes ecosystems to the next nonlinear tipping point.

So how can ecosystem services theory help answer that question? As a starting point, it seems reasonable to assume that the more people know about the economic values they derive from ecosystem services, the more they will take them into account in decision making. The market and policy demand curves for ecological conditions that provide preferred ecosystem services values thus are likely to shift to the right—that is, to increase in demand—as more and more people seek continued supply of the previously poorly understood benefits. No doubt, for example, owners of coastal properties along the Gulf of Mexico have in recent years increased their appreciation of the value of coastal wetland and dune systems. As noted above, moreover, the greatest potential for increased information about ecosystem services, and thus the greatest engine of increased demand, is through research illuminating the sources and beneficiaries of regulating and supporting services.

It is not entirely clear, however, how increased demand for ecosystem services will influence ecosystem reliability. Indeed, it could lead to decisions that intervene drastically in ecosystem processes. For example, public demand for provisioning and cultural ecosystem services on public lands led for decades to policies such as fire suppression and flood control that are now discredited as having degraded ecosystem reliability properties. Whereas under natural conditions a forest fire might have been contained to a limited area and low intensity, thus allowing the forest to repair the affected components, in some forest types the excess underbrush built up under fire suppression policies led to high-intensity fires that overwhelmed reliability capacity (Schoennagel et al. 2004). Similar missteps could follow from policies designed to meet increasing demand for regulating and supporting services. Intense management for carbon sequestration benefits, for example, could interfere not only with other services but also with ecosystem processes critical to sustained reliability (e.g., as a result of favoring species with high sequestration capacity). In other words, simply saying "we are managing to promote ecosystem services" does not necessarily advance the reliability of ecological components—it will depend on which services and how we manage for them.

On balance, however, a general shift in demand favoring regulating and supporting ecosystem services is likely to lead to decisions promoting ecological reliability, because improved information about the range of services provided by ecosystems will lead to more informed decisions about the trade-offs inevitably associated with resource management decisions. Provisioning services (corn, timber, minerals) and cultural services (rafting, hiking, visitation) focus management on narrow commodities and activities as the primary output of highly managed ecosystems, whereas regulating services (water purification, flood control) and supporting services (nutrient cycling, decomposition) depend more on the work of an intact, dynamic background of ecosystem processes and structure. An ecosystem services framework also raises awareness of need for restoration to provide a greater spectrum of services from degraded ecosystems that are characterized by high-reliability resilience.

Ecosystem services policy could therefore be shaped to promote demand for regulating and supporting services as a way of enhancing ecosystem reliability. Consider, for example, that farm policy increasingly has focused on how to shift working agricultural lands toward a multifunctional profile in which the payoff to farmers for increased land conservation practices is compensation for the enhanced flow of regulating and supporting services to surrounding lands (Ruhl 2008). The intended result of such policies is to create more complex and varied agricultural landscapes with greater integration of areas functioning closer to natural ecological conditions. With careful attention to potential downsides, therefore, ecosystem services policies can be used to promote land use practices that boost ecosystem reliability at landscape scales.

## *Efficiency*

Increased knowledge about ecosystem services benefits inevitably will lead to more economically efficient decisions about use of land and resources. If a municipal water system demands increased water supply and local farmers can increase groundwater recharge through altered land use practices, it may pay for both to sit at the bargaining table (Ruhl 2008). The question, however, is whether the potential ecosystem services theory offers to achieve *economic* efficiency translates into a capacity to also achieve *ecological* efficiency (Muradian et al. 2010; Pascual et al. 2010).

Ecological efficiency can be thought of as the capacity to maintain functions critical to sustained ecological integrity, even when essential resources such as water and nutrients become scarce. Two wetlands could be compared based on, say, biomass production and nutrient conversion during periods of drought as a way of comparing their respective efficiency attributes. The conclusion that one is more "efficient" than the other may be largely an anthropocentric perspective, but nevertheless, measuring and comparing ecological efficiency contributes to our knowledge about how ecosystems function and also how they respond to stress-inducing disturbance, including anthropogenic disturbance, that can push an ecosystem to its efficiency limits.

As with the reliability property, ecosystem services policy could affect ecological efficiency for better or worse depending on the choices made regarding which services to promote and how. Policies designed to enhance flows of provisioning and cultural services seem more likely to lead to interventions in ecosystem processes that could degrade efficiency, whereas policies designed to keep the suite of regulating and supporting services intact are more likely to keep ecological processes closer to their natural efficiency conditions. Efficiency thus is yet another potential trade-off factor that must be integrated into ecosystem services policy choices.

But ecosystem services may prove useful in this respect, not only as a policy driver but as a policy metric (Ruhl 2010a). Changes in the flow of ecosystem services from a particular landscape tell us something about conditions in the ecosystem. A sharp drop in a particular service, such as pollination, may indicate that scarcity of resources has tapped an ecosystem's efficiency capacity. Indeed, building ecosystem services measurements into policy as an ecological efficiency metric could serve to link economic and ecological efficiency in ways that will attune economic decisions to ecological conditions. If a voluntary or regulatory market exists for an economically valuable ecosystem service, a drop in the flow of the service is likely to get the attention of the sellers and the buyers, driving them to investigate the cause and invest in remedies. Even in the absence of a market, real-time information about the flow of service levels to human populations is likely to reinforce connections the public makes between their economic conditions and the ecological conditions making them possible. Where sustained economic efficiency depends on secure ecological efficiency, those interested in maintaining the former are all the more likely to be interested in maintaining the latter. Building ecosystem services metrics into policy through regulatory markets, performance standards, and other instruments can help reinforce that relationship.

### *Scalability*

Time and size matter in ecosystem dynamics, and thus temporal and spatial scales are a central focus of ecological research, particularly with the rise of the complex adaptive systems model of ecosystems (Peterson et al. 1998). Complex relationships between "small/fast" and "large/slow" ecological processes across landscapes produce feedback loops and emergent behaviors that make the study of scale properties vitally important to a more complete understanding of ecological conditions (Carpenter and Turner 2000). The scalability property of resilience captures the importance of these complex cross-scale relationships and the ability of a system to respond to disturbance or change at one scale through responses at other scales (Rodriguez et al. 2006).

Increased attention to ecosystem services flows is likely to place emphasis on scale properties in ecosystems supplying valuable services and perhaps even make scale a contentious issue in policy decisions. Consider again, for example, potential trade-offs between carbon sequestration and groundwater recharge in forest ecosystems. The service values of carbon sequestration promoted at a forest parcel are enjoyed globally, but with a long lag time and highly dispersed per capita benefits. Groundwater recharge, by contrast, is enjoyed at a far more local level with immediate and measurable payoff for local water users. Policy decisions favoring one service versus the other thus could pit one scale of human beneficiaries against another as well (Ruhl 2010b). As knowledge about such scale trade-offs increases through research, ecosystem services policy could become contentious, even if it shifts focus from provisioning/cultural services to regulating/supporting services.

Moreover, if market and regulatory policies drive private and public land managers toward services with particularized scale properties, management decisions may work toward "locking in" the ecological conditions that best optimize service flow at the scale desired by empowered users, which could undermine overall scalability of the ecosystem. If the money is in carbon, for example, land managers will drive decision making to favor carbon sequestration. It is not at all clear, however, what managing to sustain a high sequestration vegetative regime means at other ecosystem scales and for overall ecosystem scalability. Increased carbon sequestration, for example, may reduce landscape heterogeneity and complexity in the short term but increase fuel loads, leading to wildfires that increase landscape heterogeneity and scalability in the long term (Turner 2010). It will be imperative, therefore, for ecosystem services research to focus on improving knowledge about cross-scale relationships and trade-offs and for policy to be attentive to how choices among service flow alternatives affect ecosystem-scale properties and scalability capacities.

Here again there is reason to believe at a general level that a shift in demand from provisioning/cultural services to regulating/supporting services is likely on balance to enhance ecological scalability. Provisioning and cultural services are most easily visualized, managed, and delivered at the scale of a particular management unit (e.g., forest stand, watershed, estuary). By contrast, regulating and supporting services link ecosystems in one place with those in another (e.g., forest with distant stream; coastal wetland with protected inland area; carbon sequestration in one stand with the atmosphere experienced by people globally). Therefore, managing to enhance regulating and supporting services necessarily requires greater consideration and maintenance of the scalable properties of linked ecosystems.

### *Modularity*

Although disturbance regimes promote ecological resilience by introducing periodic restructuring of ecosystem components, human intervention in ecosystem dynamics has frequently overwhelmed resilience by undermining modularity properties. The introduction of nonnative species, for example, has often led to ecological restructuring that radically shifts an ecosystem into a new regime. On a much larger scale, climate change will lead to species migrations and altered hydrological conditions of magnitudes that surely will test ecological modularity across the landscape (Fox 2007; Williams et al. 2007). Some ecosystems will "absorb" the structural changes and others will not.

Ironically, ecosystem services policy could lead to decisions that reduce ecosystem modularity, or at least de-emphasize it as a desired resilience property. This is not because ecosystem services policy is likely to be a source of ecological restructuring, but rather because it could lead resource managers to resist restructuring forces. A farm, for example, can be thought of as a highly structured and maintained ecosystem-services production unit—its just that it is producing provisioning/cultural services rather than regulating/supporting services—and farmers quite rationally resist forces that threaten to restructure components of the production unit. One concern, therefore, is that even with a shift in policy emphasis toward greater production of regulating and supporting services, land managers will continue to behave as farmers do—they will resist any restructuring of the production unit. The ecological implications of intervening in disturbance regimes, even when the reason for doing so is to enhance or maintain supplies of regulating and supporting services, must be studied and taken into account in ecosystem services policy.

### *Evolvability*

A resilient ecosystem is not necessarily an unchanging ecosystem. The evolvability resilience property captures the idea that an ecosystem is a dynamic system that adapts to its exogenous context over time (Levin 1999). Even with no help from humans, species migrate, climates change, and volcanoes erupt. A resilient ecosystem uses reliability, efficiency, scalability, and modularity to respond to these outside forces, and depending on the mix and success of the blend of these properties, some ecosystems will remain relatively static (high-reliability resilience), some will evolve over the course of time, and some will "flip" into vastly different dynamic states (low-reliability resilience). With countless human interventions, of course, the evolvability property of ecosystems has been tested time and again at unnatural frequencies and extremes, with anthropogenic climate change posing the most global and intense of such tests ever.

How ecosystem services concepts interact with the evolvability property is a complex question rife with potential policy controversy. The theme that has built in the foregoing sections is that ecosystem services policy, at least in some applications, could develop into what might be described as an advanced version of farm policy. That is, just as farmers have traditionally managed agricultural lands to optimize provisioning and cultural ecosystem services, so too might land managers of the future respond to ecosystem services market and regulatory policy by managing lands to optimize regulating and supporting services. The question is whether those land managers will treat their resources as "farms" for ecosystem services. Will they, in other words, manage for resilience of the engineering type by focusing on enhancing reliability and efficiency of the service production unit to keep it "intact," as in unchanged over time, while placing less or no emphasis on the ecological resilience properties of scalability, modularity, and most holistically, evolvability?

Particularly with climate change impacts looming, ecosystem services policy could take one of two overarching directions in this regard. One would be to manage ecosystems for familiar policy goals, such as biodiversity, recreation, wilderness, or water quality, and let ecosystem services profiles of the managed areas follow from there. Another would be to manage ecosystems for the most economically valuable sets of ecosystem services, particularly regulating and supporting services, treating other policy goals as incidental benefits. The more policy favors the latter approach, the more likely it is that the "farm" model will creep into ecosystem services decision making. Of course, this may be a profoundly positive development in terms of the economic value derived from ecosystems, and it may very well lead to greater conservation of lands and resources, but it may also lead to management decisions focused on "protecting" ecosystem services production units from disturbance regimes and evolving conditions. Just as a row crop farm trades off evolvability for efficiency, will a pollination "farm" or a carbon sequestration "farm" do the same, and if so, is that the kind of conservation policy that achieves our desired quality and quantity of resilience in ecosystems?

The more likely scenario is that neither of these extremes will predominate and ecosystem services concepts will be integrated into policy decisions in varying ways and intensities across the spectrum of environmental management programs. Maintaining some areas as wilderness or habitat for an endangered species are narrow but time-tested policy goals unlikely to be supplanted by ecosystem services concepts, yet land managers pursuing those goals can make note of their incidental ecosystem services benefits. Maintaining other areas as ecosystem services "farms," such as high-yield carbon sequestration forest projects, may become a practical necessity that policy makers cannot avoid despite the trade-offs. For the many public and private land management contexts between those extremes, ecosystem services concepts can help inform and facilitate decisions about how to balance the five resilience properties depending on the broad set of goals established for a particular resource area. Because we cannot know the future value of different balances of ecosystem services, a portfolio approach that manages different portions of the landscape for different balances of ecosystem services maximizes evolvability resilience (Bormann and Kiester 2004). Most profoundly, greater awareness of the value of regulating and supporting services is likely on balance to lead to greater appreciation of the scalability, modularity, and evolvability resilience properties in ecological systems.

In sum, one thing can be concluded with confidence about the ecosystem services concept and resilience of ecosystems: ecosystem services theory expands our understanding and articulation of what an ecosystem is useful for, even in the absence of precise monetization of service values, and this new knowledge base undoubtedly will be relevant and useful for making decisions about the design and implementation of ecosystem management goals. How we use that new knowledge will make all the difference in how ecosystem services concepts contribute to ecological resilience. This observation thus leads directly to the other side of the coin—how can ecosystem services concepts contribute to the resilience of ecosystem management policy?

## Ecosystem Services and Ecosystem Management Policy Resilience

Law and policy is a complex social system that, one would reasonably hope, productively incorporates resilience properties such as reliability, efficiency, scalability, modularity, and evolvability (Ruhl 2011). The previous section pointed directly at ecosystem management policy as the gatekeeper for how ecosystem services concepts will influence ecosystem resilience. This assumes, of course, that ecosystem management policy will embrace ecosystem services concepts and use them in decision making. But that leads to the questions this section takes on—how will using ecosystem services concepts in ecosystem management policy influence the resilience of ecosystem management policy, and what kind of policy resilience are we seeking? Just as with resilience of ecosystems, for example, environmental policy resilience is not always desirable. Environmental policies that are locked in place (high-reliability resilience) by special interests to support extractive use of ecosystems often degrade regulating and supporting services and undermine overall ecosystem resilience (Trosper 2003; Beier et al. 2009). Rather than work through the five resilience properties to address those questions, but nevertheless taking them into account, this section assembles our holistic conclusions about how ecosystem services concepts could enhance or impede ecosystem management policy resilience.

### *Factors Enhancing Policy Resilience*

Let us start by assuming a world in which all the information desired about ecosystem services in any setting and scale is inexpensively and readily at our fingertips. How might this ideal state of information access make ecosystem management more resilient? Several key effects rise prominently to mind.

First, knowledge about ecosystem services is likely to make clearer and more direct the connections people have with ecosystems and with one another. For many people, policies designed to protect habitat for an endangered fly would be a hard sell, but if those policies are also framed as protecting the ecosystem services humans receive from the conserved habitat, they are more likely to receive public support. Similarly, if two landowners in an agricultural district learn that one's property supplies groundwater recharge services to the area and the other's property supplies pollination services, they are more likely to take an interest in each other's land management practices. An ecosystem management policy that builds on and reinforces these connections strengthens its legitimacy among regulated and other affected entities.

Second, knowledge about ecosystem services can improve the capacity of decision methods such as cost–benefit analysis and environmental impact assessment to conduct fully informed analyses of policy options and their trade-offs. If ecosystem management decisions are to be based on cost–benefit analysis, then better information about costs and benefits should lead to better decisions. Likewise, if environmental impact assessments are to inform action agencies and the public about the effects of different agency alternatives, profiles of how each alternative alters ecosystem services flows should improve the alternatives comparison analysis. Indeed even raw political discourse over ecosystem management policies can make use of improved information about the consequences of different political positions and interest group demands.

Knowledge about ecosystem services can also enhance the information available to voluntary and regulatory markets and other market-based instruments. Compensatory wetland mitigation, for example, is supposed to compensate for resource losses suffered at development impact sites, but such compensation is incomplete without accounting for ecosystem services losses (Ruhl et al. 2009). By requiring compensation for those losses, regulatory wetland mitigation markets will more accurately value ecosystem services and integrate them into the participants' decisions. Similarly, payment programs, such as farm conservation subsidies, can use ecosystem services concepts to more efficiently allocate subsidy payments to areas where agricultural conservation has economic as well as ecological value (Yang et al. 2010). And regulators need not always be involved, as knowledge about ecosystem services has begun to open up pure voluntary market transactions, such as payment for enhanced groundwater recharge (Ruhl 2008).

Finally, knowledge about ecosystem services can serve to build important metrics for ecosystem management policy. For example, measuring ecosystem services outputs from particular resource units could serve as a performance metric for public land management, water quality protection, and other environmental regulation programs (Ruhl 2010a). Indeed, because ecosystem services practice ultimately works toward the universally fungible metric of market values, it can serve as a metric for cross-program comparisons.

These and similar effects of integrating ecosystem services concepts into ecosystem management policy are likely to enhance the resilience of management decision processes. The reliability of and confidence in regulatory decisions will be improved by tapping into this previously undeveloped (or ignored) knowledge base. Market outcomes will produce greater efficiency as more complete information about resource values is made available. Policy itself can become more evolvable as research on ecosystem services improves the breadth and depth of detail about ecosystem services over time. In short, given the importance that people and policy makers place on knowing the economic impacts of their decisions, it is reasonable to expect that integrating ecosystem services concepts into ecosystem management decision making will contribute to decisions that enjoy greater precision and legitimacy and from which resilience is likely to be enhanced. Now, however, we must return to the real world, where all the information desired about ecosystem services in any setting and scale is *not* inexpensively and readily at our fingertips.

### *Factors Impeding Policy Resilience*

For any of the resilience-enhancing effects discussed above to take hold, knowledge about ecosystem services must be robust. But as the discussion of ecosystem services theory opening this chapter explained, the very nature of ecosystem services is that they are highly varied across time and space. This is true both ecologically and economically. Thus, a wetland parcel in Florida and one in Maine may both meet the general criteria for what makes a wetland, but they may be ecologically quite different. And even two ecologically indistinguishable wetland parcels in the same area could present vastly different economic profiles. Several studies have shown, for example, that wetlands credited for mitigation of losses at urban development impact sites often are located in areas with sparse human population densities, meaning they necessarily do not deliver the same levels of ecosystem services as did the destroyed wetlands (Ruhl and Salzman 2006). Moreover, even two wetland parcels that are ecologically and economically similar could be providing their respective service values (e.g., groundwater recharge) at different times (i.e., depending on when it rains). The upshot is that knowledge about ecosystem services is profoundly place and time specific, and this quality could substantially restrain policy resilience in several ways.

First, the idea that decision makers will have robust ecosystem services information at their fingertips for any context seems unrealistic, if not fantasy. We simply cannot afford it, and even if the financial support were deep, the scientific capacity to produce highly textured data about ecosystem services is not nearly developed, at least not now or anytime soon. This means that decision making using ecosystem services will have to employ shortcuts and generalizations, such as proxies and partial measurements. For example, based on representative samples and studies, an agency might adopt a set formula for service valuation, such as one acre of wetland type X located within 5 miles of population density Y has a value of $Z. To the extent such techniques rely on coarser and more general inputs, their reliability, and thus their resilience-enhancing effects, necessarily diminish. At some point, such proxies and other shortcuts, practically necessary as they may be, are so poor in quality that decisions using them suffer from the familiar garbage in–garbage out effect. To be sure, environmental policy is rife with dilemmas of this sort—what to measure, how much to measure, how to pay for it—the point here being that ecosystem services practice will be no exception.

The second drag effect using ecosystem services concepts could pose for policy resilience has to do with the dark side of how people behave when they have more knowledge. Knowing more about how one is connected to ecosystems may have the resilience-enhancing effect posited above, but it also may lead to greater conflict and competition between people. Carbon sequestration serves again as an example: If new knowledge about ecosystem services trade-offs shows that forest sequestration capacity comes at the expense of some groundwater recharge capacity, will local and national interests conflict over where to locate sequestration projects? Similarly, as new knowledge about the off-site benefits of public lands becomes available, will yet more interest groups enter the fray of contested public land management decisions? Surely ecosystem management policy is no stranger to conflict and competition between people over resource allocation decisions, and new information can also lead to new alliances of cooperation and coordination, but knowledge about ecosystem services will inevitably be a two-way street in terms of resolving and fueling controversy.

None of this is meant to suggest that using ecosystem services concepts in ecosystem management decision making is any more problematic than using any other form of information about the environment. But ecosystem services literature tends to paint a very rosy picture, whereas the hard reality is that the policy world can be a rough and tumble place even for what in many respects is a positive take on ecosystems and their value to humans. It is quite possible, in other words, that some interests and institutions will use knowledge about ecosystem services in ways that gum up the policy works, so to speak, and thereby impede ecosystem management policy resilience. That effect may be outweighed by the resilience-enhancing effects outlined above, but whether it will or not remains to be seen.

## Conclusion

Will using ecosystem services concepts enhance ecological resilience, ecosystem management policy resilience, or both? It all depends. That is not a particularly satisfying answer, but the theme built through this chapter is that much will depend on how robust knowledge about ecosystem services becomes and, perhaps most important, how decision makers use that knowledge. There is nothing inherent about ecosystem services theory and practice that will lead inexorably to resilience-enhancing outcomes for ecosystems or for policy. Rather, several overarching effects will drive the direction in which ecosystem services concepts move ecological and policy resilience.

First, ecosystem services theory adds a new dimension to ecosystem management decision making, principally with respect to regulating and supporting services, that expands the economic understanding of ecosystems. This broadened economic conception of ecosystems is likely to also expand the range of ecological management criteria decision makers take into account. For example, whereas a forest land manager might previously have focused predominantly on timber value and thus also on resilience of the ecosystem processes supporting timber, now other service values might become important to the management goals, which in turn may make resilience of other ecosystem processes important. As the resilience of more ecosystem processes becomes more important, overall ecological resilience is likely on balance to be enhanced.

On the other hand, the flip side is the scenario in which new knowledge about ecosystem services drives a land manager to focus even more on one particular service—i.e., the one that laid the golden egg. Even if this service is a regulating or supporting service, such as carbon sequestration or groundwater recharge, the focus on one service is likely to drive management decisions toward enhancing the resilience of ecosystem processes supporting that service, possibly to the detriment of other processes. Overall ecosystem resilience could suffer.

Ecosystem management goals for different ecosystems are likely to run the gamut from promoting broad ecosystem services profiles to single services, with everything in between. In all such contexts the quality of the management decision will depend on the quality of the information available about the service or services being managed. Poor quality information will make poor decisions more probable. High quality information, if even attainable, will be costly. Regardless of the information quality, moreover, any form of new information will open up new fronts for interest group conflict and competition (as well as cooperation and coordination). Policy resilience will be tested in some contexts and supported in others.

From this mixed bag of possibilities comes one conclusion that is clear and definitive: ecosystem services theory and resilience theory will inevitably meet through ecosystem management decision making. Ecosystem management is the birthplace of both theories and the testing ground for their practice and implementation. How each performs when they intersect is the question that motivated this chapter, and its answer is just now beginning to materialize in the field.

## References

Ackerman, F., and L. Heinzerling. *Priceless: On Knowing the Price of Everything and the Value of Nothing.* New York: New Press, 2004.

Alderson, D. L., and J. C. Doyle. "Contrasting Views of Complexity and Their Implications for Network-Centric Infrastructures." *IEEE Transactions on Systems, Man and Cybernetics, Part A: Systems and Humans* 40 (2010): 839–852.

Balvanera, P., G. C. Daily, P. R. Ehrlich, T. H. Ricketts, S. A. Bailey, S. Kark, C. Kremen, and H. Pereira. "Conserving Biodiversity and Ecosystem Services." *Science* 291 (2001): 2047.

Beier, C. M., A. L. Lovecraft, and F. S. Chapin, III. "Growth and Collapse of a Resource System: An Adaptive Cycle of Change in Public Lands Governance and Forest Management in Alaska." *Ecology and Society* 14, no. 2 (2009): 5. http://www.ecologyandsociety.org/vol14/iss12/art15.

Bennett, E. M., G. D. Peterson, and L. J. Gordon. "Understanding Relationships among Multiple Ecosystem Services." *Ecology Letters* 12 (2009): 1394–1404.

Benson, M. H., and A. S. Garmestani. "Can We Manage for Resilience? The Integration of Resilience Thinking into Natural Resource Management in the United States." *Environmental Management* 48 (2011): 392–399.

Bhaskar, V., and W. M. Adams. "Ecosystem Services and Conservation Strategy: Beware the Silver Bullet." *Conservation Letters* 2 (2009): 158–162.

Bjorklund, J., K. E. Limburg, and T. Rydberg. "Impact of Production Intensity on the Ability of the Agricultural Landscape to Generate Ecosystem Services: An Example from Sweden." *Ecological Economics* 29 (1999): 269–291.

Bormann, B. T., and A. R. Kiester. "Options Forestry: Acting on Uncertainty." *Journal of Forestry* 102 (2004): 22–27.

Brooks, R. O., R. Jones, and V.A. Ross. *Law and Ecology: The Rise of the Ecosystem Regime*. Burlington: Ashgate, 2002.

Carpenter, S. R., and M. G. Turner. "Hares and Tortoises: Interactions of Fast and Slow Variables in Ecosystems." *Ecosystems* 3 (2000): 495–497.

Chapin, F. S., III, et al. "Earth Stewardship: Science for Action to Sustain the Human-Earth System." *Ecosphere* 2, no. 8 (2011): article 89; doi:10.1890/ES1811–00166.00161.

Christensen, N .L., et al. "The Report of the Ecological Society of America Committee on the Scientific Basis for Ecosystem Management." *Ecological Applications* 6 (1996): 665–691.

Costanza, R., et al. "The Value of the World's Ecosystem Services and Natural Capital." *Nature* 387 (1997): 253–260.

Costanza, R., S. C. Farber, and J. Maxwell. "The Valuation and Management of Wetland Ecosystems." *Ecological Economics* 1 (1989): 335–361.

Daily, G. C., ed. *Nature's Services: Societal Dependence on Natural Ecosystems*. Washington, D.C.: Island Press, 1997.

Dasgupta, P. *Human Well-Being and the Natural Environment*. Oxford: Oxford University Press, 2001.

Duarte, C. M. "Marine Biodiversity and Ecosystem Services: An Elusive Link." *Journal of Experimental Marine Biology and Ecology* 250 (2000): 117–131.

Endangered Species Act of 1973, 16 U.S.C. §§ 1531–1544.

Fischman, R. L. "The EPA'S NEPA Duties and Ecosystem Services." *Stanford Environmental Law Journal* 20 (2001): 497–536.

Folke, C. "Resilience: The Emergence of a Perspective for Social-Ecological Systems Analysis." *Global Environmental Change* 16 (2006): 253–267.

Fox, D. "Back to the No-Analog Future?" *Science* 316 (2007): 823–824.
Freeman, J., and C. D. Kolstad, eds. *Moving to Markets in Environmental Regulation.* New York: Oxford University Press, 2007.
Gomez-Baggethun, E., R. de Groot, P. L. Lomas, and C. Montes. "Reconciling Theory and Practice: An Alternative Conceptual Framework for Understanding Payments for Environmental Services." *Ecological Economics* 69 (2010): 1209–1218.
Grumbine, R.E. "What Is Ecosystem Management?" *Conservation Biology* 8 (1994): 27–38.
Gunderson, L. H., and C. S. Holling, eds. *Panarchy: Understanding Transformations in Human and Natural Systems.* Washington, D.C.: Island Press, 2002.
Gunderson, L. H., C. S. Holling, and C. R. Allen. "The Evolution of an Idea—The Past, Present, and Future of Ecological Resilience." In *Foundations of Ecological Resilience,* eds. L. H. Gunderson, C. R. Allen, and C. S. Holling, 423–444. Washington, D.C.: Island Press, 2010.
Hobbs, R. J., and V. A. Cramer. "Restoration Ecology: Interventionist Approaches for Restoring and Maintaining Ecosystem Function in the Face of Rapid Environmental Change." *Annual Review of Environment and Resources* 33 (2008): 39–61.
Holling, C. S. "Resilience and Stability of Ecological Systems." *Annual Review of Ecology and Systematics* 4 (1973): 1–23.
Holling, C. S., and L. H. Gunderson. "Resilience and Adaptive Cycles." In *Panarchy: Understanding Transformations in Human and Natural Systems,* eds. L. H. Gunderson and C. S. Holling, 25–62. Washington, D.C.: Island Press, 2002.
Jackson, R. B., E. G. Jobbágy, R. Avissar, S. B. Roy, D. J. Barrett, C. W. Cook, K. A. Farley, D.C. le Maitre, B. A. McCarl, and B. C. Murray. "Trading Water for Carbon with Biological Carbon Sequestration." *Science* 310 (2005): 1944–1948.
Kinzig, A. P., C. Perrings, F. S. Chapin III, S. Polasky, V. K. Smith, D. Tilman, and B. L. Turner II. "Paying for Ecosystem Services—Promise and Peril." *Science* 334 (2011): 603–604.
Kosoy, N., and E. Corbera. "Payments for Ecosystem Services as Commodity Fetishism." *Ecological Economics* 69 (2010): 1228–1236.
Lant, C. L., J. B. Ruhl, and S. E. Kraft. "The Tragedy of Ecosystem Services." *BioScience* 58 (2008): 969–974.
Lazarus, R. J. *The Making of Environmental Law.* Chicago: University of Chicago Press, 2004.
Levin, S. *Fragile Dominion: Complexity and the Commons.* Santa Fe: Helix, 1999.
McCauley, D. J. "Selling out on Nature." *Nature* 443 (2006): 27–28.
Millennium Ecosystem Assessment. *Ecosystems and Human Well-Being: A Framework for Assessment.* Washington, D.C.: Island Press, 2003.
Milne, B. T. "Motivation and Benefits of Complex Systems Approaches in Ecology." *Ecosystems* 15 (1998): 449–456.
Mooney, H. A., and P. R. Ehrlich. "Ecosystem Services: A Fragmentary History." In *Nature's Services: Societal Dependence on Natural Ecosystems,* ed. G. C. Daily, 11–19. Washington, D.C.: Island Press, 1997.

Muradian, R., E. Corbera, U. Pascual, N. Kosoy, and P. H. May. "Reconciling Theory and Practice: An Alternative Conceptual Framework for Understanding Payments for Environmental Services." *Ecological Economics* 69 (2010): 1202–1208.

Pascual, U., R. Muradian, L. C. Rodriguez, and A. Duraiappah. "Exploring the Links Between Equity and Efficiency in Payments for Environmental Services: A Conceptual Approach." *Ecological Economics* 69 (2010): 1237–1244.

Peterson, G., C. R. Allen, and C. S. Holling. "Ecological Resilience, Biodiversity and Scale." *Ecosystems* 1 (1998): 6–18.

Pimm, S. L. "The Complexity and Stability of Ecosystems." *Nature* 307 (1984): 321–326.

Rodriguez, J. P., T. D. Beard, Jr., E. M. Bennett, G. S. Cumming, S. J. Cork, J. Agard, A. P. Dobson, and G. D. Peterson. "Trade-offs across Space, Time, and Ecosystem Services." *Ecology and Society* 11, no. 1 (2006): 28. http://www.ecologyandsociety.org/vol11/iss1/art28.

Ruhl, J. B. "Who Needs Congress? An Agenda for Administrative Reform of the Endangered Species Act." *New York University Environmental Law Journal* 6 (1998): 367–410.

Ruhl, J. B. "Agriculture and Ecosystem Services: Strategies for State and Local Government." *New York University Environmental Law Journal* 17 (2008): 424–459.

Ruhl, J. B. "Ecosystem Services and the Clean Water Act: Strategies for Fitting New Science into Old Law." *Lewis & Clark Law School Environmental Law* 40 (2010a): 1381–1399.

Ruhl, J. B. "Ecosystem Services and Federal Public Lands: Start-up Policy Questions and Research Needs." *Duke Environmental Law & Policy Forum* 20 (2010b): 275–290.

Ruhl, J. B. "General Design Principles for Resilience and Adaptive Capacity in Legal Systems—With Applications to Climate Change Adaptation." *North Carolina Law Review* 89 (2011): 1374–1403.

Ruhl, J. B., and R. J. Gregg. "Integrating Ecosystem Services into Environmental Law: A Case Study of Wetlands Mitigation Banking." *Stanford Environmental Law Journal* 20 (2001): 365–392.

Ruhl, J. B., and J. Salzman. "The Effects of Wetland Mitigation Banking on People." *National Wetlands Newsletter* 28 (2006): 8–14.

Ruhl, J. B., and J. Salzman. "The Law and Policy Beginnings of Ecosystem Services." *Journal of Land Use & Environmental Law* 22 (2007): 157–172.

Ruhl, J. B., S. E. Kraft, and C. L. Lant. *The Law and Policy of Ecosystem Services.* Washington, D.C.: Island Press, 2007.

Ruhl, J. B., J. Salzman, and I. Goodman. "Implementing the New Ecosystem Services Mandate of the Section 404 Compensatory Mitigation Program—A Catalyst for Advancing Science and Policy." *Stetson Law Review* 38 (2009): 251–272.

Ryszkowski, L. "Managing Ecosystem Services in Agricultural Landscapes." *Nature and Resources* 31 (1995): 27.

Salzman, J. "Ecosystem Services and the Law." *Conservation Biology* 12 (1998): 497–501.

Schoennagel, T., T. T. Veblen, and W. H. Romme. "The Interaction of Fire, Fuels, and Climate across Rocky Mountain Forests." *BioScience* 54 (2004): 661–676.

Seppelt, R., C. F. Dormann, F. V. Eppink, S. Lautenbach, and S. Schmidt. "A Quantitative Review of Ecosystem Services Studies: Approaches, Shortcomings and the Road Ahead." *Journal of Applied Ecology* 48 (2011): 630–636.

Trosper, R. L. "Policy Transformations in the US Forest Sector, 1970–2000: Implications for Sustainable Use and Resilience." In *Navigating Social-Ecological Systems: Building Resilience for Complexity and Chang*, eds. F. Berkes, J. Colding, and C. Folke, 328–351. Cambridge University Press: Cambridge, 2003.

Turner, M. G. "Disturbance and Landscape Dynamics in a Changing World." *Ecology* 91 (2010): 2833–2849.

Vitousek, P. M., H. A. Mooney, J. Lubchenco, and J. M. Melillo. "Human Domination of Earth's Ecosystems." *Science* 277 (1997): 494–499.

Williams, J. W., S. T. Jackson, and J. E. Kutzbach. "Projected Distributions of Novel and Disappearing Climates by 2100 AD." *Proceedings of the National Academy of Sciences USA* 104 (2007): 5738–5742.

Yang, W., B. A. Bryan, D. H. MacDonald, J. R. Ward, G. Wells, N. D. Crossman, and J. D. Connor. "A Conservation Industry for Sustaining Natural Capital and Ecosystem Services in Agricultural Landscapes." *Ecological Economics* 69 (2010): 680–689.

# Maintaining Resilience in the Face of Climate Change

ALEJANDRO E. CAMACHO AND T. DOUGLAS BEARD

Climate change, when combined with more conventional stress from human exploitation, calls into question the capacity of both existing ecological communities and resource management institutions to experience disturbances while substantially retaining their same functions and identities (Zellmer and Gunderson 2009; Ruhl 2011). In other words, the physical and biological effects of climate change raise fundamental challenges to the resilience of natural ecosystems (Gunderson and Holling 2002). Perhaps more importantly, the projected scope of ecological shifts from global climate change—and uncertainty about such changes—significantly stresses the capacity of legal institutions to manage ecosystem change (Camacho 2009). Existing governmental institutions lack the adaptive capacity to manage such substantial changes to ecological and legal systems. In particular, regulators and managers lack information about ecological effects and alternative management strategies for managing the effects of climate change (Karkkainen 2008; Camacho 2009), as well as the institutional infrastructure for obtaining such information (Peters 2008).

A number of recent initiatives have been proposed to address the effects of climate change on ecological systems. However, these nascent programs do not fully meet the needs for developing adaptive capacity. A federal, publicly accessible, and system-wide portal and clearinghouse will help regulators at all levels of government manage the effects and uncertainty from climate change (DiMento and Ingram 2005; Farber 2007). Such an information infrastructure, combined with a range of incentives that encourage regulators to engage in adaptive management and programmatic adjustment over time (Baron et al. 2009), will help governmental and private institutions become more resilient and capable of managing the physical and human institutional effects of changing climate (Camacho 2009).

## The Projected Effects of Climate Change

Substantial and mounting evidence exists supporting the conclusion that anthropogenic climate change has already had significant adverse effects on ecological and human environments throughout the world (Parmesan 2006; Worm et al. 2006; Intergovernmental Panel on Climate Change [IPCC] 2007; Staudinger et al. 2012). Such climatic shifts will increasingly raise fundamental challenges to the resilience of both ecological and human systems (Rahel et al. 2008). Perhaps more significantly, considerable uncertainty about the projected scope of ecological shifts from global climate change raises an unprecedented challenge to the existing natural resource governance system (Ruhl 2008; Camacho 2011b).

### *The Physical and Biological Effects*

Climate-related changes have already had impacts on wildlife and ecological resources (IPCC 2007; Staudinger et al. 2012). For example, biodiversity losses in many ecosystems have been driven by climate change (Millennium Ecosystem Assessment [MEA] 2005; Malcolm et al. 2006; IPCC 2007; Staudinger et al. 2012). Further, we know that timing of many of the natural events (phenology) has been affected by changes in climate (Davis et al. 2010), with often unknown consequences to the future resilience of the affected populations or species. The increased risk of extinction of plants and animals due to climate-driven events (IPCC 2007; Staudinger et al. 2012) will change the underlying biodiversity of the ecosystems and their subsequent resilience. Among its many benefits, biodiversity typically brings redundancy in ecosystem function, while loss of biodiversity and subsequent loss of redundancy as a result of climate change will likely have significant effects on the resilience of systems (Gunderson and Holling 2002).

Perhaps as importantly, climate drivers are expected to have major impacts on the health and function of habitat necessary for resilient ecosystems. The warming atmosphere alters precipitation patterns and soil moisture, leading to longer, prolonged droughts; increases the destructive nature of hurricanes; and increases mean sea level (IPCC 2007; Staudinger et al. 2012). Perhaps the biggest changes in ecosystems will be as a result of changes in the hydrologic cycle and the availability of water, further exacerbating already water-stressed regions.

### *Increased Uncertainty*

In addition to its considerable ecological effects, climate change also magnifies the uncertainty that exists for addressing environmental problems due to the many complex and confounding variables inherent in climate (Ruhl 2008). Projections of future climate-driven changes on ecosystems have high levels of uncertainty, because the underlying dynamics of global climate modeling systems are uncertain, and the dynamics of ecosystem responses are still highly variable and uncertain. Perhaps the biggest uncertainty in projecting climate-driven changes is how humans will react to changes in their environment, changes in policies and procedures, and changes to underlying norms. In many ways, uncertainty is the greatest challenge raised by climate change (Camacho 2009).

Given the highly uncertain nature of projected changes in ecosystems as a result of climate change, scenarios provide one tool for understanding projected climate-driven change and the impacts to ecosystems (MEA 2005). The use of scenarios for communicating uncertainty allows managers and policy makers the opportunity to compare and contrast alternative views of the future and craft management strategies that build resilience into the system. Given the ability of scenarios to identify common tipping points, the results of scenarios should allow decision makers to define management strategies around key drivers of change. Effective management into the future of climate-driven changes will require full adoption of an adaptive framework (Gunderson and Holling 2002; Camacho 2009) that allows policy makers to adapt to change via a systematic, learning approach incorporating rapidly changing knowledge of climate-driven impacts.

To combat the fragmented regulatory structure for most ecological systems in the United States, incorporate resilience into management approaches, and adapt to the long-term and highly uncertain climate-driven changes, it will be necessary to adopt more adaptive approaches. Adaptive capacity should reside not only in the traditional ecological resources management agencies, but adaptive approaches should work in an integrated fashion with the legal and administrative institutions responsible for ecological systems. The challenge will be, as with most adaptive management systems (Camacho 2009), to develop robust monitoring systems that integrate not only ecological monitoring, but monitoring of legal, management, and administrative approaches. Development of robust monitoring systems should allow the adaptive systems to be accountable for changes and should provide a mechanism to integrate monitoring results into management planning.

### *Existing Federal Learning Infrastructure for Climate Change Adaptation*

Despite the growing evidence for climate change–related harm and the considerable uncertainty about the precise manifestation of such effects (Salzman and Thompson 2010), the existing regulatory infrastructure in the United States does not effectively deal with the effects and uncertainty that are expected to arise due to global climate changes (Camacho 2009). Most governmental programs do not sufficiently promote learning by government officials (Gregory et al. 2006; Camacho 2009) or incentivize such managers to be more effective at achieving regulatory goals (Baron et al. 2009). Natural resources governance in the United States is also largely fragmented (Buzbee 2005; Camacho 2009), which hinders the capacity for interjurisdictional information sharing and collaboration directed toward reducing uncertainty about the effects of climate change and effectiveness of management strategies.

More specifically in the context of climate change adaptation, few resource management or regulatory agencies at the federal, state, or local level adopted any strategies for adapting to climate change or for managing the uncertainty climate change adaptation produces until recently (Stutz 2009). The U.S. Congress still has not established any programs expressly directed at climate change adaptation, although the U.S. Global Change Research Program has some limited ability to work on adaptation measures. Moreover, many regulatory authorities continue to rely on strategies premised on historically customary conditions (i.e., assuming stationarity in climate) that even agency officials concede are not likely to apply under projected climate change scenarios (Camacho 2009).

However, as detailed below, a small but growing number of adaptation planning initiatives are being developed that are attempting to address the effects of climate change on ecological systems. Though modest in scope and funding (U.S. Government Accountability Office 2009; Smith et al. 2010), several of these programs do make limited progress toward the development of processes for monitoring ambient conditions, fostering information sharing and cooperation, or encouraging adaptive management. These nascent programs are certainly improvements on preceding resource governance; nonetheless, they are unlilkely to completely provide a comprehensive learning infrastructure that cultivates interjurisdictional information sharing and agency learning.

### *The Limitations of Existing Institutions for Adapting to Climate Change*

Because natural resource governance in the United States is fairly static and fragmented, it is poorly equipped to foster agency learning, to tap the potential experimentation benefits of largely dispersed regulatory authority, or to otherwise manage the uncertainties of climate change (Buzbee 2005; Camacho 2009). Though a number of mostly federal programs have recently been created to at least in part address these shortcomings, American resource governance still lacks a more fundamental learning framework for managing the strain and uncertainty accompanying climate change.

## ADAPTIVE LEARNING

Evidence from the literature suggests that most natural resource programs in the United States are not designed to foster programmatic learning within an agency; they lack the mechanisms to adjust and improve management decisions over time, fix management missteps born from limited information and uncertainty, and promote consideration of such lessons in future management decisions (Gregory et al. 2006; Camacho 2009). Evidence now exists that climate systems are not acting in a stationary manner (Coumou and Rahmstorf 2012). However, many agencies often adopt measures that subsequent experience reveals are deficient or imperfectly tailored to current conditions. One reason for this is because environmental regulators characteristically have limited information about ambient conditions and the effects of potential strategies (Karkkainen 2004; Camacho 2007), and ambient conditions inevitably change.

Yet few agencies have developed a systematic, rigorous approach to reducing such uncertainty over time and learning from the past performance of adopted strategies. Most programs are subject to requirements to monitor agency actions or approvals, but agencies commonly neglect such obligations (Camacho 2009), and ambient monitoring is typically underfunded (Biber 2011). Perhaps most importantly, agencies are not required to assess the effectiveness of prior decisions or adjust those decisions over time (Williams et al. 2007) and as a result, it is very difficult to promote such adaptive learning institutionally. Unsurprisingly, assessments of the accuracy of prior assumptions or the efficacy of adopted decisions are rare. Similarly, adjustments of decisions to incorporate new information or changed conditions are uncommon (Walters 1997; Gregory et al. 2006). By not encouraging managers to systematically learn from prior decisions, natural resource governance remains too static (Biber 2011), rendering its capacity to reduce and manage uncertainty unnecessarily weak.

Though a number of encouraging adaptive management experiments have proliferated in an attempt to address uncertainty in the regulatory process, these approaches have left substantial room for improvement. Adaptive management was developed to help resource managers deal with uncertainty in the regulatory process through periodic monitoring and adjustment of management decisions (Walters 1986; Dorf and Sabel 1998; Freeman and Farber 2005). Such an approach can increase the regulatory process's adaptive capacity by allowing for decisions to regularly account for new information or changes in circumstances. Yet recent adaptive management experiments have not *required* periodic adjustment of agency actions (Freeman 1997) and have not sufficiently provided incentives and resources for managers to monitor and adaptively manage (Biber 2011). Most attempts at adaptive management also do not apply adaptive management principles at the program level; that is, they fail to systematically monitor and adjust the program to more effectively achieve the program's goals (Camacho 2007).

## INTERJURISDICTIONAL LEARNING

The fragmentation and limited coordination of regulatory authority over natural resources exacerbates the problem by making interjurisdictional learning very difficult. Authority over natural resources in the United States is allocated typically based on the environmental component to be regulated (e.g., species, air quality, water quality) and the level of government (e.g., local, state, federal) (Buzbee 2005). Most of the authority for management of ecological systems and water resources resides with state governments. Each state has its own approach to managing these systems, and there are few incentives to work across borders, which is especially problematic when climate-driven impacts to ecosystems occur at a scale that will almost always be larger than individual states. Even issues that are the responsibility of federal agencies, such as the management of threatened and endangered species under the federal Endangered Species Act of 1973, often suffer from fragmented jurisdictional authority. For example, management of endangered anadromous salmonids is split between the National Oceanic Atmospheric Administration (NOAA) when the salmon reside in marine waters and the U.S.

Fish and Wildlife Service (USFWS) when they reside in freshwater. This fragmentation makes management of long-term climate-driven impacts difficult.

On the other hand, there are considerable advantages to relying on decentralized and overlapping authority to manage ecological resources. Such decentralized regulatory authority is in part premised on allowing for a diversity of focused, localized strategies, thus promoting regulatory experimentation and allowing the opportunity for interjurisdictional learning about the relative efficacy of different strategies (Camacho 2011b). In fact, the need for dispersed but overlapping governance of natural resources may be even stronger with the onset of climate change. Though climate change may provide some impetus for more centralized authority due to the likely increase in interjurisdictional spillovers (Adler 2005), the considerable uncertainty raised by climate change adaptation also heightens the need for regulatory experimentation and innovation that decentralized governance is designed to provide (Adelman and Engel 2009; Ruhl and Salzman 2010). In addition, given the local variation in how climate change is likely to affect resources, the substantial tailoring benefits of more local or specialized decision making are likely to persist. Key to maximizing the diversity and experimentation benefits of decentralized and overlapping governance, however, is an infrastructure that collects the disparate information about ambient conditions and management strategies, disseminates it broadly, and otherwise encourages regulators to learn from the data and experiences of other authorities (Karkkainen 2008).

However, to date this capacity to promote interagency learning and thus help reduce uncertainty is largely untapped, because there is insufficient emphasis on coordination and information sharing between jurisdictions (Karkkainen 2004; Adler 2005). Other than through ad hoc or anecdotal opportunities, resource managers and regulators have little ability or incentive to learn from the lessons of other agencies. Even when a manager collects information about ambient conditions or the performance of adopted strategies, such information is too often not broadly accessible, because there is no comprehensive infrastructure collecting and disseminating it.

To be sure, increasing agency collaboration has been a goal of a host of regional federal regulatory initiatives (Council on Environmental

Quality [CEQ] 2010). Such ecosystem- or landscape-based networks typically are established to provide some opportunity for communication and synchronization of decision-making authority. Though such venues may provide some coordination benefits, unfortunately many of these largely focus on developing regional institutions and too often pay insufficient attention to reducing uncertainty through interjurisdictional information sharing (Camacho 2011b). Most do not adopt any shared infrastructure for managing information or reducing uncertainty and continue to leave managers with limited capacity or tools for developing or accessing data about ambient conditions and the past performance of potential management strategies used by their own agencies or other authorities. Unfortunately, too often, such initiatives simply serve as yet another level of regulation that exacerbates existing regulatory fragmentation.

## *Recent Initiatives for Climate Change Adaptation*

Though a number of initiatives have recently been established that help improve the adaptive capacity of the existing natural resource governance system, a more comprehensive commitment to promoting adaptive management and information sharing is needed in the face of climate change.

### ENVIRONMENTAL PROTECTION AGENCY

In 2011, the U.S. Environmental Protection Agency (EPA) issued a policy statement committing to complete an agency-wide adaptation plan by June 2012 (U.S. EPA 2011), and as of February 2013, a draft plan has been issued for public comment. However, the EPA adopted a National Water Program Strategy in 2008 that identifies impacts of concern from climate change to water programs in the United States, defines goals for responding to such impacts, and provides specific proposed adaptation actions for drinking water systems, water quality, and effluent standards; watershed protection; the National Pollutant Discharge Elimination System program; water infrastructure; and wetlands protection (Cruce and Holsinger 2010). In 2012, the EPA published a new, longer-term National Water Program Strategy that "describes an array of important actions that should be taken to be a 'climate ready' national water program," though the strategy "does not outline commitments to act within a specific time frame" (U.S. EPA 2012). Among other initiatives, the National Water Program Strategy helped formulate the Climate Ready Estuaries (CRE) and Climate Ready Water Utilities (CRWU) programs.

The CRE program was created in 2008 to provide estuary communities that participate in the National Estuaries Program (NEP) various tools and financial and technical assistance for assessing climate change vulnerabilities, engaging and educating stakeholders, developing and implementing adaptation strategies, and, encouragingly, sharing lessons learned with other coastal managers (Cruce and Holsinger 2010). Estuaries are among the locations most vulnerable to the effects of climate change. Most prominently, the CRE program includes a "coastal toolkit," a portal of collected data, tools, and databases on climate change, coastal vulnerability, smart growth options, adaptation planning, and financing opportunities. In addition, EPA holds occasional workshops that bring together similarly situated officials to discuss adaptation planning (CRE 2010). The goal is for this infrastructure to improve the adaptive capacity of NEP communities to more effectively identify risks and adapt to the effects of climate change.

Likewise, the CRWU program was established to provide technical resources and tools for water utilities to engage in adaptation planning (Cruce and Holsinger 2010). Promisingly, a CRWU working group developed a report recommending a framework for increasing the adaptive capacity of water utilities through increased information generation and dissemination and agency coordination (National Drinking Water Advisory Council Report 2010). The EPA has begun to implement many of the report's recommendations, seeking to promote the application of emergency management principles and sustainable infrastructure practices by utilities for assessing risk, determining vulnerability, evaluating consequences, and developing effective adaptation strategies (Cruce and Holsinger 2010). Akin to the CRE program, is an EPA-created, searchable "toolbox" containing water sector climate change information on government and utility activities, workshops, publications, funding, and tools. It also has developed a risk assessment and scenario-based tool it calls Climate Resilience Evaluation and Awareness Tool, as well as a Tabletop Exercise Tool for Water Systems to help utilities assess their vulnerability to climate change and consider potential adaptation options.

In addition to these programs, through its Office of Research and Development's Global Change Impacts and Adaptation Program and its Water Resources Adaptation Program, the EPA is conducting studies and developing various decision-support tools for resource managers about the effects of climate change and adaptation options pertaining to air and water quality, aquatic ecosystems, human health, and socioeconomic systems in the United States (Cruce and Holsinger 2010).

### COUNCIL ON ENVIRONMENTAL QUALITY

Under Presidential Executive Order 13514, the CEQ has been charged with co-chairing (along with the Office of Science and Technology Policy and NOAA) a federal interagency task force and coordinating adaptation planning across all federal agencies (CEQ 2011a). The CEQ and the participating federal agencies have adopted an approach that seeks to ensure that adaptation is integrated throughout all agency planning efforts. The CEQ has issued guidance requiring federal agencies to submit information to CEQ demonstrating that the agency is engaging in adaptation planning by a series of deadlines (CEQ 2011c). Further, CEQ is working broadly with all state, local, and other partners to promote resilience thinking in community-level planning activities (CEQ 2010). These efforts are meant to lead and support international efforts in climate change adaptation.

A cornerstone of these efforts is improving accessibility and coordination of science for decision making, and assuring that the "best available" science is available. According to CEQ, a science-based approach to adaptation planning should use integrated approaches and vulnerability assessments to identify the most vulnerable ecosystems, and then risk management approaches to prioritize adaptation responses (CEQ 2011a). The CEQ has already produced a national action plan for freshwater resources and is working closely with the USFWS, NOAA, and state and tribal agencies to develop an adaptation strategy for fish and wildlife resources (CEQ 2011b).

In addition, CEQ has led the federal agencies in creation of regional-based interagency coordination efforts on climate information, focused initially on ecological systems. The regional coordination efforts are meant to create consistency in the use of climate information across the federal government, reduce redundancy in management and science efforts, and provide state and local authorities access to information helpful for climate change adaptation planning.

## DEPARTMENT OF THE INTERIOR

The U.S. Department of the Interior (DOI) is tasked with managing one-fifth of the land in the United States, handling trust responsibilities for 562 Indian tribes, managing water supplies for 30 million people, and conserving fish and wildlife and their habitats (U.S. DOI 2011). The DOI developed an adaptive management guidebook (Williams et al. 2007) and adopted an adaptive management implementation policy in 2008 that focused on applying adaptive approaches on a landscape-wide level. The DOI is often only one of many management agencies responsible for ecological systems on any given landscape, so the application of adaptive management requires a full partnership with multiple agencies and jurisdictions. The DOI recognized early on that climate-driven management issues are far more complex and occur at a scale much larger than any one agency could handle on its own.

Therefore, DOI's approach to implementing adaptive management in response to climate change is intended to engage the entire science and management community by working with science and management partners to form twenty-two landscape conservation cooperatives (LCCs) and eight regional climate science centers (CSCs). The CSCs are operated through the U.S. Geological Survey's (USGS) National Climate Change and Wildlife Science Center (NCCWSC), established by Secretarial Order 3289 (Secretary of the Interior 2009). The CSCs are partnership-driven science centers focused on providing science in support of the various DOI resource management programs. Science support can range from development of models and better understanding of ecological processes, to development of monitoring frameworks and protocols that allow tracking of climate-driven changes. The LCCs are public–private partnerships focused on bringing applied science approaches to the management of ecological systems. LCCs, working closely with CSCs, primarily work on development of syntheses and assessments of the ecological systems, and development of decision-support tools for active resource managers. The LCCs work closely with the various partners to actually develop strategic approaches, implement consistent regulations and policies, and monitor progress toward goals.

An example of the approach being pursued by DOI is the management of western native trout. Working through the NCCWSC, and ultimately through the Northwest CSC, the agency has provided initial projections of climate-driven risk to future persistence of trout in the Rocky Mountains (Muhlfeld et al. 2011) to management partners within the Great Northern LCC. Partners in the Great Northern LCC are now working to integrate this information into applied approaches, strategic plans, and development of management strategies that will allow adaptation to forecasted changes in trout persistence.

## NATIONAL OCEANIC AND ATMOSPHERIC ADMINISTRATION

The National Oceanic and Atmospheric Administration's (NOAA) main climate responsibility resides with the physical climate system and its impacts on marine ecosystems. NOAA has extensive climate activity ongoing, ranging from the National Climatic Data Center (NOAA n.d.) to the Climate Prediction Center (NOAA 2013) and NOAA's Regional Integrated Sciences and Assessments (RISA). NOAA has focused on providing information, science, and data that broadly characterize the physical climate system, and other aspects, generally in marine systems, that help NOAA meet its missions. NOAA's broad science mission in support of climate activities includes development of global-level climate models; assessment of natural climate variability, anthropogenic change, and the global carbon cycle; research in support of policy; and other decisions in managing for and adapting to climate impacts (NOAA 2011). For example, RISAs will contribute information and work collaboratively with LCCs and CSCs organized by the DOI by performing interdisciplinary research for local private and public decision makers (CEQ 2011b).

NOAA has made a push to consolidate its various climate activities into a series of regional climate service centers (NOAA, "Proposed Climate Service in NOAA," n.d.). This budget-neutral approach would have allowed NOAA to provide access to a number of climate science products and data through a singular interface. However, during the Fiscal Year 2011 Congressional budget debates, language was inserted into the adopted budget bill that forbade NOAA from using any resources to establish the climate services (Strain 2011). If ultimately started, NOAA's climate services are intended as "one-stop shopping" for authoritative science and data about climate and climate impacts across the nation.

## UNITED STATES FOREST SERVICE

The U.S. Forest Service (USFS) is among the other large federal land management agencies, having land management responsibility for 193 million acres within the United States. The USFS is located in the U.S. Department of Agriculture and has a mission directed at sustaining the health, diversity, and productivity of its lands for a wide range of ecosystem services (USFS 2010). The USFS has management and research components focused on better understanding the impact of climate-driven changes on forests and actively working to incorporate them into forest management planning, using an adaptive management framework (USFS 2010). The USFS developed a strategic framework for responding to climate change in 2008, which led to the development of a national approach for responding to climate change across all USFS-managed lands (USFS 2010). Currently, each individual management unit of the USFS is adapting strategic plans to incorporate the long-term implications of climate on forest management strategies. Each USFS national forest and grassland will monitor progress toward climate change goals through a standard Climate Change Performance Scorecard (http://www.fs.fed.us/climatechange/advisor/scorecard.html). The USFS 2012 budget requests additional resources to support the climate-driven planning and development of adaptation strategies in response to regulations changes across the USFS (USFS 2012).

The USFS research and development program has developed a research plan that focuses on enhancing ecosystem sustainability and carbon sequestration, developing decision-support tools, and working collaboratively across the entire research infrastructure of the federal government (USFS 2009). The USFS research components contributed to the overall long-term goals of the U.S. Global Change Research Program (USGCRP) and work collaboratively, especially with the land management bureaus located in the DOI. As part of the USFS activities for integrating climate change activities into their planning activities, a climate change resource center has been developed (USFS 2013) that addresses managers' questions about what they can do about climate change (USFS 2010). The Fiscal Year 2012 omnibus budget, however, eliminates a direct line item focused on climate change activities in the USFS budget and redistributes the funding to other programs, with an implicit assumption that there will still be a focus on climate activities.

## UNITED STATES GLOBAL CHANGE RESEARCH PROGRAM

The USGCRP program was created by Congress with the passage of the Global Change Research Act of 1990 (http://www.globalchange.gov/about). The USGCRP is tasked with developing a coordinated research plan that includes development of shared information management and public participation strategies among fourteen federal agencies. Among the chief responsibilities of the USGCRP is to produce a national assessment of the effects of climate change on a number of natural, agricultural, and other resources. Pursuant to the Global Change Research Act, these assessments are to be completed at least every four years; however, only two assessments have been produced to date, and there is no ongoing sustained process for completing this legal requirement.

The federal agencies named as part of the Global Change Research Act are required to coordinate their annual budget requests through USGCRP, along with their research activities. Additionally, under the statute, all agencies are expected to make their research results available to EPA for promulgation of any rules or policies regarding climate change impacts. Further, the USGCRP and its member agencies are expected to participate in international activities focused on climate, such as the IPCC.

Given the lack of an ongoing process for producing the national assessment in a timely fashion, USGCRP, as part of its third assessment, is attempting to create an ongoing commitment and process among federal agencies to simplify the assessment process. Working closely with the NOAA RISAs, DOI CSCs, and other federal agencies, USCRP has teamed with CEQ to create regional coordinating bodies among federal agencies that would, among other tasks, create a long-term commitment to completing the national climate assessment. Whether this approach will succeed is unknown, but clearly completing a series of one-off assessments that do not build upon previous work has not been an effective approach to meeting the mandates of the Global Change Research Act. Nor does the random approach provide timely vital information about the effects of and adaptation options for addressing climate change.

## Assessment of Recent Initiatives for Climate Change Adaptation

Though a number of the recent federal agency climate change adaptation programs have promising features that are likely to help reduce uncertainty and increase agency coordination, they nonetheless fail to sufficiently incorporate a comprehensive adaptive learning infrastructure that requires, encourages, and maximizes the capacity for managers to learn from their own endeavors and those of others. A few federal initiatives do propose integrating adaptive management more fully into decision processes as an important component of climate adaptation. For example, the Federal Interagency Climate Change Adaptation Task Force's recently proposed strategy requiring adaptation planning by all federal agencies, recommending the incorporation of adaptive management and interagency cooperation and information sharing, should help emphasize the importance of agency monitoring, assessment, and adjustment for managing the effects of climate change (CEQ 2010).

Other initiatives, such as the DOI's LCCs, do aim to increase coordination between agencies that share jurisdictional authority over particular landscapes (Secretary of the Interior 2009). As climate change causes ecological shifts, it will likely increase resource scarcity and conflict and the interaction of regulatory authorities, as actions by one authority will increasingly have effects on resources regulated by others (Camacho 2011b). Endeavors such as the LCCs should help reduce interjurisdictional spillovers from adaptation activities and allow for more harmonization of management activities or creation of coordinated climate goals that hopefully will lead to more effective climate change adaptation.

Similarly, a number of new Federal programs are working diligently to contribute vital information and decision-support tools that should help reduce the dearth of information about the effects of climate change and adaptation. Research funded and undertaken by the U.S. Global Change Research Program, regional NOAA climate service centers, the USGS's National Climate Change and Wildlife Center, and the EPA's Office of Research and Development's Global Change Impacts and Adaptation Program will certainly proliferate important data about climate change and help develop information about possible adaptation strategies, with agency websites serving as portals wherein local, state, and federal resource managers can access critical information. The authorization and creation of regional CSCs will help produce missing but fundamental scientific information for use by the DOI's LCCs. At the individual program level, EPA's CRE, and CRWU programs encouragingly have collected and developed information, tools, and clearinghouses for discrete issue areas (Cruce and Holsinger 2010). Each of these attempts to increase available information and tools should help reduce some of the uncertainty about the effects of climate change and make it easier for managers to attempt adaptation planning.

Though certainly an improvement on conventional resource management (Milly et al. 2008), these few initiatives are not sufficiently directed at requiring either a more adaptive process or developing a comprehensive apparatus across multiple jurisdictions for private parties or government officials to more effectively manage uncertainty and the effects of climate change on ecological systems. As with past adaptive management experiments, despite the fact that a few of these adaptation initiatives espouse the need for increased reliance on adaptive management, few are developing systematic protocols that rigorously require the monitoring, assessment, and adjustment of agency decisions. A more comprehensive commitment to learning also requires scrupulous evaluation and adjustment not only of individual management decisions but of the individual programs and agencies as well (Doremus 2007). Moreover, these adaptation initiatives do not heed the lessons of prior attempts at adaptive management that point to the need to focus on providing concrete objectives and incentives for learning for managers (Walters 1997; Baron et al. 2009). Without clear goals, timelines for assessment and modification, resources, and other performance incentives, managers are not likely to strongly commit to adaptive management (Susskind et al. 2010). Though statements calling for integration of adaptive management are laudable, a commitment to learning and reducing uncertainty requires sustained emphasis on manager incentives as well.

Furthermore, various initiatives such as the LCCs, seek to promote better coordination, but they tend to focus on place-based or interagency dialogue and pay little attention to a broader commitment to information coordination. The creation of place-based forums for dialogue can be helpful for harmonization of management strategies for particular resources (Bardach 1998; Karkkainen 2008), and more such coordinating venues could be developed. Similarly, the Federal Interagency Climate Change Adaptation Task Force provides an important forum for preliminary discussion among federal natural resource agencies about climate change adaptation and could be productive in allowing for adaptation goals to be coordinated. However, as recently found by the U.S. Government Accountability Office (2011), such forums have yet to yield a shared understanding of strategic adaptation priorities or integration as "climate change programs and activities are set across the federal government" (86).

Perhaps more importantly, most managers are left to engage in fairly isolated adaptation planning in narrowly defined jurisdictional areas and with varying degrees of interaction with other managers from their regions. This, combined with the fact that most agencies do not generate and/or gather information about the effectiveness of their management strategies, leaves regulators with a limited capacity to manage for uncertainty, yet managing for uncertainty is critical for effective climate change adaptation. The massive uncertainty that accompanies climate change requires a more comprehensive infrastructure that allows and encourages private, local, state, and federal resource managers from throughout the country to share information, communicate, and learn from one another (Camacho 2011b).

Finally, existing governmental research initiatives are limited in their capacity to link agency information gathering, translating science into management actions and providing for information exchange. The various fragmented governmental ventures seeking the production of scientific data and decision-support tools are undoubtedly useful at regional and local levels. Yet creating information is only one part of the process. Making data, reports, and tools readily and widely accessible to others is yet another step; providing opportunities for other managers or users to contribute data is yet another; and providing opportunities for managers and other users to comment and otherwise interact is still another. Though the creation of repositories of information, such as the toolkits created by EPA's CRE and CRWU programs, is a substantial improvement on conventional management's tendency to leave managers isolated, the information flow is fairly unidirectional. In both of these programs, only EPA provides the data, guidance, and models. The portals are not at all interactive; they neither allow other managers to contribute information, nor do they facilitate communication between similarly situated managers or between managers and relevant research scientists. The Interagency Task Force continues to report that the USGCRP is "exploring options for developing and maintaining an online interagency global change information portal/system to provide 'one-stop shopping' for climate-related information" and recently a beta version of an interagency portal system has been demonstrated (CEQ 2011b, 16), but it has taken several years for this limited progress. Without a more comprehensive, shared, and evolving framework for learning, agencies will continue having difficulty managing and reducing the substantial uncertainty that is becoming compounded by a rapidly changing climate.

## Possible Legal and Institutional Reforms to Increase the Legal System's Resilience

Though there may be a range of potential substantive options for increasing the adaptive capacity of natural resource management in the United States to help prepare for and respond to the effects of climate change, the most fundamental changes necessary to support the legal system's resilience are procedural. To be sure, there are perhaps many substantive adaptation strategies that could help fortify ecosystem resilience by integrating recognition of ecological change into resource management (Peterson et al. 2008). One commonly mentioned example might be the required establishment of passive wildlife migration corridors that enable movement between ecological reserves as climatic conditions change (Simberloff and Cox 1987; Simberloff et al. 1992; Williams et al. 2005). Other substantive adaptation strategies might foster ecological resilience by building flexibility into legal rights and obligations. One example is the establishment in some jurisdictions of rolling easements, publicly owned entitlements to coastal property that shift with the coastline as sea levels rise (Titus 1998; Easterling et al. 2004). Such public entitlements would establish legal arrangements that shift obligations and rights to ensure that valuable ecosystem services remain protected as ecological conditions change (Caldwell and Segall 2007).

A more fundamental, long-term, substantive change in natural resources law might be a paradigm shift in statutory goals toward a focus on minimizing ecological harm, maximizing ecological function, or building redundancy in ecosystem functions in light of climatic and other changing environmental conditions. Rather than the traditional fidelity in American natural resources law either to maintaining ecological conditions at a specific historical baseline or to ensuring minimal human management of ecological resources (Ruhl 2010), such a transformation in regulatory goals would be more compatible with an understanding of ecological dynamism and, designed properly, could help foster ecological function and resilience (Camacho 2011b). Yet strategies that accept and promote rather than resist ecological change certainly are not without risk of harm. Inevitably, the focus of management will have to be on designing standards and deciding among strategies with an eye toward safeguarding against harmful shifts and fostering shifts that promote important ecosystem services.

Perhaps the most essential reforms for increasing the legal system's capacity to manage the effects of climate change are those that seek to improve the decision-making process by integrating and incentivizing learning to manage uncertainty. Undoubtedly, as recently proposed for federal agencies by the Federal Interagency Climate Change Adaptation Task Force (CEQ 2010), developing a process and adopting requirements for widespread adaptation planning by local, state, and federal agencies is important, as is broad assimilation in all agency actions of consideration of the effects of climate change. The myriad of individual agency actions designed to engage in climate research and adaptation planning are also significant, as are the various research programs seeking to increase information about climate change and its effects.

Yet the procedural adaptation that may be the most vital for maintaining institutional resilience is the development of a comprehensive regulatory framework for learning (Camacho 2010). Though the existence of a multitude of governmental entities with authority over natural resources provides the potential for management experimentation and consequent interjurisdictional learning, resource managers are not given sufficient incentives or opportunities to learn and adapt, and authorities are not provided opportunities to learn from one another, because there is little information gathered or shared (Camacho 2011a). As a consequence, U.S. resources management is poorly designed to promote systematic regulatory experimentation and learning. Accordingly, the two foundational elements of such a learning infrastructure would be (1) the integration of more adaptive approaches to management that require and otherwise urge officials to systematically monitor, assess, and adjust regulatory strategies over time; and (2) the creation of a collaborative and interactive information-sharing apparatus. Such an infrastructure would improve natural resource management's adaptive capacity by encouraging regulators to manage and reduce uncertainty about the regulatory programs and natural systems under their jurisdiction.

The first feature would take principles of adaptive management and lessons from its implementation and seek to apply them broadly throughout the regulatory process. Such an approach would seek opportunities not only for integrating standard adaptive management but also less formal forms of adaptive regulation that incentivize monitoring, assessment, and periodic adjustment. In the context of adaptation planning, this would include requiring science- and goal-based monitoring, assessment, and periodic adjustment of proposed and adopted adaptation strategies throughout initial planning, rule making, implementation, and enforcement. Monitoring activities would include not only ambient monitoring but also assessment of the effects and efficacy of adopted strategies, as well as of agencies themselves, at achieving stated regulatory goals (Karkkainen 2002). Significantly, because of the strategic disincentives that managers have for engaging in systematic adaptive management, past regulatory experiments suggest the need for monitoring, assessment, and adjustment to each be mandated and not voluntary, with clear goals and concrete triggers, deadlines, and other thresholds for action based on new information or changes in conditions (Susskind et al. 2010). In addition to obliging agencies to assess and adjust over time, providing other incentives for learning such as incorporation in manager performance evaluations, and enlisting stakeholders and other regulatory authorities to reinforce monitoring could also serve to increase learning. Such initiatives would likely serve to foster learning, better-tailored resource management, and regulator accountability.

To allow opportunities for regulatory experimentation and to promote collaborative learning at the national level, such an adaptive governance framework in the United States would most appropriately be led by the federal government in coordination with the states (Camacho 2011a). Federal agencies might consider identifying concrete metrics and standards against which management efforts can be measured, similar to the Office of Management and Budget's high-priority performance metrics (2010). To promote adaptive monitoring, assessment, and adjustment by state authorities, a national adaptive framework might range from federal approaches that build upon existing state information programs to federally prescribed standards for information gathering and sharing. As with other cooperative federalism measures in natural resources management, federal authorities could incentivize participation by offering funding for state adaptation efforts to state agencies engaging in continued monitoring, assessing, and reporting of information congruent with federally delineated metrics.

The second feature seeks to develop a widespread and public interjurisdictional network for information coordination, sharing, and interaction (Camacho 2010). Clearinghouses such as EPA's CRE and CRWU toolkits, which provide managers of particular resources access to data developed or gathered by a particular regulatory authority, are undoubtedly useful in helping otherwise isolated resource managers engage in adaptation planning. Accordingly, more networks can usefully be created for other resources, and such networks should be linked to one another where overlap exists.

Yet as a conduit for information sharing and learning, existing approaches are largely one-directional; a single authority provides information to others to assist their decision making, with at best limited communication in the other direction and even less among the participating managers. Though agencies increasingly are focusing on reducing uncertainty, many generally do so by collecting readily available scientific data and providing introductory guidance about what adaptation options might make sense (Camacho 2011a). In short, most existing attempts to manage uncertainty in adaptation planning have not been fully embraced as adaptive approaches that bring together diverse stakeholders to develop adaptation plans (Walters 1997; Stankey et al. 2003). Drawing on the increased reliance on and growing literature promoting the use of "collaboratories" (Bly 1998), an interjurisdictional information network should foster adaptive multiparty communication and learning through an interactive cyber-infrastructure that provides not only access to information but also opportunities to upload data and comment on and interact with such data. In the United States, this interjurisdictional information network most appropriately would be housed in a federal authority (such as the Library of Congress or the CEQ), but would continue to allow for resource management decisions to remain with each agency delegated authority over a particular resource. In the context of climate change adaptation, relevant information would include not only ambient data and developed decision-support tools, but also information about potential management strategies gleaned from mandated monitoring and adaptive management. These collaboratories would harness information from participating authorities, academic scientists, and private stakeholders; offer genuine and numerous opportunities for interaction between such authorities; and provide a shared learning environment.

Such a federally maintained, publicly accessible, and system-wide collaboratory would facilitate information dissemination among similarly situated authorities, allowing for the full diversity of experience and information on the range of regulatory alternatives to be considered

(DiMento and Ingram 2005). Importantly, this transparent network provides opportunities to other management authorities and the public at large to review, contribute to, and challenge the efficacy of proposed and adopted adaptation strategies, facilitating deliberation and debate regarding existing uncertainty and the comparative value of alternative management strategies. Perhaps of equal importance, this learning infrastructure would create opportunities for more substantive collaboration and coordination of adaptation strategies between those with overlapping jurisdiction. As such, it would help reduce some of the undesirable effects of regulatory fragmentation that lead to regulatory inefficiencies and hinder interjurisdictional learning (Buzbee 2005). Finally, this information infrastructure would help managers at all levels of government manage the effects and uncertainty accompanying climate change and engage in adaptation planning. When combined with a process mandating sustained monitoring and correction by resource managers of adopted strategies, such a cyber-infrastructure would help promote resilience in the legal system by reducing uncertainty and allowing for more nimble adjustment of management strategies over time.

## Conclusion

Global climate change brings with it not only substantial change to natural resources, but also considerable uncertainty about the precise type and magnitude of such effects on any particular location or resource. This uncertainty exacerbates existing gaps in knowledge about ambient conditions and the efficacy of strategies in managing resources and resource conflict. Maintaining the resilience and effectiveness of natural resource management in the face of climate change necessitates the development of a learning infrastructure that helps managers reduce and manage uncertainty over time.

Congress and state legislatures could provide authoritative direction in this regard. Though recently proposed and adopted federal research and regulatory initiatives are likely to help increase the adaptive capacity of natural resource agencies, they neglect key sources of uncertainty, do not provide clear prioritization or goals for resource managers, fail to consistently require (or otherwise provide incentives for) adaptive learning and management by managers throughout the regulatory process, and do not provide opportunities for interactive information sharing among similarly situated managers. Federal and state legislatures may wish to establish clear goals and priorities for resource management and concrete benchmarks, resources, and incentives for monitoring, assessment, and periodic adjustment of strategies and programs in furtherance of such goals. In addition, legislatures or agencies could assist managers in harnessing the experience of others by establishing a cyber-infrastructure that not only collects and disseminates information on ambient conditions and potential management strategies, but also provides meaningful opportunities for managers, independent scientists, and other interested parties to interact and collaborate. A focus on developing an adaptive regulatory system that encourages interactive information sharing and tailoring of management in furtherance of identified regulatory priorities will help resource managers cope with uncertainty and work toward promoting the resilience of ecological systems as those systems are increasingly taxed by climatic change.

## References

Adelman, D. E., and K. H. Engel. "Adaptive Environmental Federalism." In *Preemption Choice: The Theory, Law, and Reality of Federalism's Core Question*, ed. William W. Buzbee, 277–300. New York: Cambridge University Press, 2009.

Adler, J. H. "Jurisdictional Mismatch in Environmental Federalism." *New York University Environmental Law Journal* 130 (2005): 130–178.

Bardach, E. *Getting Agencies to Work Together: The Practice and Theory of Managerial Craftsmanship*. Washington, D.C.: Brookings Institution Press, 1998.

Baron, J. S., L. Gunderson, C. D. Allen, E. Fleishman, D. McKenzie, L. A. Meyerson, J. Oropeza. and N. Stephenson. "Options for National Parks and Reserves for Adapting to Climate Change." *Environmental Management* 44 (2009): 1033–1042.

Biber, E. "The Problem of Environmental Monitoring." *University of Colorado Law Review* 83 (2011): 1–82.

Bly, S. "Special Section on Collaboratories." *Interactions* 5 (1998): 31.

Buzbee, W. W. "The Regulatory Fragmentation Continuum, Westway and the Challenges of Regional Growth." *Journal of Law & Politics* 21 (2005): 323–363.

Caldwell, M., and C. H. Segall. "No Day at the Beach: Sea Level Rise, Ecosystem Loss, and Public Access along the California Coast." *Ecology Law Quarterly* 34 (2007): 533–578.

Camacho, A. E. "Can Regulation Evolve? Lessons from a Study in Maladaptive Management." *UCLA Law Review* 55 (2007): 293–358.

Camacho, A. E. "Adapting Governance to Climate Change: Managing Uncertainty Through a Learning Infrastructure." *Emory Law Journal* 59 (2009): 1–77.

Camacho, A. E. "Assisted Migration: Redefining Nature and Natural Resource Law under Climate Change." *Yale Journal on Regulation* 24 (2010): 171–255.

Camacho, A. E. "A Learning Collaboratory: Improving Federal Climate Change Adaptation Planning." *BYU Law Review* 2011 (2011a): 1821–1862.

Camacho, A. E. "Transforming the Means and Ends of Natural Resource Management." *North Carolina Law Review* 89 (2011b): 1405–1450.

Coumou, D., and S. Rahmstorf. "A Decade of Weather Extremes." *Nature Climate Change* 2 (2012): 491–496.

Council on Environmental Quality. *Progress Report of the Interagency Climate Change Adaptation Task Force: Recommended Actions in Support of a National Climate Change Adaptation Strategy.* Washington, D.C.: CEQ, 2010. http://www.whitehouse.gov/sites/default/files/microsites/ceq/Interagency-Climate-Change-Adaptation-Progress-Report.pdf.

Council on Environmental Quality. *Climate Change Adaptation Task Force.* Washington, D.C.: CEQ, 2011a. http://www.whitehouse.gov/administration/eop/ceq/initiatives/adaptation.

Council on Environmental Quality. *Federal Actions for a Climate Resilient Nation: Progress Report of the Interagency Climate Change Adaptation Task Force.* Washington, D.C.: CEQ, 2011b. http://www.whitehouse.gov/sites/default/files/microsites/ceq/2011_adaptation_progress_report.pdf.

Council on Environmental Quality. Instructions for Implementing Climate Change Adaptation Planning in Accordance with Executive Order 13514. Washington, D.C.: CEQ, 2011c. http://www.whitehouse.gov/sites/default/files/microsites/ceq/adaptation_final_implementing_instructions_3_3.pdf.

Cruce, T., and H. Holsinger. *Climate Change Adaptation: What Federal Agencies Are Doing.* Arlington, Va.: Pew Center on Global Climate Change, 2010.

Davis, S. J., K. Caldeira, and H. D. Matthews. "Future CO2 Emissions and Climate Change from Existing Energy Infrastructure." *Science* 329 (2010): 1330–1333.

DiMento, J. F. C., and H. Ingram. "Science and Environmental Decision Making: The Potential Role of Environmental Impact Assessment in the Pursuit of Appropriate Information." *Natural Resources Journal* 45 (2005): 283–309.

Doremus, H. "Precaution, Science, and Learning while Doing in Natural Resource Management." *Washington Law Review* 82 (2007): 547–579.

Dorf, M. C., and C. F. Sabel. "A Constitution of Democratic Experimentalism." *Columbia Law Review* 98 (1998): 267–473.

Easterling,W. E., B. H. Hurd, and J. B. Smith. *Coping with Global Climate Change: The Role of Adaptation in the United States.* Arlington, Va.: Pew Center on Global Climate Change, 2004.

Endangered Species Act of 1973, 16 U.S.C. §§ 1531–1544.

Farber, D. A. "Bringing Environmental Assessment into the Digital Age." In *Taking Stock of Environmental Assessment: Law, Policy and Practice*, eds. J. Holder and D. McGillivray, 219–258. New York: Routledge-Cavendish, 2007.

Freeman, J. "Collaborative Governance in the Administrative State." *UCLA Law Review* 45 (1997): 1–98.

Freeman, J., and D. A. Farber. "Modular Environmental Regulation." *Duke Law Journal* 54 (2005): 795–912.

Gregory, R., D. Ohlson, and J. Arvai. "Deconstructing Adaptive Management: Criteria for Applications to Environmental Management." *Ecological Applications* 16 (2006): 2411–2425.

Gunderson, L. H., and C. S. Holling. *Panarchy: Understanding Transformations in Human and Natural Systems*. Washington, D.C.: Island Press, 2002.

Intergovernmental Panel on Climate Change. *Climate Change 2007: Synthesis Report Contribution of Working Groups I, II and III to the Fourth Assessment Report of the Intergovernmental Panel on Climate Change*. Geneva, Switzerland: IPCC, 2007.

Karkkainen, B. C. "Toward a Smarter NEPA: Monitoring and Managing Government's Environmental Performance." *Columbia Law Review* 102 (2002): 903–972.

Karkkainen, B. C. "New Governance in Legal Thought and in the World: Some Splitting as Antidote to Overzealous Lumping." *Minnesota Law Review* 89 (2004): 471–497.

Karkkainen, B. C. "Bottlenecks and Baselines: Tackling Information Deficits in Environmental Regulation." *Texas Law Review* 86 (2008): 1409–1444.

Malcolm, J. R., C. Liu, R. P. Neilson, L. Hansen, and L. Hannah. "Global Warming and Extinctions of Endemic Species from Biodiversity Hotspots." *Conservation Biology* 20 (2006): 538–548.

Millennium Ecosystem Assessment. *Ecosystems and Human Well-being: Scenarios*, vol. 2. Washington, D.C.: Island Press, 2005.

Milly, P. C. D., J. Betancourt, M. Falkenmark, R. M. Hirsch, Z. W. Kundzewicz, D. P. Lettenmaier, and R. J. Stouffer. "Stationarity Is Dead: Whither Water Management?" *Science* 319 (2008): 573–574.

Muhlfeld, C. C., et al. "Seasonal Movements of Non-native Lake Trout in a Connected Lake and River System." *Fisheries Management and Ecology* 18 (2011): 224–232.

National Drinking Water Advisory Council. 2010. Climate Ready Water Utilities Final Report of the National Drinking Water Advisory Council.

NOAA. "National Climatic Data Center" (n.d.). www.ncdc.noaa.gov/oa/ncdc.html.

NOAA. "Proposed Climate Service in NOAA" (n.d.). http://www.noaa.gov/climateresources/resources/ProposedClimateServiceinNOAA_Feb15rev.pdf.

NOAA. "Climate Research" (2011). http://www.oar.noaa.gov/climate.

NOAA. "Climate Prediction Center" (2013). http://www.cpc.ncep.noaa.gov.

NOAA. "Regional Integrated Sciences and Assessments Program." http://cpo.noaa.gov/ClimatePrograms/ClimateandSocietalInteractions/RISAProgram/AboutRISA.aspx.

Obama, Barack. "Federal Leadership in Environmental, Energy, and Economic Performance." Executive Order 13154 (2009). Office of Management and Budget. *High-Priority Performance Goals*. Washington, D.C.: OMB, 2010. http://www.whitehouse.gov/sites/default/files/omb/performance/high-priority-performance-goals.pdf.

Parmesan, C. "Ecological and Evolutionary Responses to Recent Climate Change." *Annual Review of Ecology, Evolution, and Systematics* 37 (2006): 637–669.

Peters, R. L. *Beyond Cutting Emissions: Protecting Wildlife and Ecosystems in a Warming World.* Washington, D.C.: Defenders of Wildlife, 2008.

Peterson, C. H., et al. "National Estuaries." In *Preliminary Review of Adaptation Options for Climate-Sensitive Ecosystems and Resources: A Report by the U.S. Climate Change Science Program and the Subcommittee on Global Change Research*, eds. S. H. Julius and J. M. West, 215–274. Washington, D.C.: U.S. Environmental Protection Agency, 2008.

Rahel, F. J., B. Bierwagen, and Y. Taniguchi. "Managing Aquatic Species of Conservation Concern in the Face of Climate Change and Invasive Species." *Conservation Biology* 22 (2008): 551–561.

Ruhl, J. B. "Climate Change and the Endangered Species Act: Building Bridges to the No-Analog Future." *Boston University Law Review* 88 (2008): 1–62.

Ruhl, J. B. "Climate Change Adaptation and the Structural Transformation of Environmental Law." *Environmental Law* 40 (2010): 363–435.

Ruhl, J. B. "General Design Principles for Resilience and Adaptive Capacity in Legal Systems – With Applications to Climate Change Adaptation." *North Carolina Law Review* 89 (2011): 1373–1401.

Ruhl, J. B., and J. Salzman. "Climate Change, Dead Zones, and Massive Problems in the Administrative State: A Guide for Whittling Away." *California Law Review* 98 (2010): 59–120.

Salzman, J., and B. H. Thompson, Jr. *Environmental Law and Policy*. New York: Foundation, 2010.

Secretary of the Interior. *Addressing the Impacts of Climate Change on America's Water, Land, and Other Natural and Cultural Resources*. Secretarial Order No. 3285 (2009).

Simberloff, D., and J. Cox. "Consequences and Costs of Conservation Corridors." *Conservation Biology* 1 (1987): 63–71.

Simberloff, D., J. A. Farr, J. Cox, and D. W. Mehlman. "Movement Corridors: Conservation Bargains or Poor Investments?" *Conservation Biology* 6 (1992): 493–504.

Smith, J. B., et al. *Adapting to Climate Change: A Call for Federal Leadership*. Arlington, Va.: Pew Center on Global Climate Change, 2010.

Stankey G. H., B. T. Bormann, C. Ryan, B. Shindler, V. Sturtevant, R. N. Clark and C. Philpot. . "Adaptive Management and the Northwest Forest Plan: Rhetoric and Reality." *Journal of Forestry* 101 (2003): 40–46.

Staudinger, M. D., N. B. Grimm, A. Staudt, S. L. Carter, F. S. Chapin III, P. Kareiva, M. Ruckelshaus and B. A. Stein. "Impacts of Climate Change on Biodiversity,

Ecosystems, and Ecosystem Services: Technical Input to the 2013 National Climate Assessment." (2012). http://downloads.usgcrp.gov/NCA/Activities/Biodiversity-Ecosystems-and-Ecosystem-Services-Technical-Input.pdf.

Strain, D. "House Science Panel to Investigate NOAA Climate Service." *Science* [online] September 22, 2011. http://news.sciencemag.org/scienceinsider/2011/09/house-science-panel-to-investigate.html.

Stutz, B. "Adaptation Emerges as Key Part of Any Climate Change Plan." *Yale Environment 360* May 26, 2009. http://e360.yale.edu/content/feature.msp?id=2156.

Susskind, L., A. E. Camacho, and T. Schenk. "Collaborative Planning and Adaptive Management in Glen Canyon: A Cautionary Tale." *Columbia Journal of Environmental Law* 35 (2010): 1–55.

Titus, J. G. "Rising Seas, Coastal Erosion, and the Takings Clause: How to Save Wetlands and Beaches Without Hurting Property Owners." *Maryland Law Review* 57 (1998): 1279–1399.

U.S. Department of the Interior. "Who We Are" (2011). http://www.doi.gov/whoweare/index.cfm.

U.S. Environmental Protection Agency. *Policy Statement on Climate Change Adaptation*. Washington, D.C.: EPA, 2011. http://www.epa.gov/climatechange/Downloads/impacts-adaptation/adaptation-statement.pdf.

U.S. Environmental Protection Agency. *National Water Program Strategy 2012: Response to Climate Change*. Washington, D.C.: EPA, 2012. http://water.epa.gov/scitech/climatechange/upload/NWP_Draft_Strategy_03–27–2012.pdf.

U.S. Forest Service. "Forest Service Global Change Research Strategy, 2009–2019." Washington, D.C.: U.S. Department of Agriculture, Forest Service Research and Development, FS-917a (2009).

U.S. Forest Service. *National Roadmap for Responding to Climate Change*. Washington, D.C.: USFS, 2010. http://www.fs.fed.us/climatechange/pdf/roadmap.pdf.

U.S. Forest Service. *Fiscal Year 2012 Budget Overview*. Washington, D.C.: USFS, 2012. Available at: http://www.fs.fed.us/aboutus/budget/2012/justification/FY2012-USDA-Forest-Service-overview.pdf.

U.S. Forest Service. "Information and Tools for Land Managers" (2013). http://www.fs.fed.us/ccrc.

U.S. Global Change Research Act of 1990, Public Law 101-606(11/16/90) 104 Stat. 3096-3104.

U.S. Government Accountability Office. *Climate Change Adaptation: Strategic Federal Planning Could Help Government Officials Make More Informed Decisions*. Report No. GAO-10–113. Washington, D.C.: GAO, 2009.

U.S. Government Accountability Office. *Climate Change: Improvements Needed to Clarify National Priorities and Better Align Them with Federal Funding Decisions*. Report No. GAO-11–317. Washington, D.C.: GAO, 2011.

Walters, C. *Adaptive Management of Renewable Resources*. New York: Macmillan, 1986.

Walters, C. "Challenges in Adaptive Management of Riparian and Coastal Ecosystems." *Conservation Ecology* 1, no. 2 (1997):1. http://www.consecol.org/vol1/iss2/art1.

Williams, B. K., R. C. Szaro, and C. D. Shapiro. *Adaptive Management: The U.S. Department of the Interior Technical Guide.* Washington, D.C.: Adaptive Management Working Group, DOI, 2007.

Williams, P., L. Hannah, S. Andelman, G. Midgley, M. Araujo, G. Hughes, L. Manne, E. Martinez-Meyer, and R. Pearson. "Planning for Climate Change: Identifying Minimum-Dispersal Corridors for the Cape Proteaceae." *Conservation Biology* 19 (2005): 1063–1074.

Worm B., et al. "Impacts of Biodiversity Loss on Ocean Ecosystem Services." *Science* 314 (2006): 787–790.

Zellmer, S., and L. Gunderson. "Why Resilience May Not Always Be a Good Thing: Lessons in Ecosystem Restoration from Glen Canyon and the Everglades." *Nebraska Law Review* 87 (2009): 893–949.

# Matching Scales of Law with Social-Ecological Contexts to Promote Resilience

JONAS EBBESSON AND CARL FOLKE

The scales of social-ecological contexts and interactions—including environmental degradation, ecological processes, societies, and societal development—are changing, and these changes challenge our thinking on the scales of law. Increasing exploitation and trade, altered migration patterns, intensified harvesting, and new scientific breakthroughs transform issues that were previously seen as national or subregional to global or regional in scope. Social-ecological contexts expand from local to regional and from regional to global or leap from local to global. Cross-scale interactions play out in new ways; new linkages and feedbacks are created that increasingly connect distant peoples, places, and parts of the biosphere. Examples include emerging markets of urban lifestyles and demands that reshape whole landscapes and regions and may tip them from sinks to sources of greenhouse gases, some examples being the explosion in demand for palm oil and widespread deforestation (Folke et al. 2011) and the lack of adaptive capacity among institutions to deal with fast-moving global markets for seafood (Berkes et al. 2006). Accordingly, not only ozone layer depletion and climate change are perceived as global features, but also ocean acidification, atmospheric aerosol loading, freshwater use, chemical pollution, and biodiversity loss. These are either systemic processes global in scope or global implications resulting from the aggregated local and regional processes

(Rockström et al. 2009a, 2009b). Societies are interconnected globally through political, economic, and technical systems, and these complex connections and cascades challenge any ambition for sustainable governance (Walker et al. 2009; Biermann et al. 2012).

Equally challenging for such ambitions are the complex global interconnections of societies through legal systems. Law—including legal structures, principles, institutions, and conceptions—is a fundamental element of society, and one of many integrated parts that, through individuals and societies, affect the biosphere and its functioning and life-support. Law guides and constrains human actions and has played an important role for the globalized and interconnected world of seven billion people. This expansion of the human dimension on Earth, especially since World War II, is now reaching the planetary boundaries for human well-being (Steffen et al. 2011). When Earth is viewed from outside the atmosphere, state borders can hardly be observed, and of course they do not mark out adequately the scales of social-ecological contexts. Nevertheless, state borders coupled with state sovereignty are fundamentals of the current paradigm of international law and relations and for the anachronistic divide of environmental issues and concerns as something "national" or "international." This last perception is changing, but law remains a slow driver, which in the international context is much shaped around state sovereignty and the nation-state rather than, say, network societies (Castells 1996) or polycentric conceptions (Ostrom 2010).

Established legal concepts on scales and jurisdictions do not align nicely with the often unclear boundaries and complex interactions, loops, and interdependences of social-ecological systems, and we believe that this mismatch affects the resilience of these systems. We define resilience of social-ecological systems as their capacity to absorb disturbance and reorganize while undergoing change so as to retain essentially the same functions, structures, identities, and feedbacks.

The pursuit of *legal resilience building* is more than adjusting requirements on polluting activities, planning provisions, or fishing quotas; it also involves concerns for law's validity, legitimacy, and scope of application; for standard setting and regulation in general; for notions of responsibility and liability; and for implementation, compliance control, and enforcement—and reflections on scales. Cross-scale social-ecological interactions require effective transboundary legal regimes that are capable of and mandated to integrate concerns for the environment, natural resources, and health, as well as in such areas as trade, investment protection (Romson 2012), and energy distribution.

In this chapter, we limit our attention to three structural issues of legal resilience building and the matching of legal scales with social-ecological scales:

1. The notion of who is entitled to participate in decision making and management of social-ecological concern;
2. The impact of state sovereignty for effective management of social-ecological systems;
3. The prospect of legal structures for inclusive multilevel governance.

The scales of law pertain essentially to its *geographical* reach, which is apparent in the jurisdictional divisions and allocations within and between states. Municipalities, counties, internal states, and territories within states often have legislative and enforcement powers on issues related to the local environment; in these settings, property rights and other sets of rights and obligations of individuals and corporations loom large. In the international setting, states' jurisdiction to adopt and apply laws and policies affect human health and the environment inside as well as outside the state borders. International environmental agreements, mostly developed as responses to transboundary environmental effects, span from bilateral to genuinely global instruments and reveal the wide range of spatial legal scales.

As to the first issue, new understandings of the scales of social-ecological processes and interactions put to question the legal framing of who has "sufficient interest," who is "affected," and whose right is impaired in a specific case, to the extent that these persons (and organizations) may have the right to be informed, participate in decision making, and initiate enforcement and legal reviews of decisions, acts, and omissions with a bearing on human health and the environment. An expanded social-ecological context requires expanding the understanding of who may be qualified, in legal terms, to act in or initiate administrative and judicial procedures. This framing matters for the legitimacy, fairness, and effectiveness of the decision-making processes and for institutional structures when managing ecosystems.

The second issue, the impact of state borders and state sovereignty, involves different aspects of self-determination, in particular to the delimitation of jurisdictions for legislative capacity, courts' adjudicative competence, and effective enforcement of laws by public authorities and members of the public. Jurisdictional borders also appear within states, for example, in federal systems and in systems in which regional and local authorities are given administrative and sometimes legislative capacities. Whether or not the social-ecological context is sufficiently embraced by a specific jurisdiction depends in part on its geographical size; whereas Lake Baykal is within the jurisdiction of one country, Lake Michigan is shared by two countries, and Lake Victoria by three. Many rivers are shared by even more countries, and the atmosphere of the earth obviously by all. Moreover, components such as lakes, which may appear to be distinct, interact with larger dynamics, for example, through watercourses, the atmosphere, migrating species, and human trade patterns. Accordingly, numerous social-ecological contexts cross numerous jurisdictions, and we are witnessing an increasing number of processes that even cross and challenge the biophysical boundaries of the entire planet (Rockström et al. 2009a, 2009b). These observations require a fresh thinking of sovereignty and self-determination.

On the third issue, the described legal and institutional dimensions cannot be reduced to just micro-level versus macro-level (Young 1994) or to national versus international. Rather, increasing cross-scale interactions in social-ecological systems are likely to require new legal and institutional interactions of different levels, scales (local, subnational, national, regional, and global law), and forms. Such interactions can be observed in the emergence of new forms of collaboration, such as the Global Partnership on Climate, Fisheries and Aquaculture (PaCFA) initiative, for new types of problems, including climate change, ocean acidification, and loss of marine biodiversity (Galaz et al. 2012), and in the increasing activity of nonstate actors in international contexts and even global regimes.

## Law and Resilience

Scale is but one of many features of law and institutions that impinge on the resilience of social-ecological systems, and matching scales is not enough for effective management. Other factors with a bearing on law are the flexibility of systems and institutions, multilevel governance, openness of institutions, public participation in decision making, and the potential for learning and adaptation (Folke et al. 2005; Ebbesson 2010). Legal rules, principles, and institutions define rights, obligations, and responsibilities; who has the power to do what, who has the right to participate in decision making, and who can request reviews; and what are the consequences of different acts, omissions, and situations. The extent to which legal structures, concepts, principles, and institutions adequately address the relevant cross-scale dynamics and complex interplays of societies and ecosystems not only depends on environmental laws in a narrow sense. The degree of social trust, experience, and distributive justice, and the legitimacy of rules and institutions more generally affect the resilience of social-ecological systems (Ebbesson 2010).

The legitimacy and trustworthiness of social institutions and decision making depends on the notion of *the rule of law* as such, with implied legal certainty and predictability and constraints on the power of government. Although the integrity and rule of law and legal certainty may curb flexibility in different situations, it also promotes trust in social relations, decision making, and public institutions in national as well as international contexts. Thus, in different ways, legal resilience building requires a combination of legal certainty and flexibility. Moreover, the legitimacy of legal frameworks is affected by the way they are created, and this must be taken into account when appraising and developing legal structures and institutions, especially in transboundary contexts, when several states are involved. In large-scale international contexts, legal structures are critical for determining who may make binding decisions on the content and application of general norms in different jurisdictions. Thus, for global legal arrangements (as in the case of climate change) to be legitimate, all states and peoples concerned must sense that they have a say in the law-making procedures and that the applicable law is worthy of being accepted (Ebbesson 2010).

This brings us back to the scales and reach of legal norms. A specific law may be intended to apply only for a certain area (a country or part of a country or a set of countries, as with multilateral environmental agreements and the European Union [EU]), but its legal and other implications may still reach well beyond the scale defined by such formal territorial limits. This is the case when national courts, authorities, and private actors apply national laws also in transboundary situations, so as to give the laws extraterritorial effect (e.g., when corporations are held responsible in their home countries for harm caused abroad). Laws may also have extraterritorial effects by influencing corporations and corporate structures outside a country or region when corporations wish to market their products and establish themselves in that country or region. Setting high safety standards or restrictions on products for health or environmental reasons can thus create new patterns of corporate behavior outside the given country or region. The early ban on certain products with ozone-depleting substances in the United States and other countries is one such example; EU legislation on chemicals, with quite far-reaching obligation on importers to provide risk information and so on is another. Trade-related provisions in international environmental agreements can also reach beyond formal borders of jurisdiction, because trade restrictions on, say, ozone-depleting substances or endangered species, prevent the parties from trading with nonparties as well.

On the other hand, the difficulty in measuring or defining the scales of the social-ecological context, including complex adverse environmental changes, potential thresholds, and tipping points, complicates the development of adequate law and policy responses. Still, there are examples wherein social-ecological systems have emerged locally to later connect to international environmental institutions and agreements for improved stewardship of natural resources and ecosystems. A striking one is the development of a global adaptive governance system to curb illegal, unregulated, and unreported fishing in the Southern Ocean, now embedded within the Commission for the Conservation of Antarctic Marine Living Resources (CCAMLR; Österblom and Bodin 2012). While the scale of the social-ecological context should be somehow matched legally, global and regional legal regimes and normative frameworks must be implemented, supported, and even supplemented by regional and local measures, including legislative and judicial acts—thus creating different forms of multilevel regime complexes (Dunoff 2007; Keohane and Victor 2011). Merely matching formally the scales of social-ecological processes with institutions and legal regimes of the same size is too simple and not adequate. At the end of the day, even global agreements and law making must reach to and influence the conduct of persons, either acting individually or collectively (e.g., as corporations, civil society organizations, and administrative bodies). Effective management requires multilevel structures and regimes into which multiple actors and different dimensions and scales can be included and integrated (Young 2011).

## Scales and Public Participation in Decision Making

### *Framing Interests Geographically*

The scales of law are reflected in the degree to which members of the public can participate in environmental decision making. In many jurisdictions, only persons in the vicinity or with a certain proximity to the activity or installations in question have the right to be part of decision making, for example, on whether to permit new installations and activities with adverse environmental impact, whether to adopt city plans, and whether to approve the marketing of new chemicals. Usually, the legal requirements are even stricter to be entitled to appeal or otherwise challenge such decisions.

Different states take different approaches and define the persons with rights to participate and to request legal reviews of decisions by different criteria. A common and traditional approach is to limit these rights to persons with a qualified interest in the case, thus excluding from decision making or judicial review anyone who cannot show that he or she is sufficiently affected by the decision in question or that his or her rights are sufficiently impaired. Spatial scales play a part when determining whether this interest test is met, so the larger the impact of projects and activities, the larger the scope of persons with participatory rights.

If such geographical criteria apply consistently, as new cross-scale interactions are identified, with new linkages and feedbacks connecting distant peoples, places, and parts of the biosphere and as the scales of the environmental issues expand, then so should the notion expand of who is affected by a decision in the legal sense. For example, if it is shown that a decision to allow the marketing of a chemical product or to increase a fishing quota in one region of the world affects other parts of the world, the scope of persons concerned expands dramatically. This understanding should influence the determination of which members of the public are considered sufficiently affected to be admitted relevant information and allowed to participate in the decision making in question. This should also transcend state borders, to allow persons concerned in other countries to be part of the decision-making process and challenge the decision made.

In many countries the scope of persons with the right to have access to environmental information and to participate in environmental decision making, for example, in environmental impact assessment procedures, has expanded, reflecting the understanding of expanded social-ecological contexts. One way of broadening the group of persons with such rights is to relax the link between the cause and the effect, so the person interested in the information or decision making does not have to provide much, if any, evidence to show that he or she is affected. Another, more radical approach, is to allow anyone who so wishes to access information and to be part of the decision-making process. Yet another approach is to acknowledge environmental organizations and other civil society groups in the decision-making process. In all, such tendencies reflect a loosening of the conceptual dichotomy between *private* and *public* interests, which has, and still is, commonly applied with regard to standing. More fundamentally, this reveals that many relevant social-ecological issues cannot be reduced to either private or public interests; they pertain to both matters, and environmental decision making has collective as well as individual repercussions (Ebbesson 1997). This understanding should make it easier for persons and actors who previously could not engage in environmental matters to do so. It should also imply that members of the public, when participating in decision-making processes, may invoke not only their individual concerns but also may refer to public interests, including, for instance, such disparate issues as protection of biodiversity and public health and precautions against flooding. While there is no clear trend or development, to different degrees, these approaches can be found in quite a few jurisdictions.

The situation is different with respect to "access to justice," that is, the right of members of the public to access courts or administrative authorities in order to challenge decisions, acts, and omissions by public authorities and private actors. Standing for judicial review procedures usually remains limited to persons with some interest or impaired rights in the case, for example, those being affected by the decision in question. There has been some broadening in this respect as well of the scope of actors with standing in court or before administrative review bodies, yet in a more restricted way than for participatory rights. One approach has been to relax the distinction between private and public interests for access to judicial review procedures. We believe that this may be appropriate when considering the multiple facets of environmental and health issues and the difficulty in clearly drawing a line between those who are affected, and thus granted standing, and those who are not. Another approach has been to grant civil society organizations devoted to health and the environment standing in judicial and administrative review procedures. A few jurisdictions do not request any specific link between the subject and the interest pursued to be granted access to justices, implying that anyone can initiate a judicial review procedure of an environmental decision, act, or omission (*actio popularis*). Although we do not see a very clear legal trend here, changes to expand the group of persons, including civil society organizations, have taken place in numerous jurisdictions in response to changed notions of interests and scales of social-ecological contexts.

Adequate public participation in environmental decision making adds to the legitimacy of the decisions made and the trust in the administration, and thus fosters the resilience of the systems. Resilience is also enhanced by the information and knowledge provided by members of the public and by the learning opportunities for all stakeholders, including corporations, when engaging in decision-making procedures. This should lead to higher-quality decisions and better outcomes. Even so, the argument is sometimes made, particularly in times of economic recession, that expanded opportunities for participation create huge transaction costs and make decision making ineffective and time-consuming. For sure, procedures and institutions to foster democracy and resilience involve costs, and there are most certainly cases in which participatory rights are abused. Even so, if indeed legitimacy and resilience is enhanced, the cost argument is not very strong. Moreover, the effects of large-scale decisions on health, the environment, and infrastructure remain for a long time, so in that sense allowing time for proper decision making and participation may also be justified. In fact, there is little evidence that expanding participatory rights, including giving standing to civil society organizations, has resulted in dramatic abuses of such opportunities. On the contrary, studies on the standing rights for environmental organizations indicate that they do not create floods of cases in the judicial system (de Sadeleer et al. 2005).

New cross-scale linkages and cascading dynamics in social-ecological contexts challenge the described ways of framing the group of actors who may access information; participate in decision making; and request review of decisions, acts, and omissions with a bearing on health and the environment. It puts to question the very notion of who is concerned or affected, and whether and how such criteria should be used by authorities and courts. Climate change is a pertinent example. While decisions regarding oil refineries and the building of new highways, which add emissions of greenhouse gases, affect not only people and organizations in the vicinity, but also people living far from the facility, indeed even at the other side of the world, this is not yet recognized in the legal context so as to create matching participatory rights.

The case of *Kivalina v. Exxonmobil Corporation* (2009), decided by a U.S. federal court, illustrates the challenge. The native Inupiat village of Kivalina, in Alaska, sued Exxonmobil Corporation and 23 other oil, energy, and utility companies for damage under a U.S. federal common law claim of nuisances. The federal district court dismissed the case in part by ruling that there were no "particular judicially discoverable and manageable standards" to guide the balance of interest, in part because the plaintiffs lacked standing under Article III of the U.S. Constitution. The first ground for dismissing the case, the lack of applicable standards for deciding the case, is a result of the lack of adequate U.S. federal standards for reducing greenhouse gas emissions. However, our discussion of framing scales and transboundary dimensions refers rather to the question of standing for the Kivalina community. The plaintiffs conceded that they were unable to trace their alleged injuries to any particular defendant and argued that they did not have to. On this point, the court held that, because there were no federal standards limiting the discharge of greenhouse gases, there was no presumption that the defendants' conduct harmed the plaintiff, and as a result "it [is] entirely irrelevant whether any defendant 'contributed' to the harm because a discharge, standing alone, is insufficient to establish injury." The court concluded that the causation between the discharges of greenhouse gases by the defendants and the injuries was too weak.

When claiming compensation for damage due to effects of increased emissions of greenhouse gases, it is not possible to link a specific injury to a specific activity, and it is likely that courts of other countries would reason in the same vein as the U.S. federal court. Nevertheless, the *Kivalina* case also shows the link between the lack of standing and the lack of effective regulation at an adequate level. Had there been standards related to the emission of greenhouse gases in place that were not complied with by the corporations, the issue of standing might have taken a different route. Despite the outcome in the given case, climate change reveals how the geographical expansion of social-ecological contexts may involve larger groups of persons who are affected and concerned. In these contexts, new forms of participation, possibly through multiple and polycentric governance, and through representation, may develop in transboundary contexts.

### *Transboundary Dimensions*

Realizing that social-ecological contexts increasingly transcend state borders and that opportunities for public participation may enhance resilience, can we somehow find this reflected in legal structures? Can we see any expanded opportunities for public participation across state borders or international efforts to promote public participation? To some extent we can (Ebbesson 1997, 2007). Public participation in environmental decision making was addressed in various national contexts before and around the 1972 Stockholm Conference on the Human Environment, although the opportunities provided varied considerably from one country to the other. This was reinforced at the international level at the time of the 1992 Rio Conference on Environment and Development. Principle 10 of the 1992 Rio Declaration on Environment and Development (Rio Declaration 1992), agreed upon by more than 150 countries, prescribes that each individual shall have the opportunity to participate in decision-making processes relating to the environment and have access to environmental information and to judicial and administrative proceedings with redress and remedy (Rio Declaration 1992). According to Agenda 21 (Agenda 21 1992), the action program for the twenty-first century, also adopted at the Rio Conference, "[o]ne of the fundamental prerequisites for the achievement of sustainable development is broad public participation."

The Rio Conference helped spur a global debate on public participation in environmental decision making, but the development of international law has not matched the dimension of the debate. Rather, it has developed mainly in some regions. The most advanced treaty is the 1998 United Nations (UN) Economic Commission for Europe Convention on Access to Information, Public Participation in Decision-making and Access to Justice in Environmental Matters (Aarhus Convention 1998), which sets minimum standards for the participatory rights described for the EU and almost fifty countries in Europe and Central Asia. The Aarhus Convention requires that anyone shall have access to information held by public authorities, without having to show a specific interest. It also requires that a broad category of members of the public, including nongovernmental organizations engaged in environmental matters, have access to decision-making procedures and effective review procedures. Some global environmental treaty regimes endorse public participation, and important developments have also taken place in international human rights bodies (courts, commissions) in Europe, the Americas, and Africa, to the extent they recognize such participatory rights as fundamental human rights (Ebbesson 2009). Indeed, these signs that public participation is increasingly becoming an issue of international law reflect a shift of scales and perception. Concerns that were previously only national have become international.

Allowing members of the public concerned, also across state borders, to be part of decision making and management concerning natural resources reduces the risk of ignoring adverse effects, even outside the country of origin. It also promotes learning across state borders and legitimacy of decisions made. In the same vein, resilience is promoted if members of the public can initiate judicial review procedures to the extent that polluters and others responsible for harm to health and the environment cannot hide behind state borders to escape responsibility. That pushes for change and adaptation, and also promotes learning by those involved and responsible. The case of *Kivalina v. Exxonmobil Corporation* (2009) is again a useful example. The social-ecological context involving the responsibility for the defending oil, energy, and utility companies obviously does not end at U.S. borders; these emissions of greenhouse gases affect fishing and coastal communities in other parts of the world. Not only environmental justice, but also resilience in general, would be promoted by expanding opportunities for persons affected across the state border to bring polluters and others to court for injunctive measures or compensation, in order to cover costs for adaptation and restoration (Ebbesson 2009).

Climate change is but one of many examples of social-ecological contexts in which everyone is in everyone else's backyard, and multiple acts affect multiple people, communities, countries, and regions (Folke et al. 2011). Fisheries, forestry, and the marketing of hazardous chemicals are other examples. The legal—and indeed political and moral—question for resilience and participatory rights in decision making is how to adequately define, through spatial or other criteria, who is sufficiently affected or concerned to be part of decision making and granted standing to challenge acts and omissions by pubic authorities and private actors with adverse impact on the social-ecological system. It is increasingly acknowledged in international law that members of the public should be able to participate in environmental decision making and management of natural resources across borders *without discrimination* (Ebbesson 1997, 2007). Accordingly, state borders should not prevent members of the public, including civil society organizations, from being ensured the same opportunities to participate in decision-making procedures as members of the public in the country where the decision is to be made. In addition to some regional conventions, the International Law Commission, with the mandate of the UN to progressively develop and codify international law, has promoted the principle of nondiscrimination in transboundary environmental decision making (International Law Commission 2001). If properly implemented, state borders could thus be transcended by members of the public requesting information, participating in decision making, and accessing the judiciary across state borders—and not only transcended via the offices of central governments. While such transboundary participation in environmental decision making remains quite unusual in practice, expanded social-ecological contexts will make it more relevant. As regards formalized transboundary cooperation in environmental matters of administrations below the level of governments, there are still few such arrangements. Such direct coordination and cooperation of subunits would in many situations support resilience, but this remains an underdeveloped area in international law and governance. Moreover, transboundary rights implied by nondiscrimination are not enough, because they would not as such ensure participatory rights in transboundary contexts. A state with restrictive rules on public participation and access to justice would still only need to ensure equally limited rights across the border. Therefore, more is required than simply ignoring state borders in decision-making processes.

We are not arguing that transboundary *actio popularis*, whereby any individual at anytime can initiate judicial procedures on any environmental issue, in pace with expanded social-ecological contexts, would be the most appropriate approach to promote resilience. For practical reasons, legal resilience building must enable representative units, such as local and national governments, intergovernmental bodies, and civil society organizations, to take effective actions, and cannot rely only on the participation of individual members of the public. Yet an updating of the notion of who is sufficiently concerned to have access to information and to legal reviews and to be part of environmental decision making is essential to promote resilience, as the scales of social-ecological contexts are expanding. Despite some positive developments in the last twenty years, international law remains rather fragmentary and patchy in supporting participatory opportunities in environmental decision making. Clearer and more robust international legal frameworks of regional and/or global scope remain to be developed. This transnational perspective on decision making, accentuating participatory rights for individuals and civil society organizations across large scales and state borders, is paralleled by the interstate dimension and the challenge to state sovereignty and state borders.

## Scales, State Sovereignty, and International Dimensions of Law

### *One Regime or Many?*

Law is part of the complex global interconnections of societies. International law, in parallel to the set of international norms that apply globally to all states (customary law and general principles), consists of numerous treaty regimes that define the rights, obligations, and competence of states and nonstate actors. These treaty regimes, covering anything from two to some 200 parties, create legal and institutional overlaps, but also plenty of lacunae, so international law only inadequately or asymmetrically deals with important issues.

The creation of adequate regimes is an important element in matching the scales of law with social-ecological contexts. Our hypothesis is that legal resilience building requires scale-matching, so that the geographical scale of the regime reflects the social-ecological issue at stake, but this does not mean that the matching should consist of only one comprehensive regime at one level for each concern. Most such matching and efforts to avoid and adapt to changes will involve multiple legal acts, international policy documents, declarations, and decisions of different normative status, thus creating regime complexes (Keohane and Victor 2011) and providing for multilevel governance. Still, resilience would gain from some form of overriding "umbrella treaty" or framework convention, reflecting the entire scale of the social-ecological context, such as the 1992 UN Framework Convention on Climate Change (UNFCCC), or the 1982 UN Convention on the Law of the Sea, to which other regimes can relate.

While general international law, in particular customary international law, applies to all states, it is often not distinct or adequate enough to ensure effective action in social-ecological contexts. Moreover, to a great extent it does not address truly global concerns, but rather focuses on interstate relations. On the other hand, international treaty arrangements, intended to establish intergovernmental forums for cooperation, and more specifically in defining norms of conduct in order to obtain an environmental objective, only bind the parties to the treaties.

### *General International Law: The Principle of No Harm*

The divide between "national" and "international" and the paradigm of state sovereignty in international environmental matters is plainly revealed by the most quoted principle of the 1972 Stockholm Declaration on the Human Environment (Stockholm Declaration 1972):

> States have, in accordance with the Charter of the United Nations and the principles of international law, the sovereign right to exploit their own resources pursuant to their own environmental and developmental policies, and the responsibility to ensure that activities within their jurisdiction or control do not cause damage to the environment of other States or of areas beyond the limits of national jurisdiction. (Stockholm Declaration 1972, Principle 21)

The quoted principle has been confirmed at subsequent Earth summits on sustainable development (Rio Declaration 1992; Johannesburg Plan of Implementation 2002) and in treaties and international court decisions as an acknowledged norm of international law. Although the principle says nothing about scales, it reflects a balance between the right to self-determination and the duty not to cause external harms, thus confirming the impact of state borders in addressing environmental concerns.

Leaving aside the debate on the extent to which this principle allows any balancing of interests or considerations (such as environmental, demographic, social, or economic conditions) for the states involved (Handl 2007), the principle formulates the fundamental responsibility of states to ensure that activities within their jurisdiction and control do not cause significant harm across state borders.

The principle of no harm (also referred to as the principle of good neighborliness) is particularly relevant in bilateral and rather local conflicts, and it has been influential for decisions by international courts and arbitration tribunals in such disputes as, for instance, the Trail Smelter arbitration award in the dispute between the United States and Canada (Trail Smelter case 1941) and the judgment by the International Court of Justice in the Paper Pulp dispute between Argentina and Uruguay (Paper

Pulp case 2010). When it can be shown that an activity under the control of a state has caused or may cause significant harm outside the territory of that state, the state may be held responsible and obliged to stop the cause of the harm/risk and to repair or compensate for the harm caused.

On expanding the scale, the principle of no harm is important also in regional and global contexts. Indeed, as a customary principle of international law, the principle of no harm applies globally and to *all* states. This feature in itself fits nicely with resilience thinking, because no state can "escape" from the application of the principle in transboundary contexts by means of state sovereignty or lack of state consent. The principle provides for flexibility in environmental governance, because each state has considerable discretion in deciding how *not* to cause transboundary harm. The principle of no harm is also a fundamental principle, on which other, more detailed rules and principles can be established. For instance, the obligations to require environmental impact assessments when activities may cause transboundary harm is today considered a part of general international law (International Law Commission 2001; Paper Pulp case 2010), but is also codified in various international treaties.

If complied with so as to avoid transboundary harm in local contexts, the given restriction of causing transboundary harm would also contribute to avoiding some harms of a larger scale. Nevertheless, the more complex the social-ecological context gets, the less adequate and effective is this principle for allocating responsibility to the states concerned. The principle generally falls short in effectively preventing harms in cases in which activities under a state's control do not cause significant harm when seen in isolation, but they nevertheless contribute to the harm together with other activities in other states. Many of the diffuse cases of pollution and environmental degradation are such, including the adding to the atmosphere of ozone-depleting substances, greenhouse gases, and aerosols. In these contexts, in which there is no clear causal link between the harm and the activity, the principle of no harm becomes inadequate, or at least insufficient, to guide and control states in avoiding harms. In some such situations, activities may even have little impact locally, in neighboring states, while adding to global harms. Moreover, by referring to the adverse effect elsewhere as a criterion for proper state conduct, the principle is often too vague, in itself, to effectively guide states and other actors.

The principle of no harm, with its moral underpinning, is not without legal relevance, for example, when assessing the responsibility of major contributors to climate change, such as the United States and China. Still, in these more complex contexts of large scales, the principle of common but differentiated responsibility is much more influential, for example, as the basis for negotiations to define the responsibility of each state (Cullet 2003; Brunnée 2009).

### *International Institutions and Treaty Regimes*

The principle of no harm should thus be supplemented, and the legal approach to defining states' duties to controlling harmful activities within each state should take other forms and be further defined. So should the responsibility of states to cooperate and coordinate their measures. Examples of such approaches are found in international treaty regimes on pollution of/through the atmosphere, such as acidification, ground-level ozone pollution, depletion of the stratospheric ozone layer, and climate change. States' duties and responsibilities are not defined in terms of the possible harm caused or through the general notion of common but different responsibilities, but in more measurable and verifiable terms, for example, through national caps, restrictions and prohibitions, technical standards, implementation plans, and strategy documents.

These treaties range significantly in the material scope of application, from focusing on very specific concerns, such as the 1973 International Agreement on the Conservation of Polar Bears (Polar Bears Agreement 1973) and the 1987 Montreal Protocol on Substances That Deplete the Ozone Layer, as amended (Montreal Protocol 1987), to covering almost all aspects of a given social-ecological context, such as the 1992 Convention on Biological Diversity, the 1994 UN Convention to Combat Desertification, and the 1992 UNFCCC. Narrowing down the scale or scope of application can be motivated in terms of effectiveness, because it is easier to negotiate and agree on far-reaching measures with respect to a specific substance, such as CFCs, DDT, and mercury, or specific species, such as the fur seal, than to agree on adequate measures to protect biodiversity or prevent climate change in general. On the other hand, if the focus is narrowed down too much, the regime may fail to effectively address the entire relevant social-ecological context.

Although there is no immediate geographical scale implied by the exact number of treaty parties, the challenges consist in including as many states as possible—if not all—covered by the social-ecological contexts, while not diluting the agreement with the increase in the number of parties. Here again, state sovereignty plays in, with the possibility of states staying out of such regimes, even when contributing significantly to the problem at stake, thus complicating the matching. In an optimistic account of the "new sovereignty" of the mid-1990s, Abram Chayes and Antonia Handler Chayes argue that "for all but a few self-isolated nations, sovereignty no longer consists in the freedom of states to act independently, in their self-interest, but in membership in reasonably good standing in the regimes that make up the substance of international life. To be a player, the state must submit to the pressures that international regulations impose." Chayes and Chayes continue: "Sovereignty, in the end, is status—the vindication of the state's existence as a member of the international system. In today's setting, the only way most states can realize and express their sovereignty is through participation in the various regimes that regulate and order the international system" (Chayes and Chayes 1995, 27). Their argument, that countries have been increasingly participating in various international regulatory regimes, is supported by the numerous international treaties and institutions established with the task of dealing with different transboundary and international issues, including health, the environment, natural resources, and emergency situations. In this respect, international law *did* change significantly during the twentieth century, even before the expansion of environmental law, from a system of international coexistence to a system of international cooperation (Friedmann 1964). Despite their obvious shortcomings in addressing and resolving environmental issues, the numerous international treaties and institutions relating to transboundary environmental matters, and changed notions of what is internationally relevant, show the possibility of changing and adapting law to the scope and scale of the issue at stake.

Contrary to the principle of no harm, the rationale underpinning most multilateral treaty regimes is that the common objective can be achieved (only) by the totality—the combined impact—of all measures and cooperation by the parties. Still, as we point out in the next section, it is too simple to think of treaty regimes as jigsaw puzzles (Ebbesson 1996), wherein each party provides its piece. We think that the establishment of adequate international regimes will remain crucial for legal resilience building across state borders, but we need better metaphors to understand how regimes operate.

We also agree that when states enter into international agreements, they generally tend to alter their mutual expectations and actions over time in accordance with the agreements (Henkin 1979) and that the propensity to comply with international law increases with new strategies to deal with the large bulk of compliance problems by means of ensuring transparency, dispute settlement, capacity building, and the use of persuasion (Chayes and Chayes 1995). Even though these factors mainly address compliance with what is already there, they are highly relevant for the effectiveness of international law, as well as for scales and resilience in larger contexts. Compliance with international commitments creates trust, which in itself is crucial for international regimes. Compliance also means that even if the applicable international norms reflect too low ambitions on the protection of human health or the environment, at least these weak ambitions are complied with.

The scale and reach of law is also affected by the extent to which states apply their national laws extraterritorially, whether as part of treaty commitments or on their own account. Some such application is controversial, yet in many cases it is not only accepted, but necessary. For example, effective protection and management of global commons and areas beyond exclusive state jurisdiction, such as the high seas and Antarctica, requires that states adopt, apply, coordinate, and enforce their laws outside their territory as well. In many such situations, this is required by international law. There are also situations wherein a state's extraterritorial application of its national laws for the protection of health and the environment *in the territory of other states* can be justified, even though contested by other states as infringing on their sovereignty. One such situation occurs when multinational corporations cause harm to environmental health in countries with weak authorities, poor or no legislation to protect these interests, and no remedy for the persons and interests harmed. The only effective remedy may be found in the company's home country. An illustrative example is the export by the major Swedish mining company Boliden of large quantities of highly toxic wastes to Chile in the mid-1980s, during the Pinochet era (Johansson and Edman 2009), with the obvious intention of avoiding responsibility and lowering its costs. In such cases, when corporations hide behind the state veil and abuse weak state structures, and when justice for those affected cannot be pursued in the state where the harm occurred, it should be possible to bring the corporations to justice elsewhere, presumably in the home state of the corporation (Ebbesson 2009). Both these situations of normative overlap reveal complex dimensions of scales in law, highly relevant as new transboundary social-ecological contexts are identified. They also reveal why state sovereignty and state borders should not prevent, in some cases, a state applying its laws to areas outside its borders, sometimes even to activities in the territories of other states, in pursuit of environmental justice and the protection of the commons.

Solving this mismatch of scales requires fresh thinking on sovereignty and jurisdiction not only in these two respects. State sovereignty must be constrained, so as not to allow the causing of extraterritorial harm, but also so as not to be an excuse for remaining outside international cooperative institutions related to problems to which the state in question contributes and is a part. That is, the claim for legitimate self-determination should never imply an escape from the duty to be part of international cooperation to solve common problems and to sustainably manage common resources when the state is part of the social-ecological context.

## Scales, Law, and Multilevel Governance

As social-ecological systems expand, intricate cross-scale interactions play out in novel ways and present new challenges for international law. Law and governance must meet the demands both of incremental change, when things develop in roughly continuous and predictable ways, and of sudden abrupt change, when past experience is often insufficient for understanding and reacting, consequences of actions are ambiguous, and the future of system dynamics is often uncertain. Social-ecological resilience in such situations draws on scale interactions (Folke et al. 2011), and multilevel governance will be required to avoid constraining options for societal development and future capacity for adaptation (Galaz et al. 2008).

Urban fads, lifestyle changes, emergent markets, and flows of resources, people, and information create new cross-scale linkages and feedbacks that increasingly connect distant peoples and places, in often fast and unpredictable patterns, and shape the capacity of the biosphere to sustain human well-being in new ways. New markets can develop so rapidly that the speed of resource exploitation often overwhelms the capacity of local institutions to respond (Folke et al. 2011). Rising human numbers, pervasive urbanization, new migration patterns, emerging markets, diffusion of new technologies, and social innovations may combine with shocks such as ecological crises, rapid shifts in fuel prices, and volatile financial markets. Such new interactions present a range of institutional and political leadership challenges, which have been insufficiently elaborated by crisis management researchers, institutional scholars (Galaz et al. 2010), and lawyers. For such governance to be effective, joint understanding of ecosystems and social-ecological interactions is required, as is building adaptive capacity to harness exogenous institutional and ecological drivers that might pose possibilities or challenges to social actors.

How, then, can and should law stimulate multilevel and adaptive governance and promote pathways of development that operate in synergy with the biosphere and frame creativity for sustainability (Folke et al. 2005)? The first step would be, again, to demote the mental and formal divide between notions of what is "national" and "international," which still dominates transboundary cooperation in environmental matters (Ebbesson 2010). The routine of centralizing interstate cooperation means that all activities and communication in transboundary contexts must pass through the central government rather than being entrusted to the regional or local units concerned in the respective state. Such centralization precludes potential dynamics of cooperation across state borders between municipalities and also limits the opportunities for members of the public to engage in transboundary decision making and management of common resources.

There are signs of change in the legal structures and forms of governance across borders relevant to multilevel governance and the resilience of social-ecological systems. In interstate relations, increasing regionalism with supranational elements reflects the need to go beyond this divide. The most apparent example is the EU, which is a unique case of transboundary law, institution building, and cooperation, not least in the field of health and the environment. With its quasi-federal structure, the EU has managed to loosen the stiff distinction between international and national law. Through formalized procedures, the EU member states and the European Parliament have adopted a large volume of legislation for the protection of health and the environment and thereby improved the implementation of international treaty obligations in the member states.

Despite considerable internal conflicts, clashes of interests, compromises, and self-interests among the member states, the EU has been able to develop legislation on environmental issues that promotes public involvement, takes into account the ecological features in management and administration of resources (Water Framework Directive 2000), and perpetuates compliance by the member states with international obligations relating to the environment. Moreover, even though its engagement in global environmental matters is not really escalating, the EU stands out as being able to push for a more active, and sometimes radical, transnational and international environmental policy than most industrialized countries and regions. We do not argue that the EU does enough to ensure social-ecological resilience or sustainable management of resources (and currently its global engagement may be weakening). But, while the EU appears as a bureaucratic and stiff organization in different respects, rather than as a flexible social institution, it nevertheless provides a relatively advanced form of transboundary cooperation for managing common resources and promoting multilevel governance compared with most international organizations.

The quasi-federal structures of the EU also highlight the fact that multilevel governance is not limited to interstate relations but also pertains to domestic structures. Although countries are formally considered equal in international law, they have very different domestic structures. Most states are somehow divided internally with respect to judicial, legislative, and/or administrative capacities of public bodies and institutions. Such divisions also affect the prospects for effective governance and management of natural resources and of dealing with environmental and health issues, and the legal and administrative divides often do not align with the relevant social-ecological formations. Administrative tasks and mandates are geographically defined in ways that often cut through rivers, forests, coastal areas, and other ecosystems, which may lead to deficits in the coordination and implementation of environmental and other laws.

The traditional geometrical notion (metaphor) of law, that international law is horizontal in structure—essentially state to state and government to government—while national law is vertical with different levels (central, regional, municipal) does not capture all dimensions of the normative directions or the relations between the actors involved. We are not suggesting that "polycentricity" provides a perfect spatial metaphor to cover all aspects of law, including the legitimacy of law. Nevertheless, the notion of polycentricity addresses not only the multiple and asymmetric institutions and norm-making in transboundary contexts, but also the diversity of actors and stakeholders in a way that the traditional notions of law do not. It reflects the existing "regime complex" (Keohane and Victor 2011) with respect to, for instance, climate change, biodiversity, and marine environments. It also stresses the fact that governance and initiatives are not a one-way, top-down issue and that many efforts and initiatives stem from the bottom and move up. When used in this way, polycentricity also grasps the many situations of international environmental law wherein the obligations may formally be addressed to state parties, while the implementation measures and the legal context required also relate to rights, duties, and competencies of private persons, corporations, and civil society organizations. Moreover, it pertains to transboundary situations in which transnational rather than international law dominates, for example, when states apply their national laws extraterritorially, when corporate codes of conduct apply across state borders, and when members of the public engage across state borders in decision making to object to harmful activities or policies.

In parallel with the increasing attention to public participation before national forums, including in transboundary settings, members of the public, not least civil society organizations, are increasingly active with regard to international governance (Ebbesson 2007). An ambitious approach to promoting such engagement of nonstate actors in international contexts is the 2005 Almaty Guidelines on Promoting the Application of the Principles of the Aarhus Convention in International Forums, adopted by the Aarhus Convention parties. Another example of multilevel governance is the notion of partnership projects, promoted by the Johannesburg Plan of Implementation of the 2002 Earth Summit on Sustainable Development (Johannesburg Plan of Implementation 2002), which essentially refers to projects involving states and the corporate sector. These examples not only confirm the polycentric dimensions of international law and governance, they also reveal the spatial expansion and complexity of governance, in the sense that civil society organizations, as well as corporate entities, are increasingly involved in areas and contexts far from their "home area"; again, everyone is in everyone else's backyard.

## Conclusion

Societies are not only interconnected globally through legal, political, economic, and technical systems, but also through the earth's biophysical life-support systems. Globalizing human–environment interactions are characterized by increasing connectivity, speed, mobility, and scale, and new forms of abrupt change through these interactions (Folke et al. 2011). It is time for the law, whether national or international, and for *lawyers* to swiftly shift perspective and start tackling the fundamental challenges facing humanity. These include expanding across contexts and borders, respecting planetary boundaries, allowing for social innovation in relation to stewardship of ecosystems and the environment, and supporting building capacity to deal with both new globally interconnected interactions and abrupt change. In this context, we have introduced the concept of *legal resilience building* for shifting human development into new pathways of collaboration with the biosphere, the very foundation for societal development and of which we are part and from which we have evolved. This includes rethinking scales of law to better match those of social-ecological contexts.

## References

Aarhus Convention. "United Nations Economic Commission for Europe Convention on Access to Information, Public Participation in Decision-making and Access to Justice in Environmental Matters. 25 June 1998." *International Legal Materials* 38 (1998): 515.

Agenda 21. "Report of the United Nations Conference on Environment and Development. 13 June 1992." *International Legal Materials* 31 (1992): 881.

Almaty Guidelines. Almaty Guidelines on Promoting the Application of the Principles of the Aarhus Convention in International Forums. Annex to Decision II/4. Report of the Second Meeting of the Parties. UN Doc ECE/MP.PP/2005/2/Add.5. 20 June 2005.

Berkes, F., et al. "Globalization, Roving Bandits, and Marine Resources." *Science* 311 (2006): 1557–1558.

Biermann, F., et al. "Navigating the Anthropocene: Improving Earth System Governance." *Science* 335 (2012): 1306–1307.

Brunnée, J. "Climate Change, Global Environmental Justice and International Environmental Law." In *Environmental Law and Justice in Context*, eds. J. Ebbesson and P. Okowa, 316–332. Cambridge: Cambridge University Press, 2009.

Castells, M. *The Rise of the Network Society*. New York: Wiley, 1996.

Chayes, A., and A. H. Chayes. *The New Sovereignty: Compliance with International Regulatory Agreements*. Cambridge, Mass.: Harvard University Press, 1995.

Convention on Biological Diversity. 5 June 1992. *International Legal Materials* 313 (1992): 818.

Cullet, P. *Differential Treatment in International Environmental Law*. London: Ashgate, 2003.

de Sadeleer, N., G. Roller, and M. Dross. "Access to Justice in Environmental Matters and the Role of NGOS: Empirical Findings and Legal Appraisals." Groningen, Netherlands: Europa Law, 2005.

Dunoff, J. F. "Levels of Environmental Governance." In *The Oxford Handbook of International Environmental Law*, eds. D. Bodansky, J. Brunnee, and E. Hey, 85–106. Oxford: Oxford University Press, 2007.

Ebbesson, J. *Compatibility of International and National Environmental Law*. London: Kluwer Law International, 1996.

Ebbesson, J. "The Notion of Public Participation in International Environmental Law." *Yearbook of International Environmental Law* 8 (1997): 51–97.

Ebbesson, J. "Public Participation." In *The Oxford Handbook of International Environmental Law*, eds. D. Bodansky, J. Brunnee, and E. Hey, 681–703. Oxford: Oxford University Press, 2007.

Ebbesson, J. "Piercing the State Veil in Pursuit of Environmental Justice." In *Environmental Law and Justice in Context*, eds. J. Ebbesson and P. Okowa, 270–293. Cambridge: Cambridge University Press, 2009.

Ebbesson, J. "The Rule of Law in Governance of Complex Socio-Ecological Changes." *Global Environmental Change* 20 (2010): 414–422.

Folke, C., T. Hahn, P. Olsson, and J. Norberg. "Adaptive Governance of Social-Ecological Systems." *Annual Review of Environment and Resources* 30 (2005): 441–473.

Folke, C., et al. "Reconnecting to the Biosphere." *Ambio* 40 (2011): 719–738.

Friedmann, W. *The Changing Structures of International Law*. London: Stevens & Sons, 1964.

Galaz, V., T. Hahn, P. Olsson, C. Folke, and U. Svedin. "The Problem of Fit Among Biophysical Systems, Environmental Regimes and Broader Governance Systems: Insights and Emerging Challenges." In *Institutions and Environmental Change: Principal Findings, Applications, and Research Frontiers*, eds. O. Young, L. A. King, and H. Schroeder, 147–186. Cambridge, Mass.: MIT Press, 2008.

Galaz, V., F. Moberg, E.-K. Olsson, E. Paglia, and C. Parker. "Institutional and Political Leadership Dimensions of Cascading Ecological Crises." *Public Administration* 89 (2010): 361–380.

Galaz, V., B. Crona, H. Österblom, P. Olsson, and C. Folke. "Polycentric Systems and Interacting Planetary Boundaries: Emerging Governance of Climate Change–Ocean Acidification–Marine Biodiversity." *Ecological Economics* 81 (2012): 21–32.

Handl, G. "Transboundary Impacts." In *The Oxford Handbook of International Environmental Law*, eds. D. Bodansky, J. Brunnee, and E. Hey, 531–549. Oxford: Oxford University Press, 2007.

Henkin, L. *How Nations Behave: Law and Foreign Policy*. New York: Columbia University Press, 1979.

International Law Commission. *International Liability for Injurious Consequences Arising out of Acts not Prohibited by International Law (Prevention of Transboundary Harm from Hazardous Activities)*. Report of the Work of Its Fifty-Third Session. UN Doc. A/56/10/suppl.10: 366–436 (2001).

Johannesburg Plan of Implementation. *Resolution 2*. Report of the World Summit on Sustainable Development. 26 August–4 September 2002. UN Doc. A/CONF.1999/20: 6–72 (2002).

Johansson, W., and L. Edman. *Blybarnen* [Toxic Playground]. DVD. Laika Film & Television/Folkets Bio, 2009.

Keohane, R. O., and D. G. Victor. "The Regime Complex for Climate Change." *Perspectives on Politics* 9 (2011): 7–23.

Kivalina v. Exxonmobil. Native Village of Kivaklina, and City of Kivalina vs. Exxonmobil Corporation et al, N.D. Cal. Case No: C 08–1138 SBA, Order granting defendants' motion to dismiss for lack of subject matter jurisdiction, September 30, 2009.

Montreal Protocol. "Protocol on Substances That Deplete the Ozone Layer. 16 September 1987." *International Legal Materials* 26 (1987): 1550.

Österblom, H., and Ö. Bodin. "Global Cooperation Among Diverse Organizations to Reduce Illegal Fishing in the Southern Ocean." *Conservation Biology* 26 (2012): 638–648.

Ostrom, E. "Beyond Markets and State: Polycentric Governance of Complex Economic Systems." *American Economic Review* 100 (2010): 1–33.

Paper Pulp case. International Court of Justice. Pulp Mills on the River Uruguay (Argentina v. Uruguay). Judgment 20 April 2010. International Court of Justice Reports 2010: 14.

Polar Bears Agreement of 1973. "International Agreement on the Conservation of Polar Bears. 15 November 1973." *International Legal Materials* 13 (1974): 13.

Rio Declaration. United Nations Declaration on Environment and Development. 13 June 1992. *International Legal Materials* 31 (1992): 876.

Rockström, J., et al. "A Safe Operating Space for Humanity." *Nature* 461 (2009a): 472–475.

Rockström, J., et al. "Planetary Boundaries: Exploring the Safe Operating Space for Humanity." *Ecology and Society* 14, no. 2 (2009b): 32. http://www.ecologyandsociety.org/vol14/iss2/art32.

Romson, Å. *Environmental Policy Space and International Investment Law.* Sweden: Stockholm University, Acta Universitatis Stockholmiensis, 2012.

Steffen, W., et al. "The Anthropocene: From Global Change to Planetary Stewardship." *Ambio* 40 (2011): 739–761.

Stockholm Declaration. United Nations Declaration on the Human Environment. 16 June 1972. *International Legal Materials* 11 (1972): 1416.

Trail Smelter case. Trail Smelter Arbitration. 11 March 1941. *United Nations Reports of International Arbitral Awards* 3 (1941): 1905.

United Nations. "Convention on the Law of the Sea. 10 December 1982." *International Legal Materials* 21 (1982): 1261.

United Nations. "Framework Convention on Climate Change. 9 May 1992." *International Legal Materials* 31 (1992): 849.

United Nations. "Convention to Combat Desertification in Those Countries Experiencing Serious Drought and/or Desertification, Particularly in Africa. 17 June 1994." *International Legal Materials* 33 (1994): 1328.

Walker, B. H., et al. Looming global-scale failures and missing institutions. *Science* 325 (2009): 1345–1346.

Water Framework Directive. "Directive 2000/60/EC of the European Parliament and of the Council of 23 October 2000 Establishing a Framework for Community Action in the Field of Water Policy." *Water Framework Directive Official Journal* L327 (2000): 1–73.

Young, O. R. "The Problem of Scale in Human/Environment Relationships." *Journal of Theoretical Politics* 6 (1994): 429–447.

Young, O. R. "Effectiveness of International Environmental Regimes: Existing Knowledge, Cutting-Edge Themes, and Research Strategies." *Proceedings of the National Academy of Sciences USA* 108 (2011): 19853–19860.

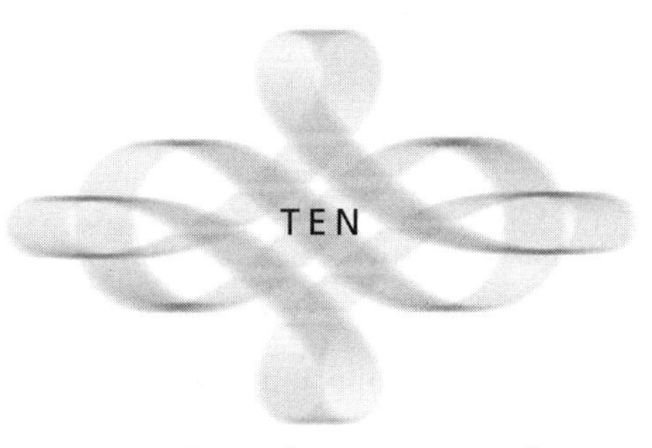


# Incorporating Resilience and Innovation into Law and Policy

## *A Case for Preserving a Natural Resource Legacy and Promoting a Sustainable Future*

TARSHA EASON, ALYSON C. FLOURNOY, HERIBERTO CABEZAS, AND MICHAEL A. GONZALEZ

The concept of sustainability has been widely embraced by society and in environmental law and policy as a measure to ensure a heritage of economic viability, social equity, and environmental stewardship. In a large number of statutes, Congress and many state legislatures have begun to adopt the goals of protecting a natural resource legacy and promoting sustainable use of the nation's valuable natural resources. However, many of the statutes enacted have been virtually unenforceable due to lack of standards and guidance on reconciling complex and often competing priorities. Moreover, reports continue to surface regarding such problems as diminishing natural resources, freshwater supplies, and biodiversity (World Wildlife Fund 2008). These undesirable impacts illustrate that contrary to the stated goals of existing law, the way we do business and consume resources remains unsustainable (Flournoy and Driesen 2010). Hence, while it is clear that the ideals of sustainability are widely supported, the shift toward this paradigm is essentially unrealized.

One key aspect of sustainability is ensuring that the resources and ecosystems on which we depend continue to support human existence from generation to generation; thus, it inherently is a dynamic concept. However, this ideal is in conflict with contemporary environmental law and policy, which have traditionally assumed that systems are predictable and that change is linear, incremental, and generally slow.

For the past several decades, ecologists have noted the dynamic nature of ecosystems and sought to account for this reality. In turn, legal scholars have begun to grapple with the challenge of developing policy for systems that change and evolve through time (Bosselman and Tarlock 1994; Wiener 1996). The importance of these efforts is highlighted by such irreversible ecological impacts as may occur due to climate change. Therefore, it is critical that laws and policy be developed and adapted to ensure the patterns of human interactions with the ecosystems on which we depend are sustainable. Because many of the present natural resource management laws still embody the goal of preserving ecosystems in some ideal or "natural" state, it is a goal that demonstrates tension with the dynamic reality of these systems. As such, the question is: How do we bridge the gap between the goals we espouse, the nature of what we want to protect, and current practice?

Resilience has particular promise as a concept that can help us to design laws that account for ecosystem dynamics. Resilience refers to the ability of a system to withstand perturbation and continue to function, thus acknowledging the variable nature of ecosystems over time. In the context of sustainability, Mayer et al. (2004) stated that resilience "focuses on the degree to which human activities increase or decrease the [persistence] of a particular dynamic regime that provides desirable goods or services" (420). This concept embodies pertinent aspects of sustainability and helps define a path toward sustainability. Further, it highlights the fact that preserving a resource legacy requires the protection of the ecosystems that provide it.

Complementing the focus on using resilience to achieve better stewardship of public natural resources, this paper also explores efforts to harness technological innovation and competitive pressures of the market to encourage industry to innovate in ways that promote sustainability, thus reducing risks to human health and the environment. Just as static measures are inadequate to ensure sustainability of natural resources, static regulatory approaches are also inadequate to engage industry's innovative power in the quest for operational, human health, and environmental benefits. The question this raises is how to create an economic incentive for those creating and selling goods and services to improve the environment, when the improvement benefits society as a whole rather than their consumers.

In response to the questions posed, this chapter articulates two strategies. One applies to decision making affecting management of publicly owned natural resources and the second to decision making in the private sector. The implementation of these strategies is illustrated by drawing on two proposals for new legislation and a recent public–private initiative. The first strategy is to incorporate resilience into policy and laws that govern the management of public natural resources to help reach the broader goals of sustainability—protecting ecosystems and natural processes, as well as the goods and services they provide. To date, few if any laws take direct, explicit account of resilience, although managers in many cases have discretion to, and already may, consider resilience in implementing decisions under these laws. To illustrate the potential embodied in the concept of resilience, we explore the implementation of a piece of model legislation: the National Environmental Legacy Act (Legacy Act) proposed by Flournoy et al. (2010), which is designed to promote sustainable resource use and decision making for public natural resources. Further, we investigate how the incorporation of metrics and indicators of resilience and sustainability into the Legacy Act framework may aid in target setting.

The second strategy is to harness technological innovation and competition to promote sustainable outcomes in private sector decision making. To illustrate the untapped potential of these forces, we explore how the Environmental Competition Statute (ECS) proposed by Driesen (2009) could advance sustainability if applied within the chemical sector of the economy. This second strategy is designed to overcome regulators' timidity and to create a market for innovation that benefits the environment. We then also draw on the example of a recent public–private cluster initiative that the U.S. Environmental Protection Agency (EPA) has embraced as a tool for maximizing economic development and environmental protection.

## The Legacy Act

The concept underlying the Legacy Act is to give meaning to the often invoked goal of sustainable use of publicly owned natural resources and, therefore, to effectuate our desire to protect a stock of these resources for future generations. This requires a proactive, adaptive management strategy with associated monitoring tools and metrics that can react in something close to real time. Adoption of the Legacy Act would first require us to confront a key question: What resource legacy do we wish to leave our children and grandchildren?

Despite the stated goals, and in some cases mandates, of existing laws to achieve sustainable resource use, it is clear that they have not achieved their aspirations (Flournoy et al. 2007). Many public natural resources are currently managed under statutes with notoriously open-ended standards that require federal agencies to "balance" a variety of often-incompatible uses, many of which degrade or deplete relevant resources. Many of these statutes contain no enforceable standard mandating protection of any particular quality or quantity of a resource (Flournoy et al. 2007). The on-the-ground results of implementation of statutes such as the Federal Land Policy Management Act of 1976, the Multiple-Use Sustained Yield Act, and the Magnuson-Stevens Fishery Conservation and Management Act demonstrate failure to accomplish even laudable stated mandates of sustainable use of publicly managed rangeland and fishery resources. The Legacy Act seeks to address the problem of death by a thousand cuts, the ongoing incremental loss of resources that leads to a shifting baseline and public acceptance of degradation of resources as inevitable (Ankersen and Regan 2010).

Building on the goals already expressed in numerous laws, the Legacy Act incorporates some key concepts. First is the demand that legislators articulate the legacy of resources that we seek to leave the next generation, covering a broad and diverse array of publicly owned and managed resources. Second, the law must define what type of degradation of the resources' quality or quantity is compatible with preserving that legacy. Therefore, if the legacy is to leave forest resources of the same quality and quantity as we currently have, then the permissible standard of uses of forests would be to prevent any degradation. Rather than authorizing use as long as it is sustainable, a Legacy Act shifts the burden and mandates that the stewardship agency not permit activity that would cause degradation in quality or quantity of the relevant resource. Further, actions inconsistent with this enforceable mandate would be expressly prohibited (Flournoy and Driesen 2010).

Thus, the Legacy Act would require management of public resources to conserve some defined stock of resources for future generations. Embracing the Legacy Act concept would demand that we identify our long-term goals, which would then help us chart and maintain a course to achieve our shared goals. It would also improve our decisions over the long term by generating the information base needed to support adaptive learning. This type of clearly defined limitation and prohibition on degradation beyond statutorily defined limits has proven successful in several statutory schemes. The Endangered Species Act's prohibition on taking of endangered species and the Clean Air Act's prevention of significant deterioration mechanism have both provided mandates that are specific and amenable to monitoring and that have been successfully enforced. The Legacy Act would seek to bring similarly effective management to a broad array of public resources.

A key issue to be resolved by Congress, through its democratic legislative process, would be to specify the contours of the legacy we commit to preserve. We could decide to commit to preserving a quantity and quality of all renewable resources equivalent to those we have today. Alternatively, we might decide to permit a specified degree of depletion or degradation. For nonrenewable resources, such as fossil fuels, we would very likely want to allow some specified pace of depletion or degradation, but for the first time we would actually consider what pace is a responsible one in consideration of future generations. The choices made in a Legacy Act would be intentional, and there would be greater accountability and transparency about the choices—both of which are frequently lacking under current law.

Under a Legacy Act, agencies already charged with the responsibility as trustees for federal lands and natural resources would now have a new mandate: to manage these lands and resources to ensure no impairment of the designated legacy, whatever that legacy may entail. This would require that the agencies translate the general standard of permissible degradation and depletion (if any) into operational terms. Design and implementation of such a statute poses two key challenges that resilience and sustainability metrics could help to meet. These related challenges are the dynamic and interdependent nature of ecosystems and the need to avoid creating excessively costly or impractical data demands.

The dynamic nature of ecosystems means that we cannot preserve ecosystems in a static state—whether this is the current state or some ideal state. Thus, to operationalize the Legacy Act, resource-specific goals (e.g., a prohibition of any degradation of water quality) are supplemented with the goal of retaining the resilience of overall ecosystems. This allows a more holistic and flexible approach, complementing the resource-specific mandates that serve as a backstop. It also addresses the reality that it is not feasible to monitor and set enforceable mandates for every specific resource. The overall mandate to maintain resilience provides broader protection and, as described below, there are analytic tools and data available that make this a feasible goal.

Implementation of a Legacy Act also depends upon identification of metrics that allow us to track degradation of publicly owned natural resources without imposing an unrealistic data demand. The concept of resilience and recent developments in sustainability metrics could have tremendous power and facilitate our efforts to achieve our stated goals. Adopting such methods into a Legacy Act could provide managers of federal public natural resources with a workable tool that would ensure federal public natural resources are managed to facilitate continuous progress toward sustainability.

## Managing Ecosystems for Sustainability and Resilience

Experts have amassed an abundance of evidence on important properties of ecosystems, enabling the delineation of key features of ecosystem function and structure. For example, many ecosystems are characterized by gradual accumulation of biomass and nutrients. These processes typically cycle and change along usual patterns driven by the daily and yearly cycles of light, tides, and seasons, along with longer-term decadal, centurial, and other temporal and spatial variations. Although complex, these processes often follow orderly patterns and are sufficiently consistent to be studied and understood (Odum 1971; Maurice and Phillips 1992).

If managed systems are resilient, they can withstand periodic fluctuations and still maintain self-organization and function through time (Eason and Cabezas 2012). However, it is possible for a dynamic system to reach a threshold and abruptly shift from one set of system conditions (i.e., regime) to another. When any system undergoes a change from one characteristic pattern or set of behaviors to another, the change is generally termed a *regime shift*. Regime shifts have been demonstrated for a multitude of ecological and social systems and often have significant ecological and economic consequences. For example, a lake may shift from oligotrophic to eutrophic due to the inflow of phosphorus, resulting in algae overgrowth, lack of oxygen needed for fish species survival, and consequently, a reduction in biodiversity and water quality. Hence, ecosystem change typically tends to be episodic rather than constantly erratic. Further, unlike engineered systems, ecosystems have multiple equilibria (i.e., multiple stable regimes) and may transition from one set of conditions to another with different underlying structure and behavior. Management of such systems must be flexible and adaptive, because it is often very difficult to predict exactly how or to what regime a system may transition. Because no two regimes have the same observable patterns, the characteristics of these regimes may be measured using metrics and indicators of underlying system behavior.

### *Demystifying Sustainability and Resilience*

The practical application of the concepts of sustainability and resilience require a new perspective. From a systems point of view, a system may be characterized by the parameters critical to its survival. For example, a regional system that supports human populations can be described by its economic, technical, ecological, social, and legal dimensions. The behavior of a dynamic system may be depicted as a trajectory through time in a space where the dimensions are its critical parameters (dotted line labeled "system trajectory" in Figure 10.1). In this context, sustainability relates to finding a set of system conditions that can support the social and economic development of human and ecological systems without major, irreversible environmental consequences (Karunanithi et al. 2008). Although optimizing based upon one aspect alone may result in localized benefits, managing systems myopically may result in burden shifting or lead to adverse or even catastrophic events for the entire system. Hence, a system is sustainable only if it meets key criteria

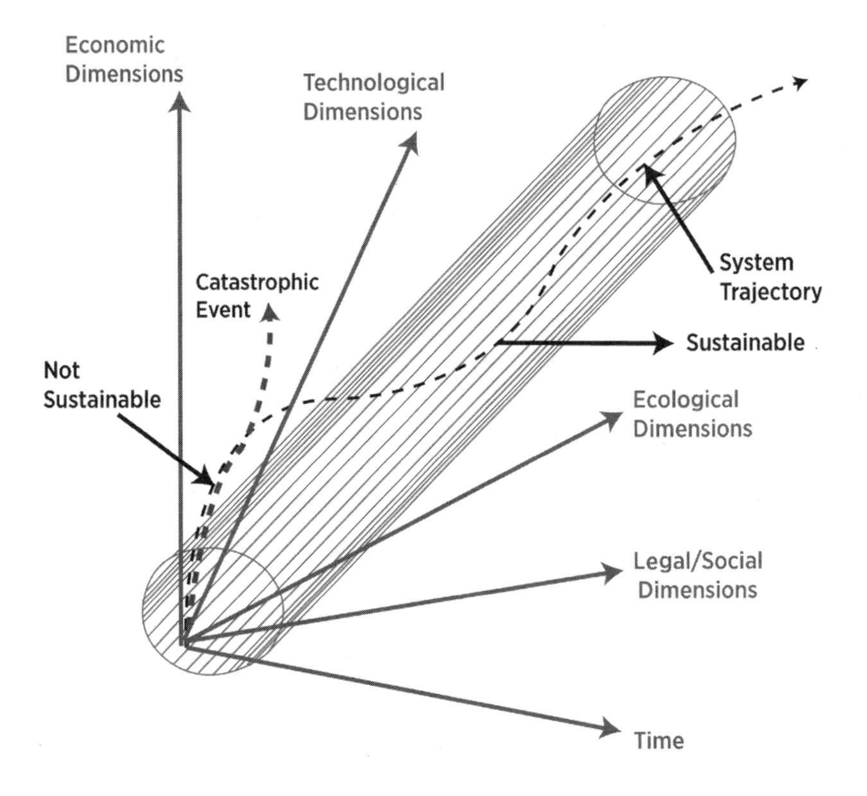


FIGURE 10.1. Sustainable systems path.

for all pertinent aspects. Accordingly, its trajectory must remain within a region of acceptable variability (tubular shape in Figure 10.1) for all of its critical dimensions. Resilience, then, is the ability of the system to remain within this region over time while under pressure from external forces and its own internal dynamics.

One important fact that must be kept in mind is that there is no explicit way to measure sustainability (U.S. EPA 2010). However, it is possible to assess whether the patterns of system behavior imply movement toward or away from sustainable conditions. Moreover, there are two principles to consider: (1) while an unsustainable system moving toward sustainability is trending in a desirable direction, a sustainable system moving away from sustainability will eventually become unsustainable; and (2) systems that are sustainable or that are moving toward sustainability must have sufficient resilience to remain on the path in the face of perturbations. Hence, sustainability is impossible to maintain or to achieve without system resilience. Metrics and indicators are a means of assessing these patterns. An indicator is a measure that provides information on specific system attributes (e.g., carbon dioxide emissions). Multiple indicators may be aggregated to form an index or metric, which may be used to assess integrated attributes of systems (e.g., the burden of human demand on land is evaluated using Ecological Footprint Analysis, as outlined below in the section on metrics and indicators). Both types of measures are used to provide understanding of human and natural systems, their corresponding linkages, and the burden of human activity on the systems that support them.

## *Metrics and Indicators*

Metrics and indicators tracked over time afford the ability to evaluate the observable behavior of systems and assess dynamic trends with respect to associated criteria. Although most widely used metrics and indicators focus on tracking specific and important concerns (e.g., the concentration of ozone or the estimated reserve in a fishery), the quantities needed for resilience and sustainability are broader concepts that represent the ability of ecosystems and societies to meet human needs. For example, assessing human burden on resources and ecosystems is pertinent for evaluating whether the burden is increasing or decreasing with time and, ultimately, whether that burden is within the capacity of the earth's system. While there is no consensus on specifically how to measure sustainability or resilience, there is a great deal of activity in this area.

### SUSTAINABILITY INDICATORS

The National Environmental Policy Act of 1969 (NEPA), a precursor to the establishment of the U.S. EPA, formalized a growing understanding of the importance of the relationship between humans and the environment, foreshadowing ideals soon to be of great importance on a global stage. Nearly two decades later, the World Commission on Environment and Development (WCED) coined the term "sustainable development" as "development that meets the needs of the present without compromising the ability of future generations to meet their own needs" (United Nations 1987). Years after the WCED definition was accepted, researchers have continued to struggle with reaching a consensus on exactly how sustainability should be measured. One key development and widely accepted convention was to establish what are termed the "three pillars of sustainability": environment, economy, and society. Each of these pillars denotes particular aspects of the product, process, or system that may be assessed via observable and measurable criteria.

Using the "triple bottom line" as a working principle for sustainability, there is a plethora of activity related to sustainability indicators throughout the world. From individual researchers to large international task groups (e.g., World Bank, Eurostat, Organization for Economic Cooperation and Development), researchers have produced a multitude of measures for assessing aspects of sustainability for various scales and topics. The International Institute of Sustainable Development (IISD) lists 895 sustainable development initiatives worldwide, which include nearly 150 indicator development activities (IISD 2011). One of the key challenges of assessing sustainability is determining indicators based on sound science and pertinent to the system under study.

The protection of natural resource inventories and ecosystems is core to the mission of the National Environmental Legacy Act. Hence, indicators such as resource consumption rate, resource availability, land use and type, freshwater availability, deforestation, and biodiversity are critically important, as these indicators may be tracked over time and used to determine suitable targets for management. However, as mentioned previously, the true question is whether the trends in these indicators are sustainable. Accordingly, not only should systems be assessed by trends in sustainability indicators, but the resilience of the system should also be measured to evaluate its propensity to experience undesirable shifts.

## MEASURES OF RESILIENCE

As previously noted, resilience relates to the ability of a system to withstand perturbation and continue to function. Even under extreme external pressure, many natural systems will adapt and shift into alternative regimes without human intervention. However, like the case of a clear, thriving oligotrophic lake that shifts to a turbid, eutrophic regime, the new regime may be undesirable and unproductive from a human perspective. One reason is that the costs of remediation and potential for unalterable consequences can be substantial. Consequently, the resilience community has been focused on identifying catastrophic shifts before they occur.

Using variance-based measures, Carpenter and Brock (2006) indicate that increases in variability signal impending regime shifts. Van Nes and Scheffer (2007), Dakos et al. (2008), and Chisholm and Filotas (2009) propose rate of recovery from perturbation (i.e., "critical slowing down") as a measure of system resilience. Biggs et al. (2009) suggested increasing variability, skewness, kurtosis, autocorrelation, and slow rates of recovery from perturbations (i.e., critical slowing down) as leading indicators of impending regime shifts. However, they noted that increases in these indicators typically occur once the regime shift is underway, which is too late to implement effective management actions (Biggs et al. 2009). Moreover, Scheffer et al. (2009) state that while these indicators show great promise in detecting regime shifts in simple and model systems, work is still needed to determine whether these indicators provide warning of imminent shifts in real complex systems. In response to this research question, Eason et al. (2013) used various model and complex real systems to compare the performance of traditional indicators to that of Fisher Information, an information theory approach. Fisher Information affords the ability to characterize the dynamic behavior of systems and to include its regimes and regime shifts. Results of this work offer great promise for resilience science and sustainability.

## USE OF METRICS AND INDICATORS FOR ENVIRONMENTAL MANAGEMENT

Arguably, the measures needed for environmental management should be those representing the most important, fundamental processes and services essential to human existence. Fortunately, the number of such critical processes is likely to be modest. For example, human existence depends on the ability of ecosystems and the environment to: (1) cycle nutrients, including carbon, nitrogen, phosphorus, and oxygen; (2) capture and distribute energy from the sun throughout the planetary ecosystem; (3) support an economy that can help to provide for human welfare over the long term; and (4) maintain the system's integrity and self-organization, which is the basis of life and societal existence. Naturally, this listing is not complete; however, it does provide a reasonable sense of the matter. In this context then, assessing resilience and sustainability is related to finding appropriate scientifically grounded measures that can be used to track and assess these various system properties, subsystems, and functions.

Researchers have recently undertaken a study to assess whether a seven-county region in south-central Colorado is moving toward or away from sustainability over time (U.S. EPA 2010). By using four metrics based on publicly available data, it was demonstrated that it is possible to assess sustainability through time as a function of basic properties of the system. These properties are the ecological impacts of human activity through Ecological Footprint Analysis (EFA); economic well-being with Green Net Regional Product (GNRP); flow of available energy through the system using Emergy Analysis (EmA); and system order and stability with Fisher Information (FI). Brief information on each metric is provided below; however, more details may be found in the work by U.S. EPA (2010).

Time-series data of variables characterizing pertinent features (e.g., demographic, production, consumption, land use) of the San Luis Basin region in Colorado were compiled and used to calculate the aforementioned sustainability metrics, each of which captures distinct aspects of the system. Ecological Footprint Analysis is a measure of the impact of a population on environmental resources and involves identifying the amount of biologically productive land (biocapacity) and determining the demand (ecological footprint) placed on the land to support the human consumption, production, and waste-generation activities. Emergy Analysis assesses energy flow through a system and is a means of estimating the value (in terms of captured solar energy) that the environment contributes to society. Roughly stated, emergy is the amount of solar energy invested by the environment into the creation of something (e.g., a living thing, a resource, a product) or in maintaining a natural process. This may include emergy created over geologic time, as is the case for fossil fuels (petroleum, coal, etc.), which are considered nonrenewable over human timescales. Green Net Regional Product is a macroeconomic measure that captures the economic well-being of a system and is equal to the difference between aggregate consumption and the depreciation of human and natural capital, i.e., (value of all market transactions) – (depreciation of human + natural capital). Fisher Information assesses dynamic order by capturing the patterns in the observable behavior in a system. Because order relates to the ability of a system to maintain a desirable steady state (i.e., regime), Fisher Information is used to characterize self-organization, regimes, and regime shifts (Karunanithi et al. 2008).

Criteria were established to interpret each metric and determine whether the system shows signs of moving toward or away from sustainability. The criteria did not seek to preserve any particular state of the system (e.g., pristine or ideal), rather they ensured the maintenance and preservation of the basic properties, products, and processes necessary for continued function and livelihood. Further, an unsustainable path was defined as one whereby human welfare, ecological balance, emergy flows into the system, or system organization (as defined by GNRP, EFA, EmA, and FI, respectively) is compromised. Accordingly, as illustrated in Figure 10.1, a violation of the criteria for any metric indicated an unsustainable path and undesirable trajectory.

The regional sustainability project described above provides key insights on evaluating the dynamic changes in a region largely comprised of publicly owned land. Although it was a pilot project, it offers a model that enables resource management agencies to track trends, which delineate movement toward or away from sustainability, and provides a scientifically credible set of measures suitable for assessment and management (U.S. Department of Housing and Human Development 2011). With increasing recognition of the need for such strategies and the ongoing interest in finding ways to assess and achieve sustainability, the study offers a practicable path for policy development in land and resource management.

## Linking Metrics and Indicators to a Legacy Act

Embedding the concept of resilience more deeply into the framework of the Legacy Act and incorporating measures like those used in the

San Luis Basin study offer potential strategies for improving on the model Legacy Act described by Flournoy et al. (2010). The core of the act is a standard that precludes impermissible degradation of the quantity or quality of all public natural resources and charges the agencies managing them to document that no such degradation is occurring over a long time horizon. As originally described, the Act would require agencies to collect data on all individual resources and monitor the change in relevant parameters over time, a potentially monumental task. For some subset of nonrenewable or threatened resources, this level of monitoring of individual resources to ensure no impermissible degradation or depletion would be appropriate. But for the broader array of resources in public ownership (and particularly for assessing impacts to ecosystems and renewable resources), a modification in the approach under the Legacy Act could streamline implementation and take better account of both the dynamic nature of ecosystems and the key role of resilience. We therefore propose a modification to the proposal for the Legacy Act. For the majority of renewable resources under public management, the statute could direct agencies to ensure that the ecosystems remain on a sustainable path and that their resilience is not impaired.

Thus, rather than requiring agencies to identify the current quantity and quality of all individual natural resources under public management and to monitor these on an ongoing basis, the Legacy Act would require agencies to focus on a key subset of resource parameters. This would involve collecting the data needed to assess the sustainability and resilience of the relevant ecosystems, thereby affording the ability to monitor trends and ensure that the system is on a sustainable path. Such an approach would not only ensure that the services and value of the ecosystems are preserved, but would also allow the agencies to use publicly available data in many cases, thus avoiding unrealistic or impractical data and monitoring demands. The resource-specific measures would remain an important backstop, ensuring that key resources are preserved according to whatever standards are set forth in the statute. This information would enable near real-time adaptive management by public resource managers and provide the same transparency, accountability, and long-term protection that the Legacy Act seeks.

## Environmental Competition Statute

As detailed in the previous sections, the roles that environmental law and science play to safeguard and protect human health and environment for not only our current generation but also for future generations are extremely important. In addition to their importance, the ability for each to complement the other has been established, and society is continuing to learn of new opportunities to expand these complementary efforts. The science of sustainability is one that is transdisciplinary in nature and operation and is continuing to evolve. Thus, there exist opportunities to bridge concepts, theories, and methodologies to understand, advance, and operationalize this new approach to environmental preservation. One concept that has been introduced demonstrates the merging of new approaches to make environmental law a mechanism for generating change for a more sustainable environmental future. This concept, the ECS (Driesen 2009), provides a mechanism that embraces dynamic and constructive change. The ECS incorporates a "triple bottom-line" concept into an environmental regulatory regime. This triple bottom line embraces economics, environment, and society rather than solely addressing environmental protection.

In the case of the ECS, this triple bottom-line approach to sustainability uses economic incentives to stimulate innovation in environmental technology. While the role of innovation and technology in environmental protection will be discussed in more depth in the next section, it is important to point out that technology has long played an important role in the remediation of environmental degradation but is now also widely employed as a means of preventing pollution. By introducing the use of competition into the mix, the ECS opens the door to advancing the development and application of novel and innovative technologies for environmental protection and overcomes the inherent limitations of traditional regulatory approaches in stimulating innovation. As with any next-generation environmental law or statute, the stimulating of movement toward a more sustainable future is the goal.

The ECS responds to a number of flaws that Driesen (2010) identifies as characteristic of first- and second-generation environmental laws. First, regulation under most pollution control laws is hampered by regulators' timidity in setting ambitious standards. Driesen describes why and how agencies are overly concerned with impacts on the most antiquated actors in an industry and too little concerned with the positive benefits of incorporating new technology. Even under statutes designed to force technology, agencies tend to demand "relatively modest improvement based on well-understood technology" (Driesen 2010, 175). He describes how this occurs not just with command-and-control regulation but can also affect the design of emissions-trading regimes and pollution taxes, limiting their efficacy to spur innovation and implementation of the best technology.

So what is the goal of an ECS? While it can be seen as advancing the role of law in environmental protection, it is also a mechanism for introducing the inclusion of competition and innovation to improve the environmental quality of a system—to create a race to the top in the development of environmentally superior technology. This is analogous to creating the dynamics of a market in a business sector. In industry, competitors vie for greater market shares within their sector by introducing new, better, or less costly products to meet consumer demand and desires. In return, they are typically rewarded by increased sales, greater profits, and an increased market share. If this type of atmosphere can be created in the environmental technology and protection market, the development of novel and innovative technologies would be promoted and nurtured. The ECS creates this type of competitive environment by providing a premium or reward (i.e., incentive) to businesses for introducing new technologies that exceed regulatory baselines. This "market" provides the ability for technology developers (innovators) to have free rein in improving environmental quality while advancing their firms' economic interests, making it possible to reduce pollution, preserve natural resources, and generate a profit from doing so.

The following scenario highlights the possible implementation of ECS. Five manufacturers of a plastic polymer product are subject to the same environmental regulations, utilize similar technologies in the manufacture of their products, and have similar emission profiles. Additionally, they comply with the relevant environmental regulations. One manufacturer, Company D, has recently been purchased and the new CEO wants to introduce a greener technology for producing the polymer. While this new process has higher operating costs, which the company is willing to absorb, it will allow Company D to lower its emissions in one category below the regulated level. Because the company is already in compliance with the environmental regulation, there may be no market incentive to reduce the emissions. As noted above, these emissions are externalities—costs borne by the public at large. Thus, the market will not inherently create incentives to minimize or eliminate the emissions. The inevitable timidity characteristic of regulators operating under a traditional regulatory regime similarly will prevent regulators from setting an ambitious regulatory standard, even though it may be within the industry's economic and technological reach. Therefore, society misses the benefit of applying an existing technology to reduce pollution even further. Under this scenario, if an ECS were in place, the other four companies would be required to pay a fee to Company D (i.e., a reward) for exceeding compliance that would cover the cost of using and developing the environmentally superior approach and provide a premium to Company D. The other companies, therefore, have an incentive to implement a new (or a similar) technology not only to avoid paying the reward to Company D but also to have the opportunity to be paid a fee for implementing a compliance-exceeding technology themselves by the remaining companies who have not exceeded regulatory requirements. From this scenario, we can see that a competition strategy could create a "domino effect" through its market and incentive approach, resulting in companies developing and implementing cleaner technologies and creating a less polluted environment.

Although this is a fictional example, it illustrates the ability to create the incentives to exceed the norm, a major facet of the proposed ECS. This "surpassing of the norm" also presents an opportunity to incorporate the concepts and practices of sustainability and innovation through environmental protection.

## Harnessing Technology and Innovation for Sustainability

As touched on briefly above, technology has played a pivotal role in the area of environmental protection. However, this role has largely been limited and relegated to the remediation of environmental degradation and to implementing strategies to allow for remaining within regulatory limits. These actions are typically associated with a reactive mindset.

Since the groundwork laid over four decades ago with the NEPA of 1969 and the definition formally introduced over twenty years ago (United Nations 1987), the concept of sustainability has begun to play a critical role in environmental protection. True adoption of such an ideal requires a paradigm shift from a reactive to a proactive approach in business, science, and society at large. This preemptive approach avoids creating an environmental challenge, which would later require a corrective action and can be demonstrated by such examples as developing a technology that no longer uses toxic chemicals (or creates a toxic product) or one that releases minimal to zero fugitive emissions. Although this green chemistry approach (discussed further in the section on green chemistry) is still relatively new, there are numerous reports and examples of successes (U.S. EPA 2011).

It is imperative that society become as proactive in considering resource needs and environmental challenges as we are reactive in handling environmental consequences. Many of the environmental challenges that our national and international partners face are the result of human activities and human-generated products, including chemicals. Chemicals, in particular, have the potential to generate environmental and human health impacts throughout their entire life cycle. With this potential for impact, it is evident that a sustainable and holistic approach to chemical design, synthesis, management, and reuse can contribute significantly to addressing current and future environmental and human health impacts created by these manufactured chemicals. By applying a proactive and holistic approach, we can begin to minimize or eliminate these impacts across the entire chemical life cycle and increase protection of the environment.

The goal of sustainability is being employed in the chemical sector to reduce negative effects on the environment and human health. To achieve sustainability across the life cycle of a chemical, we must have the ability to not only minimize or eliminate this risk across the life cycle, but we must also be able to assess and quantify any remaining risk and ensure a more sustainable path is being achieved. As the life cycle of a chemical or technology is mapped out, many opportunities exist for improvement to current technologies, as do research areas for development of novel and innovative processes. This is where a proactive approach coupled with a holistic view provides the best opportunity to increase the sustainability of a system. The ECS illustrates a novel regulatory approach that achieves this.

### *Green Chemistry and Green Engineering*

To further build on this point, examples of this proactive approach that have received tremendous support since their introduction are the areas of green chemistry and green engineering. These are not new disciplines for chemistry and engineering; they are new approaches to performing chemistry and engineering. Introducing concepts based on pollution prevention, sustainability, and industrial ecology into the disciplines of chemistry and engineering encourages the development of new technologies and methods that have environmental protection at their foundation.

The twelve principles of green chemistry described by Anastas and Warner (1998) offer a philosophical basis to identify potential areas in which the level of greenness in designing or implementing chemical technology and reactions can be increased. With these efforts in mind, the number of opportunities increases significantly upon the introduction of chemical engineering and the twelve principles of green engineering articulated by Anastas and Zimmerman (2003). These principles introduce concepts for process design, scaling up, and the use of alternative reactor configurations and geometries for influencing reaction conditions and product and emission profiles. Collectively, these twenty-four principles provide a significant foundation for utilizing technology to protect the environment and advance society along the path of sustainability. Regulatory models like the ECS offer the opportunity to build on and engage these promising foundations in the sciences by providing companies an economic incentive to implement these new opportunities for greener industrial practices.

### *Innovation*

Technology generally is central to minimizing any potential for impact to the environment and human health, but there are other contributors to the goal of increasing the sustainability of a system. One such contributor is innovation. Innovation can be described using a multitude of definitions; common in those definitions is the desire to capitalize on new ideas or technologies. This is usually driven by economic considerations, yet typically also affects societal and environmental conditions. The impact of innovation is manifest in advances ranging from the use of a cell phone and its technical capabilities to increase access to medical information in a remote region, to the development of a new technological product that creates a completely new market sector. From an environmental perspective, the value of innovation depends on how we capture a technology to satisfy a need and advance society's other interests while preserving the environment.

In recent years, the ability to incorporate innovation into our daily and professional lives has seen a dramatic increase. Within the environmental research and business communities, innovations have included the use of: (1) technology applications to make access to environmental information easier; (2) social networking and media to bring "up to the minute" information to users; (3) innovation as a means of increasing entrepreneurship and intrapreneurship; (4) the use of transdisciplinary teams for technology development; and (5) environmental protection as a means of spurring economic development. The ECS would help to promote innovation that enhances environmental performance by providing an economic incentive for such innovation in lieu of existing economic disincentives.

### *Maximizing Economic Development and Environmental Protection in the Administration of Environmental Law and Policy*

Beyond the regulatory approach embodied in the ECS, government can also play a role in promoting environmentally beneficial economic development through participation in creative public–private initiatives. Recent developments by the U.S. EPA provide an example of utilizing environmental protection to spur economic development. U.S. EPA is using the agency's research and development mission to transform some aspects of the function of the agency. In a 2010 speech to the National Press Club (http://yosemite.epa.gov/opa/admpress.nsf/a883dc3da7094f97852572a00065d7d8/70ba33a218b8f22f852576e00

06b2a53!OpenDocument), the administrator stated "that it is not the economy *or* the environment, but it is the economy *and* the environment." In other words, EPA policy and actions will be structured to stimulate economic development in the United States while preserving and improving our environment. Although the agency will continue its role in environmental remediation, it will also become much more proactive in increasing the incorporation of sustainability. To achieve this, it will work more closely with the private sector and local governments to identify practices that increase sustainability and provide alternatives derived from the development and application of innovative policies and technologies.

One example of this new approach to environmental protection is the formation of clusters. A cluster is a unified group that brings together all the expertise needed to take technologies from conceptualization into end use, while maintaining a sustainability focus. From research to demonstration and marketing to deployment, a cluster calls upon the skills of key sectors, including universities and colleges; large corporations; emerging companies; federal, state, and local government; and support groups. The cluster will identify the needs of the industries and communities they serve to produce products that help to protect human health and the environment.

An example cluster is the Confluence—Water Technology Innovation Cluster located in the Cincinnati, Ohio, metropolitan area. This regional activity encompasses southwestern Ohio, northern Kentucky and southeastern Indiana and is based on an EPA and Small Business Administration initiative that recognizes the importance of harnessing regional expertise in public utilities, research partners, and innovative business to encourage economic development and environmental and human health protection. While the concept is not new, the bringing together of partners and groups from the onset to advance the technology continuum is a novel contribution of this concept. Additionally, the cluster concept also recognizes the need to include policy makers into the discussion and executable actions to ensure newly designed and commercialized technologies take into account their environmental and human health impacts.

While this is only a subset of recent activities and successes that demonstrate the use of innovation and technologies for increasing environmental preservation and protection of a system, it is very evident that society can no longer approach problems and challenges as in the past if a sustainable future is truly desired. Further, the role of technology is more than a means of providing an innovative solution in isolation from legal, societal, and economic considerations. The ECS and the cluster initiative involving U.S. EPA and other entities show two complementary ways in which government can play a critical role in ensuring that economic activity also advances society's environmental objectives.

## Conclusion

The preservation of natural resources and ecosystems is directly tied to the ability to promote a long-term vision and to secure a commitment to implement proactive strategies for meeting human needs from generation to generation. Hence, it is critical that approaches are developed that consider the dynamic nature of systems and avoid static policies that are insufficient in this context. Further, the approaches developed must be bounded by resource supply horizons, environmental regulations, technological capabilities, and the capacity of critical supporting ecosystems to absorb human burdens while continuing to thrive. While there is still much effort needed to further develop the practical means of moving toward sustainability, this work provides key insights on the type of legal and policy instruments, market practices, and technological innovation that are critical to the success of this endeavor. As articulated in this chapter, the following mechanisms provide three possible strategies to support this effort. Model legislation such as the Legacy Act provides a foundation that is flexible and adaptive when coupled with scientifically sound measures of sustainability and resilience. The ECS offers a way to harness economic incentives to promote environmentally beneficial technological innovation. Lastly, the U.S. EPA's participation in the cluster initiative provides an example of how public and private sectors working together can advance economic and environmental goals. We therefore propose a multi-pronged approach to preserving a resource legacy, promoting intergenerational equity, and moving toward a more sustainable future.

## References

Anastas, P. T., and J. C. Warner. *Green Chemistry: Theory and Practice.* New York: Oxford University Press, 1998.

Anastas, P. T., and J. B. Zimmerman. "Design Through Twelve Principles of Green Engineering." *Environmental Science and Technology* 37 (2003): 94A–101A.

Ankersen, T. T., and K. E. Regan. "Shifting Baselines and Backsliding Benchmarks: The Need for the National Environmental Legacy Act to Address the Ecologies of Restoration, Resilience, and Reconciliation." In *Beyond Environmental Law: Policy Proposals for a Better Environmental Future,* eds. A. C. Flournoy and D. M. Driesen, 53–79. New York: Cambridge University Press, 2010.

Biggs, R., S. R. Carpenter, and W. A. Brock. "Turning Back from the Brink: Detecting an Impending Regime Shift in Time to Avert It." *Proceedings of the National Academy of Sciences USA* 106 (2009): 826–831.

Bosselman, F. P., and A.D. Tarlock. "The Influence of Ecological Science on American Law: An Introduction." *Chicago-Kent Law Review* 69 (1994): 847–873.

Carpenter, S. R., and W. A. Brock. "Rising Variance: A Leading Indicator of Ecological Transition." *Ecology Letters* 9 (2006): 311–318.

Chisholm, R. A., and E. Filotas. "Critical Slowing Down as an Indicator of Transitions in Two-Species Models." *Journal of Theoretical Biology* 257 (2009): 142–149.

Dakos, V., M. Scheffer, E. H. van Nes, V. Brovkin, V. Petoukhov, and H. Held. "Slowing Down as an Early Warning Signal for Abrupt Climate Change." *Proceedings of the National Academy of Sciences USA* 105 (2008): 14308–14312.

Driesen, D. M. "Sustainable Development and Air Quality: The Need to Replace Basic Technologies with Cleaner Alternatives." *Widener Law Journal* 18 (2009): 883–895.

Driesen, D. M. "An Environmental Competition Statute." In *Beyond Environmental Law: Policy Proposals for a Better Environmental Future,* eds. A. C. Flournoy and D. M. Driesen, 173–197. New York: Cambridge University Press, 2010.

Eason, T., and H. Cabezas. "Evaluating the Sustainability of a Regional System Using Fisher Information, San Luis Basin, Colorado." *Journal of Environmental Management* 94 (2012): 41–49.

Eason, T., A. S. Garmestani, and H. Cabezas. "Managing for Resilience: Early Detection of Regime Shifts in Complex Systems." *Clean Technologies and Environmental Policy* (2014, in press).

Endangered Species Act of 1973, 16 U.S.C. §§ 1531–1544.

Federal Land Policy and Management Act of 1976, 43 U.S.C. §§ 1701–1782.

Flournoy, A., M. C. Giblin, and M. Shudtze. "Squandering Public Resources." Center for Progressive Reform Publication No. 705, Washington, D.C., 2007. http://www.progressivereform.org/articles/Squandering_Public_Resources.pdf.

Flournoy, A., R. Feinberg, M. C. Giblin, H. Halter and C. Storz. "The Future of Environmental Protection: The Case for a National Environmental Legacy Act." Center for Progressive Reform White Paper No. 1002, Washington, D.C., 2010. http://www.progressivereform.org/articles/NELA_1002.pdf.

Flournoy, A. C., and D. M. Driesen. *Beyond Environmental Law: Policy Proposals for a Better Environmental Future*. Cambridge: Cambridge University Press, 2010.

International Institute for Sustainable Development. "Global Directory of Indicator Initiatives" (2011). http://www.iisd.org/measure/compendium/searchinitiatives.aspx.

Karunanithi, A. T., H. Cabezas, B. R. Frieden, and C. W. Pawlowski. "Detection and Assessment of Ecosystem Regime Shifts from Fisher Information." *Ecology and Society* 13, no. 1 (2008): 22. http://www.ecologyandsociety.org/vol13/iss1/art22.

Maurice, S. C., and O. R. Phillips. *Economic Analysis: Theory and Application*, 6th ed. Homewood, IL: Irwin, 1992.

Mayer, A. L., H. W. Thurston, and C. W. Pawlowski. "The Multidisciplinary Influence of Common Sustainability Indices." *Frontiers in Ecology and the Environment* 2 (2004): 419–426.

Odum, H. T. *Environment, Power, and Society*. New York: Wiley, 1971.

Scheffer, M., J. Bascompte, W. A. Brock, V. Brovkin, S. R. Carpenter, V. Dakos, H. Held, E. H. van Nes, M. Rietkerk, and G. Sugihara. "Early-Warning Signals for Critical Transitions." *Nature* 461 (2009): 53–59.

United Nations. *Report of the World Commission on Environment and Development, General Assembly Resolution 42/187*. New York: UN, 1987.

U.S. Department of Housing and Human Development. "Developing a New Approach for Measuring Regional Sustainability, Summer 2011" (2011). http://www.huduser.org/portal/periodicals/em/summer11/highlight2_sidebar.html.

U.S. Environmental Protection Agency. *San Luis Basin Sustainability Metrics Project: A Methodology for Evaluating Regional Sustainability*. EPA Number: EPA/600/R-10/182. Washington, D.C.: EPA, 2010.

U.S. Environmental Protection Agency. "Presidential Green Chemistry Awards 2011" (2011). http://www.epa.gov/opptintr/greenchemistry/pubs/pgcc/past.html.

van Nes, E. H., and M. Scheffer. "Slow Recovery from Perturbations as a Generic Indicator of a Nearby Catastrophic Shift." *American Naturalist* 169 (2007): 738–747.

Wiener, J. B. "Beyond the Balance of Nature." *Duke Environmental Law & Policy Forum* 7 (1996): 1–24.

World Wildlife Fund. *Living Planet Report*. Washington, D.C.: WWF, 2008. http://assets.panda.org/downloads/living_planet_report_2008.pdf.

ELEVEN

# Adaptive Law

CRAIG ANTHONY (TONY) ARNOLD AND LANCE H. GUNDERSON

The need for "adaptive law"—for law to be adaptive and resilient—is clear. What is not as clear, though, is what adaptive law would look like. What would be its primary features? Can we learn lessons from the panarchic dynamics of interconnected social and ecological systems that would guide legal reform to achieve adaptive structures and functions in U.S. law?

These questions must be answered, though, in the context of how law, society, and nature actually interact with one another. The study of adaptive law has an empirical imperative, not just a normative imperative. In particular, we are not interested in articulating an ideal model that works in theory but not in practice. Likewise, we are not interested in articulating a proposal for a utopian legal system that has little to no likelihood of ever actually being adopted in the United States, at least in any kind of widespread way. We are well aware that while "truth speaks to power" (Wildavsky 1987), power often talks back, usually forcefully. There are and will be substantial obstacles to making legal institutions more adaptive: powerful political and economic interests that benefit from the status quo, institutional inertia and path-dependence, the human psychology of self-confidence and certainty, and sociocultural systems embedded with maladaptive structures and norms. However, many counterforces demand a more adaptive legal system: the economic, political, and social risks of uncertainty in natural and social conditions; incidence of legal and policy failures; growing understanding of limits to human knowledge and action and the complexities of social-ecological dynamics; the human and societal costs of collapse of ecosystems and/or social institutions; applied environmental and social ethics; the human psychology of innovation and problem solving; and actual incidence of unpredictable transformations in ecological and/or social conditions. The features of adaptive law are likely to emerge over time, manifested in a variety of ways in legal institutions, rather than through a decision maker unilaterally adopting a new regime of adaptive law (Arnold 2011a).

Furthermore, we need broad, cross-cutting principles of adaptive law. There are many excellent suggestions for reforms of particular laws or aspects of the law to make them more adaptive. Some of the best include reforms of the National Environmental Policy Act (Karkkainen 2002; Farber 2009a; Benson and Garmestani 2011), the Endangered Species Act (ESA; Camacho 2007; Ruhl 2008; Doremus 2010), the Clean Water Act (CWA; Adler 2010), and the use of adaptive management methods in federal agency resource management and restoration projects (Doremus 2007; Camacho 2009; Glicksman 2009; Zellmer and Gunderson 2009). While these reform proposals offer very useful specifics about particular manifestations of adaptive law and management, they leave a void of general overarching principles to guide the legal system toward legal, societal, and ecological resilience. Even recent scholarship that offers broader principles of adaptation in law primarily focuses on resilience to particular disturbances and uncertainties, mostly climate change (Ruhl 2010, 2011).

We aim to describe a set of features of a legal system that are both internally adaptive and resilient to a wide range of possible disturbances and facilitate the resilience and adaptability of both nature and society—and their constituent systems—to disturbances. At times, we will draw on examples from environmental, natural resources, land use, and property law to illustrate the contrasts between maladaptive and adaptive features of law. Nonetheless, the principles we identify have broad application to many fields of law and many aspects of the legal system generally.

The basic features of the U.S. legal system that are maladaptive fall into four large categories: (1) systemic goals that are narrow; (2) a structure that is monocentric (i.e., centralization of authority to solve problems), unimodal (i.e., the use of single, uniform models as solutions to problems), and fragmented; (3) inflexible methods that employ rules, legal abstractions, and resistance to change; and (4) rational, linear, legal-centralist processes. In contrast, the features of an adaptive legal system are: (1) multiplicity of articulated goals; (2) polycentric, multimodal, and integrationist structure; (3) adaptive methods based on standards, flexibility, discretion, and regard for context; and (4) iterative legal-pluralist processes with feedback loops and accountability. The features of maladaptive law and adaptive law are compared in table 11.1. Subsequent sections of this chapter address the details and implications of each of these four features, followed by an assessment of the need for additional research on adaptive law principles.

Table 11.1 Comparison of features of maladaptive law and adaptive law.

| Feature | Maladaptive law | Adaptive law |
|---|---|---|
| Goals | Legal regimes aim to advance particular stability of single systems. Current regimes focus primarily on political and economic goals. Alternative (reform) regimes focus primarily on ecological goals. | Legal regimes aim for multiple forms of resilience: the resilience and adaptive capacity of both social and ecological systems, including constituent subsystems, such as institutions and communities. |
| Structure | Law is monocentric, utilizing fragmented and unimodal responses to problems. | Law is polycentric, utilizing multimodal and multiscalar responses to problems that are loosely integrated. |
| Methods | Law controls society through rules, limits on action and authority, demand for certainty, and legal abstractions that resist change. | Law facilitates social and ecological resilience through moderate/evolutionary adaptation to changing conditions, context-regarding standards, tolerance for uncertainty, and flexible discretionary decision making. |
| Processes | Law presumes rational, linear decision-making and implementation processes by a single authority and the centrality of law to the ordering and management of human affairs. | Law recognizes and embraces iterative processes, with feedback loops among multiple participants; limits to human and organizational rationality; the effects of social and ecological forces on the ordering and management of human affairs; and accountability mechanisms for the conservation of capital. |

## Adaptive Law Goals

Adaptive legal regimes foster resilience: the resilience and adaptive capacity of both social and ecological systems, including constituent subsystems, such as institutions and communities. The failure of legal institutions to value and facilitate the resilience of ecosystems, such as watersheds, wetlands, forests, deserts, and urban ecosystems, threatens the health, sustainability, and resilience of social systems that depend on ecosystems. The relationships between ecosystems and social systems are multiscalar, multifunctional, complex, and dynamic; threats to or transformations in one system can affect others (Berkes and Folke 1998; Gunderson and Holling 2002; Ostrom et al. 2007). Legal institutions are prone to give primary or sole value to the resilience of political and economic institutions, such as production of and transactions in consumer goods, private property rights, the diffusion of governmental power horizontally and vertically through separation of powers and federalism, and the facilitation of financial capital and investment (Sprankling 1996; Eagle 2006). Driesen (2003), for example, has critiqued the bias in environmental law to protect presumed static economic efficiencies and to ignore dynamic relationships between economics and the environment. At times, the legal system seems to operate as though its primary function is to promote the resilience of the legal system itself (e.g., Arnold 2011b). Systemically narrow definitions of resilience, though, are maladaptive in that they undermine the health, functioning, and resilience of ecosystems and other social institutions (e.g., local communities, diverse cultures, families, religions) that in turn affect the whole of society.

On the other hand, alternative conceptions of law, particularly environmental law, focus too narrowly on the resilience of ecosystems and of natural functions and processes, without adequate attention to the vitality and adaptability of the social systems and institutions that often seem to be at odds with the natural environment. For example, environmentalists may criticize features of the U.S. political and economic systems that are central to American culture and political structure, such as liberty, private property rights, localism in governance, quasi-free markets, consumerism, and the like (Delgado 1991; Freyfogle 2006). These critiques may call for substantial, even radical, transformations of American law and society in order to promote ecological health and resilience, biodiversity, and environmental protection. Legal changes that give primacy to ecosystems or biodiversity, particularly if they require substantial transformations to social systems and institutions, may produce a variety of unintended consequences, including political backlash, nonimplementation or underimplementation of the reforms, political and social conflict, and fiscal or economic hardships (e.g., Doremus and Tarlock 2008). Disturbances to social systems and institutions often adversely affect ecosystems and biological communities (Dutt 2009; Downey et al. 2010). Moreover, ecocentric legal reforms may fail to address the most significant pathologies of the interconnections among nature, society, and law. In contrast, an adaptive legal system aims for structures, methods, and processes that build the resilience and adaptive capacity of both nature and society—of a range of ecosystems and social systems and institutions.

Both the need for and difficulty in achieving poly-resilience in law are illustrated by examples from ESA's protection of species on the verge of extinction. On its face, ESA seems to give primacy to the survival and recovery of wildlife species in the wild and to strictly prohibit any federal, state, local, or private actions that would harm federally listed endangered or threatened species, including adverse modification of these species' habitats. The U.S. Supreme Court has expressly recognized ESA's absolute prohibitions and institutionalization of caution (*Tennessee Valley Authority v. Hill* 1978), and ESA seems to have weathered challenges to federal authority to regulate and protect purely intrastate species, at least so far.

However, the design and operation of ESA has also produced strong political backlash undermining its protections. ESA has been unpopular with private property owners, land developers, state and local governments, water users, the timber industry, and other resource users affected by its prohibitions (Burke 2004; Babbitt 2005; Doremus and Tarlock 2008). ESA has been blamed for declining industries and high costs of housing and land development. ESA has been "framed" in American politics and culture as a conflict between humans and nonhuman species, such as cave bugs (Gold 2005), blind salamanders (Gordon 2012), and desert flies (Goad and Santschi 2011) that many people do not value. This framing phenomenon has put federal regulators and conservation biologists on the defensive (Burke 2004; Babbitt 2005). In addition, listing decisions have been attacked as based on bad science or no science at all, both from those attacking a perceived proregulatory bias toward listing and those attacking a perceived antiregulatory bias toward nonlisting or delisting (Burke 2005; *Otter v. Salazar* 2012). At one point, Congress imposed a moratorium on the listing of new species (Scharpf 2000). It has expressly exempted two different river projects from ESA after courts have found that they would violate the act's provisions (Energy and Water Development Appropriation Act of 1980, 1979; Energy and Water Development Appropriations Act of 2004, 2003, § 208), and legislatively delisted the Wyoming populations of gray wolves (Associated Press 2012). Some landowners responded to the potential application of ESA to their property by removing the land's features that would support listed species (or the species themselves) before the species could be detected as present on their property, a phenomenon referred to as "shoot, shovel, and shut up" (Nelson 1995; Elmendorf 2003).

More systematically, Congress has authorized regulatory agencies to grant "incidental take permits" that allow landowners and resource users to alter listed species' habitats if they develop approved habitat conservation plans (HCPs) to minimize and mitigate the impacts of their actions on the species' survival and recovery (ESA § 1539(a)). Congress has also authorized a committee of federal agency officials, known as the "God Squad," to exempt important federal projects from ESA's prohibitions (ESA § 1536(e)–(o)). Congress has regularly underfunded ESA's implementation and has considered legislation to repeal ESA altogether (Burke 2004).

ESA has also been a target of takings litigation. In some cases, courts have held the federal government liable for compensation for ESA's taking of private property (e.g., *Tulare Lake Basin Water Storage District v. United States* 2001), whereas in other cases, they have found that no taking of property occurred (e.g., *Southview Associates v. Bongartz* 1992).

Federal regulatory officials have yielded to political pressure in some decisions to list or delist particular species and have adopted a "no-surprises" policy to provide the regulated community with regulatory certainty against new information or changes, including changed conditions of species and their habitats (Habitat Conservation Plan

Assurances "No Surprises" Rule 1998). Little attention has been given to climate change impacts on species, their likely shifting and changing habitats, or the regulatory and management functions of ESA (Camacho 2010). Many HCPs have been poorly implemented or not monitored for function and resilience over time (Camacho 2007).

Some HCPs have contemplated multispecies bioregional planning and management, but ESA is structured primarily as a species-by-species, project-by-project, parcel-by-parcel regulatory regime, not an ecosystem conservation law. While ESA has stimulated ecosystem-scale, multi-dimensional, and multi-stakeholder planning and management in some situations (Guercio and Duane 2009), its rigidity has proven relatively maladaptive to resolving many complex resource issues, such as competing uses of water in a water-stressed or water-scarce river basin.

ESA's noble goals of protecting species from extinction and preserving biodiversity have been undermined because Congress and federal agencies did not systematically integrate into the law and its implementation two critical goals: (1) the vitality and resilience of the ecosystems of which species and their habitats are parts, and (2) the vitality and resilience of the existing political, economic, and social systems that affect the legitimacy and effectiveness of environmental laws. ESA is not a total failure; it has protected species from extinction and adapted in certain ways to the conflicts over its implementation (Goble et al. 2006; Guercio and Duane 2009). However, the law's narrowness has impeded it from serving as a tool to strengthen ecosystem resilience generally. The law's narrowness has also impeded it from enhancing the adaptive capacity of political, economic, sociocultural, and legal institutions to integrate the goals of biodiversity, ecosystem vitality, and resource conservation with the goals of human freedoms and rights, economic vitality, and resource use.

One important lesson learned from ESA is that, in order for law to be adaptive, it must build and strengthen the adaptive capacity of multiple institutions, just as it builds and strengthens the adaptive capacity of multiple ecosystems. This is a fundamental feature of a poly-resilient legal system. For example, adaptive law could stimulate or aid economic institutions to value ecosystem services, the often undervalued benefits that ecosystems provide to human society and the economy (Daily 1997;

Ruhl et al. 2007). Likewise, adaptive law could facilitate strong participatory and deliberative governance of local communities adapting to extreme weather events, rising coastlines, or other environmental transformations, while also maintaining basic human freedom, dignity, and rights.

Capacity-building is not a matter solely for environmental law. To be sure, legal institutions need to help to build adaptive capacity among federal environmental regulatory agencies and federal natural-resources management agencies, as a growing literature emphasizes (e.g., Camacho 2009; Glicksman 2009; Doremus 2010). This can be done, for example, through express authority for adaptive management and planning, statutory reforms of specific standards and processes, and development of adaptive governance structures and iterative processes with robust feedback loops. However, there is also a strong need to build nonfederal adaptive capacity: the adaptive capacity of state and local government agencies (including regional, hybrid, and quasi-governmental authorities), private landowners and resource users, environmental and resource conservation groups, local communities and community groups, businesses and industries both individually and in associations and networks, universities and professional organizations, and other civil society groups and networks (Berkes and Folke 1998; Hirokawa 2012). Society's impacts on ecosystems and nature's impacts on society are simply too large, extensive, varied, and complex to be moderated solely by federal resource management or federal environmental regulation. Adaptive law is adaptive to the effects of culture, psychology, and power, including the ways in which society–nature issues are framed—risk preferences, political forces, and the culture and politics of private property rights and takings—which are often much stronger and more influential than the legal principles of private property rights and takings (Arnold 2007b). Adaptive law principles must pervade fields of law that are not narrowly defined as environmental law: property law, energy law, land use law, water law, agricultural law, disaster law, constitutional law, business law, tort law, international law, human rights law, and so forth.

Another particular aspect of the legal system's poly-resilience is its function as one of the institutions mediating among science, culture, and politics in society. The legitimacy and centrality of science as a source of knowledge and a guide to action in the physical and natural world is highly contested in contemporary U.S. society. The modernist paradigm of science as rational, objective, experimental, and self-policing is increasingly at odds with a postmodern outlook of knowledge as subjective, political, and vulnerable to many kinds of critiques (Pielke 2007). Those skeptical of science find increasing popular support for their contentions that science is hostile to religious belief, biased against minority or disempowered groups, tainted by political and policy preferences of researchers, or unreliable, as new discoveries debunk old knowledge and theories (e.g., Mooney 2005). Indeed, two central aspects of panarchy research and theory contribute to political and cultural skepticism of science: (1) knowledge is provisional, as we cannot predict *ex ante* with great certainty cause-and-effect relationships in complex interconnected systems with nonlinear dynamics; and (2) prior theories about equilibrium and stationarity in ecological conditions have been replaced with new theories of disequilibrium and transformational dynamics (Walker and Salt 2006). Legal institutions play a significant role in determining the roles of science in collective choice and action. This is true not only for judicial decision making, but also for legislation, regulation, private legal arrangements, and even informal law. It is also true not only for the use of science in environmental protection and natural resource management, but also for the use of science in human health care choices and standards, intellectual property regimes, evidentiary proof of cause-and-effect for liability purposes, and the technology–social relationships interface, among others. Adaptive law aims to help scientific institutions to adapt to the changing sociocultural and political environments in which science is evaluated and to help social institutions to adapt to changing scientific knowledge and understanding.

Poly-resilience is a systemic feature, not an easy-to-apply mechanical decision rule that naively presumes lack of trade-offs in any given legal, policy, or management decision. The lack of clear rules or principles about choices and trade-offs is both a weakness and a strength of the poly-resilience feature of adaptive law. One can imagine that it is not possible to maximize the resilience of all systems in any given decision. Ecosystem managers may have to make choices among systemic features or even between ecosystems, particularly if disturbances have transformed ecosystems. An example has been the National Forest Service's choice between managing the Shawnee National Forest for hardwood restoration or for preservation of nonnative pines supporting a group of species anchored by the pine warbler (*Glisson v. United States Forest Service* 1995). Furthermore, the resilience of existing political, social, or economic systems may adversely affect the restoration and resilience of ecosystems like the Everglades (Zellmer and Gunderson 2009). Thus, the resilience of the existing American culture and politics of private property rights arguably could be incompatible with the resilience of extinction-threatened species, their habitats, and the natural communities of which they are a part. Or the resilience of a system of federalism that allocates governance across national, state, and local scales arguably could be incompatible with the resilience of wetland ecosystems that are under constant stresses from development, farming, pollution, and the like.

Many environmental experts contend that social systems must undergo substantial transformation if ecosystems and the social systems that are integrally interconnected with ecosystems are to survive, and thus we should not strengthen the resilience of at least some of the features of our current systems (Zellmer and Gunderson 2009; Benson and Garmestani 2011). There are three major problems with the theory that resilience requires major social system transformation, though. First, social systems may simply be robustly resistant to the kind of major changes that environmental experts contend are required. For example, the decline of private property rights in the United States has not materialized as predicted (Sax 1983), and many ecology-regarding changes in U.S. consumer culture, business and agricultural practices, and land use patterns have been incremental: gradual, particularistic, and more in the nature of partial reformation than complete transformation or revolution. Second, the very phenomenon of social system resistance to major change may cause harm to ecosystems and their resilience. For example, landowners sought to eliminate species and their habitat conditions from their lands ahead of application of ESA to their lands (i.e., "shoot, shovel, and shut up"; Nelson 1995; Elmendorf 2003). In addition, the regulated community's strong political backlash against ESA has undermined its effective implementation to protect species and the ecosystems on which they depend. Third, major changes to existing social systems can have all kinds of unanticipated consequences to ecosystems and other social systems. Economic decline and civil unrest contribute to environmental degradation (Downey et al. 2010). Suppression of local authority, culture, and knowledge by a centralization of authority, culture, and knowledge can lead to loss of indigenous understanding of natural systems, place-based commitments to environmental conservation, and the functionality of local, indigenous, or minority communities (Berkes and Folke 1998; Reid et al. 2006).

One way to advance poly-resilience is to focus on policies, methods, and tools with co-benefits: multiple, ancillary benefits arising from a particular approach (Long 2010; Hirokawa 2012). The multiplicity and diversity of co-benefits aid the resilience of multiple systems and subsystems. For example, local ordinances protecting and enhancing the urban tree canopy produce many co-benefits. Urban trees mitigate urban heat island effects (thus helping to save human lives in extreme heat), sequester carbon, moderate storm-water runoff, stabilize soils and prevent erosion, shelter wildlife, maintain temperatures in urban streams, contribute to the walkability of urbanscapes, add economic value to land, improve mental health, enhance area aesthetics, build human connectivity to nature, and provide many other ecological and social benefits (Alliance for Community Trees 2011).

Another co-benefits example is the organization of resource planning, management, and/or governance at watershed scales. These watershed institutions or processes, whether formal or informal, offer opportunities to address a variety of interrelated issues involving land and water at ecosystem scales that transcend artificial political–legal boundaries: water supplies; instream flows; aquatic habitat; management of dams and reservoirs; competing uses of water; drought; flooding; stormwater runoff; water pollution/quality; erosion and sedimentation; stream channel and bed integrity; riparian vegetation; interconnections between surface waters and groundwater; estuaries; wetlands; regional land use practices and growth management; water-focused or water-dependent land development projects; and restoration projects for rivers, streams, lakes, wetlands, or even large-scale aquatic systems (e.g., the Everglades), among many other issues (Arnold 2011b). Watershed planning, management, and governance systems can also serve as vehicles for addressing issues of other ecosystems within or interrelated with the watershed, such as forests, grasslands, coastal lands, or marine ecosystems.

A third co-benefits example involves the principles of environmental justice. Environmental justice demands the fair treatment of low-income communities and communities of color in environmental laws, policies, and practices, and includes elements of distributive, procedural, remedial, and social justice (Cole and Foster 2001). The principles of environmental justice call for protecting low-income communities of color from pollution and intensive land uses: pollution prevention; remediation of lands and waters affecting these communities; re-use of vacant or abandoned properties; economic, social, and physical revitalization of neighborhoods and reservations; mechanisms to prevent gentrification of low-income neighborhoods; opportunities for equal participation in decisions that affect low-income and minority communities; equal access to natural resources and community infrastructure (e.g., parks, green infrastructure, transportation); and community empowerment, among many other goals (Mutz et al. 2002; Arnold 2007a). Environmental justice activists, for example, have insisted that local government storm-water policies preserving aquatic conditions in wealthy suburban areas be balanced with restoration of degraded, channelized, often-ignored streams in their urban neighborhoods (Karvonen 2011).

Adaptive legal systems enforce and defer to laws producing co-benefits. Adaptive legal systems recognize the authority of new planning, management, and governance structures created for co-benefits. Adaptive legal systems adopt and apply principles facilitating co-benefits. By their nature, many co-benefits laws, policies, structures, and principles do not fit neatly into pre-existing legal categories. Therefore, the legal system must adapt to recognize their multiple benefits to the resilience of multiple systems.

## Adaptive Law Structure

Adaptive law is polycentric, utilizing multimodal and multiscalar responses to problems that are loosely integrated. This feature contrasts sharply with frequent arguments that threats to the resilience of social systems and ecosystems require strong national or global governance authorities that control behavior through command-and-control regulations and the rule of law. Monocentric legal approaches are often encouraged to address complex problems with substantial multiscalar dimensions across jurisdictions, sectors, or time (Babbitt 2005; Farber 2009b). Advocates for monocentric approaches, such as federal government climate change laws and policies, contend that subnational governments and private sector actors lack sufficient incentives, authority/power, expertise, or resources to address these problems adequately. In fact, some argue that subnational governments and private sector actors enhance their competitiveness through decisions and actions that push harms—negative externalities—onto other jurisdictions or sectors, competitors, or future generations. Furthermore, monocentric structure advocates argue that centralized authority is needed to coordinate multiple responses to complex and multidimensional problems. Thus, federal laws and policies are needed to achieve climate change mitigation and adaptation, for example, when those both causing and experiencing impacts are as diverse as corn growers in Iowa, commuters in Louisville, and oil-and-gas developers in coastal Louisiana, among countless others.

Contrary to this conventional wisdom, though, resilience and adaptation throughout the legal system, society, and nature are enhanced through polycentrism: a structure in which there are multiple centers or sources of authority. Bell (1991–1992) has articulated a concept of "polycentric law" that questions the government (i.e., the state) as the sole source of legal authority and celebrates the emergence of law and legal authority in the private sector, such as customary law or private contractual arrangements. Furthermore, Roberts (2011) has identified a diverse range of private governance institutions. Certainly an adaptive law structure recognizes the importance of the private structure as a source and location of law, governance, and adaptive capacity. However, we adopt a broader definition of polycentric structure in adaptive law that embraces a multiplicity and diffusion of governmental sources of law and authority (e.g., strong roles for local and state governments; regional, hybrid, and quasi-public governance institutions; and mixed forms of dispute resolution), as well as nongovernmental sources. Thus, we link the classic work on decentralized metropolitan governance by

Ostrom et al. (1961), which defined the public sector as a polycentric system instead of a monocentric hierarchy, with the private-sector work of Bell and Roberts.

A growing number of experts on complex environmental problems (e.g., Doremus 2009; Ostrom 2009) have argued for the superiority of polycentrism. There are several reasons why polycentric legal structure promotes resilience in social and ecological systems. First, polycentric structure allows for experimentation and innovation in governance and management: "laboratories of democracy" (Galle and Leahy 2009). United States Supreme Court Justice Louis D. Brandeis famously made this point in his assertion: "It is one of the happy incidents of the federal system that a single courageous State may, if its citizens choose, serve as a laboratory; and try novel social and economic experiments without risk to the rest of the country." (*New State Ice Co. v. Liebmann* 1932, 311, Justice Brandeis dissenting).

Second, polycentrism enables risk diversification, a key element of resilience. A monocentric structure creates risk that a single approach taken by the central authority will fail, whereas multiple approaches taken by multiple authorities are not all likely to fail (Berkes and Folke 1998; Goldstein 2012). In a monocentric structure, the direct lines of governance emanating from the central authority throughout society are means by which policy/management failures or shocks to the system are transmitted throughout society or can produce cascade effects throughout society, often with adverse consequences for interconnected ecosystems. In a polycentric system, failures and shocks can move through lines of connections (networks) between the various centers of authority, but the multiplicity of power centers and lines of connection are more likely to buffer, slow, mitigate, or even resist these effects. Moreover, a crisis of legitimacy within a single centralized authority undermines the capacity of society to govern and manage itself and the environment in adaptive ways. It is unlikely that all authorities within a polycentric system would experience crises of legitimacy at the same time.

Third, redundancy of resources characterizes a polycentric system, which is also a critical feature of resilient systems. Redundant resources within a system can absorb shocks: if some are lost, others can still support the essential functions of the system (Folke et al. 2004; Gunderson et al. 2006). The very nature of polycentric systems means that each center of authority is replicating at least some of the resources and functions of the other centers and thus can continue providing these functions if others decline. On the other hand, a monocentric system faces the temptation to maximize efficiency and resolve conflicts between subunits of the system over functions and roles by eliminating redundancies.

Fourth, polycentrism is better matched to the scales, scope, and speed of problems that legal and governance institutions must address than monocentrism is. Many problems are multiscalar, requiring at least some amount of decentralization of authority to address small-scale localized aspects of the problem. Many of the problems that challenge the resilience of social systems and ecosystems today are so massive and complex as to exceed the capacity of even large, powerful, centralized authorities (such as federal government agencies) to address them adequately. These problems have been called "policy jungles" (Ruhl and Salzman 2010) because of a complex aggregation of causes and effects that can be nonlinear and spatially and temporally discontinuous. Moreover, some problems arise out of the interrelationships among multiple complex "policy jungles," creating "policy super-jungles of policy jungles" (Arnold 2011a). The seeming advantage of a single centralized authority addressing large-scope problems melts away when the problems become so massive and complex that no single entity can solve them (Freeman and Farber 2005). Instead, we need efforts and solutions from multiple authorities that are components of the complex adaptive systems that produce complex problems. Furthermore, these problems may require quick and agile responses, which are more likely to be possible among many decentralized authorities than among a single centralized authority, which may be too slow and cumbersome to respond speedily or make quick changes to adapt to new conditions or feedback-loop lessons.

Fifth, polycentrism is more adaptive to U.S. social norms, culture, and politics, which value localism and some degree of private-sector autonomy (Arnold 2007b). Some would contend that U.S. politics and culture are simply wrong, as well as maladaptive with respect to both society and nature (Freyfogle 2006). However, these critics are typically less hostile to polycentric elements of U.S. politics and culture when they involve local sustainably grown food movements pushing back against national and international agriculture and food-chain policies; grassroots environmental justice advocates decrying land use and environmental laws that either gentrify their neighborhoods or allow polluting and intensive land uses in their neighborhoods; landowners and tenants asserting rights to be free from law-enforcement monitoring and intrusion; or water conservation advocates asserting property interests in instream flows, water quality, or riparian lands against federal dams and water development projects. Normative arguments aside, the reality is that the forces favoring polycentrism are strong in U.S. society. Adaptive law recognizes these forces and harnesses them to promote ecological and social resilience.

The role of societal forces that shape law highlights another feature of adaptive law structure: the emergence of integrationist multimodality in response to the inadequacies of unimodal and fragmented structures. Integrationist multimodality is the use of multiple modes or methods of achieving a policy goal but in ways that aim to integrate or interconnect those multiple modes or methods (Arnold 2011b). Arnold (2011b) has posited that integrationist multimodality is an emerging new generation of environmental law and policy, which have typically been unimodal and fragmented and therefore maladaptive and inadequate. Unimodality is the choice of a particular mode, instrument, method, or design as "optimal," frequently characterized by advancement of a particular model or uniform—"one size fits all"—approach. For example, the classic law and policy debate as to whether command-and-control regulations or market mechanisms are more effective at achieving policy goals is about a unimodal choice. However, many examples of unimodality are more fine-grained: model statutes or ordinances, standard design or management procedures, uniform laws, the new preferred program or policy of the day, and so forth. In contrast, "multimodality" is a toolbox approach, facilitating multiple actors' selection from among a variety of instruments, methods, and tools to respond to complex problems. Moreover, these multiple modes can be and often are linked, although uniformly tight linkages—true integration—can lead to transmission of disturbances and shocks throughout the system, producing cascade effects that lead to system decline or collapse (Gunderson and Holling 2002; Folke et al. 2004). Loose connections or networks—integrationist, rather than integrated—offer coordination and synergy while reducing the risk of weakness contagion.

Nonetheless, fragmentation of legal, governance, and management actions is maladaptive. Unimodality often manifests itself in fragmented ways, because each issue/problem (e.g., sprawl, wetlands protection, forest management) and/or each decision-making entity (e.g., the city of Atlanta, U.S. Army Corps of Engineers, U.S. Forest Service) has its own "optimal" or "preferred" mode or method. In our typically fragmented system of governance, interconnected problems—such as water quality, water supply, instream flows, aquatic habitat, land use and development, and regional growth patterns—are treated as separate and distinct, each with its own standardized ways of addressing its fragment of a larger network of problems (Arnold 2011b).

Ironically, diffusion of authority across multiple centers and institutions enhances society's capacity to achieve at least loosely integrated, or interconnected, responses to interrelated problems. In their call for polycentric responses to large-scale, complex environmental and resource problems, such as climate change, Ostrom and colleagues (2009) rejected the idea that there are any "panaceas" or "optimal" solutions, thus linking integrationist multimodality concepts to polycentrism. Monocentric authorities have to break down problems into manageable categories and tend toward bureaucratic specialization, a policy or governance "silo" effect (Ostrom et al. 2009) Each "subunit" of the centralized authority is tempted to invest resources, including political capital, into a preselected standardized or optimal "solution" to each particular "subproblem." However, no single unit, no single law, and no single mode or method are adequate to tackle complex issues such as the causes and effects of climate change; the resilience of aquatic ecosystems; marine system collapse; changes in wildfire regimes; or the nexus among food supply/security, energy supply/security, and water supply/security.

As a result, the phenomenon of integrationist multimodality has emerged, particularly where environmental and natural resources law and policy intersect with other fields of law, governance, and management, such as land use, water supply, disaster preparedness and response, energy, and agriculture (Arnold 2011b). One example is the emergence of watershed institutions in the United States: the increasing conceptualization and organization of resource planning, management, regulation, advocacy, and problem-solving around watersheds, which are areas of land draining to common points on a body of water (Arnold 2011b). Watershed institutions are diverse. They can range from informal, collaborative, multistakeholder problem-solving processes to formal management initiatives within federal or state government agencies. They can encompass the extensive work of citizen-action groups focused on the health of particular waters and watersheds. They can also encompass regulatory programs created for watershed protection. Watershed institutions may engage in planning, management, restoration, allocation (or re-allocation) of resources, market incentives, regulation, dispute resolution, and public education and engagement, among other types of activities. They may address any combination of problems that affect the watershed, including water supplies, instream flows, water quality and pollution, storm-water runoff, land uses, growth patterns and management, aquatic species and habitat, riparian lands and vegetation, groundwater, and many others. They often rely on networks of experts, officials, and lay participants, pushing problem-solving outside the "silo" effect of bureaucratic structures or professional/disciplinary categories, sharing information and ideas, and providing links among possible solutions and actions. In addition, watersheds have nested scales in which smaller units are nested within larger units, and watershed institutions can be organized around any level of watershed, from large river basins down to microcatchments. Examples of watershed institutions include (Arnold 2011b)

- large-scale ecosystem restoration projects, such as the Comprehensive Everglades Restoration Plan or the Upper Mississippi River Basin ecosystem restoration project;
- an interstate river commission created by interstate compact to control water diversions and uses, such as the Delaware River Basin Commission;
- multiparticipant groups or councils created to plan, manage, or resolve disputes over competing uses of waters in a watershed, such as the Middle Rio Grande Water Assembly in New Mexico;

- federal agencies' management of public lands and resources by assessing and protecting watershed features, such as watershed analyses and management of the Gifford Pinchot National Forest under the Northwest Forest Plan;
- watershed protection advocacy groups formed around particular watersheds, such the Connecticut River Watershed Council and the Nashua River Watershed Association.

In the absence of any centralized and systematic ecosystem-based planning and management regime in the United States, a diversity of watershed governance and management has emerged, employing a wide variety of methods but in ways that aim at integrated treatment of environmental and societal problems that arise in particular watersheds.

The example of watershed institutions as emergent integrationist multimodality also highlights the importance of multiscalar structures in societal and ecological adaptation. Both society and nature are organized at multiple scales. In addition, the effects of social systems on ecosystems and ecosystems on social systems occur in varying qualities and degrees across many different geographic and temporal scales (Gunderson and Holling 2002). Even systems that seemly have hierarchical structure, such as watersheds' nested scales, do not merely aggregate smaller-scale functions and processes into larger-scale versions of the smaller scales. According to the National Research Council (NRC 1999), the smallest drainage basins are heavily influenced by soils and most susceptible to storm events and to human land development and disturbance due to their functions as source areas. Intermediate basins are more likely to be influenced by larger natural and human disturbances, such as tropical storms, large dams and channel engineering projects, and extensive changes to hillsides and channels, because they have larger storage capacity and have complex mixes of slopes, channels, and floodplains. Large river basins provide scale for alluvial and deltaic processes, store sediment, and host a diversity of ecosystems and hydrologic processes, with disturbances to large river basins rarely having uniform effect throughout the basin.

Multiple centers of authority (polycentrism) and multiple modes of action (multimodality) are not sufficient if they are not accompanied by multiple scales of organization (multiscalar organization). A central task of adaptive law is to make linkages among governance scale, governance function, ecosystem scale, and ecosystem function, which requires both multiscalar and multimodal governance (Arnold 2006). Many of our most complex problems, such as watershed protection, climate change, or energy policy, are targets of single-scale governance proposals (or "solutions" or "panaceas") that fail to consider the need for more nuanced multiscalar strategies and structures that involve both scaling up and scaling down to make linkages among the multiple scales (Osofsky 2010; Osofsky and Wiseman 2012). Multiscalar integration occurs through nonhierarchical vertical and horizontal networks that legal institutions must recognize and respect (Feldman and Ingram 2009). New concepts of federalism—sometimes called dynamic federalism or adaptive federalism—call into question traditional hierarchical choices between national power and state or local power, but instead treat the constitutional distribution of power across political–legal geography as a more variable, adaptable, problem-oriented structure of multiple governance scales with authority being replicated—not necessarily divided—among jurisdictions (Ryan 2007; Adelman and Engel 2006).

In addition, preselected and rigid legal determinations about which entity will have governance authority and which methods of governance and management will be used fail to allow new governance approaches and institutions to emerge. Recognizing that regulation will play a critical role but will need to be structurally adaptable, Freeman and Farber (2005, 795) have proposed a concept of "modular environmental regulation," which is defined by "a high degree of flexible coordination across government agencies as well as between public agencies and private actors; governance structures in which form follows function; a problem-solving orientation that requires flexibility; and reliance on a mix of both formal and informal tools of implementation, including both traditional regulation and contract-like agreements."

Modular environmental regulation is neither a panacea nor a guide to ecological and sociopolitical goals for environmental protection and management. It is a structural design proposal that merits consideration as facilitating polycentric, multimodal, multiscalar, and loosely integrated responses to problems in society and the environment.

## Adaptive Law Methods

Adaptive law facilitates social and ecological resilience through moderate/evolutionary adaptation to changing conditions, context-regarding standards, tolerance for uncertainty, and flexible discretionary decision making. In contrast, four maladaptive features currently pervade the American legal system: (1) a preference for establishing predetermined pathways of action through rigid rules and conventional planning requirements, (2) a preference for certainty and security in resources and social structures, (3) a preference for risk avoidance and liability for mistakes, and (4) a preference for decisions based on universally applicable legal abstractions. Nonetheless, the American legal system also contains several alternative adaptive features, including flexible legal standards to guide discretionary decision making, the practice of and capacity for change, protection of responsible and bounded risk taking, and context-regarding decisions. The plasticity of U.S. law is highly contested, not just about the degree to which law can or should change but also about what kind of flexibility is permissible or desired. For example, some who resist flexibility and change in property rights and arrangements may be on the opposite side of the proverbial law-reform fence than some who resist flexibility and experimentation in environmental regulations and management.

The problem for the legal system is to provide enough and the right kind of stability that helps society and ecosystems to absorb shocks and changes without going into decline or collapse, while also providing enough and the right kind of flexibility that helps society and ecosystems to adapt to shocks and changes in resilient and sustainable ways (Garmestani et al. 2009; Benson and Garmestani 2011). We use coastal land use issues to explore three particular adaptive law methods that aim for the right amount and type of flexibility: (1) discretionary decision making governed by standards and context, (2) adaptive planning and management, and (3) the evolution of property law.

The combination of coastal resource uses and coastal-change dynamics are putting tremendous stresses on coastal ecosystems and coastal communities (Mann 2000; Nordstrom 2000; Beatley et al. 2002). Intense development of coastal lands has destroyed or altered critical ecosystem features such as sand dunes and their stabilizing grasses (e.g, sea oats), wetlands, and wildlife habitats. Coastal land uses also have generated pollution, storm-water runoff, and high-concentration activities, such as human crowds on beaches (e.g., Davenport and Davenport 2006), that have nonadjacent or long-term impacts on lands, waters, and wildlife, such as degradation of biologically highly productive estuaries (Hueiwang et al. 2005) and endangerment to beach species like the piping plover, the least tern, and sea turtles from habitat disturbances (Karleskint et al. 2009). Coastal lands are naturally highly dynamic, serving as frequently shifting buffers between marine systems and terrestrial systems (Bird 2008). Attempts to protect human uses of coasts from natural and human-induced changes such as shoreline erosion, sea level rise, or storm surge often exacerbate the disturbances to coastal ecosystems and human communities or create new adverse effects (Beatley at al. 2002). For example, cutting inlets or building barriers, such as breakwaters or groins, interrupts sand flow along the shoreline and accelerates erosion downshore. Shoreline armoring, such as the building of sea wall or rock revetments, may only temporarily protect existing structures and destroys beaches by cutting off the natural flow of sands, which in turn reduces the coastline's storm- and sea-buffering functions. Beach restoration and nourishment alter critical natural features of the entire coastal system, while giving coastal residents and tourists "restored" beaches for only a short period of time. The combination of groundwater withdrawals to support coastal communities, channel dredging, and sea level rise are altering coastal aquifers (Moore 1999), which puts both interconnected aquatic ecosystems and coastal communities' water supplies at risk.

Various exercises of broad discretion regarding the use of coastal resources are blamed for these harms: government permits and variances allowing private and public entities to use regulated resources, administrative agency resource-management decisions, the broad rights and authority of private resource users in how they use these resources, and even loose legal standards that give courts room to authorize (or fail to prevent) actions that prevent environmental harm (Arnold 2011a). Arguably, stricter legal rules and regulations that limit these exercises of discretionary authority would prevent or at least reduce further harm to coastal environments.

Discretionary decision making, however, is a necessary feature of resource conservation and use. Rigid legal rules prevent the flexibility needed to adapt to changing conditions and to consider situations and knowledge not contemplated by the rule makers. Furthermore, rigid rules put pressure on the legal system to create or allow means of flexibility through nonenforcement, new behaviors that are not covered by the rules, variance-creating reforms to the rules, or invalidation of the rules altogether. The complex and uncertain dynamics of interconnected ecosystems and social systems, such as those illustrated by coastal land dynamics, require that resource regulators, managers, and users have a certain amount of discretion.

However, the problem is typically not the lack of narrow and absolute rules but the lack of appropriate and relevant standards to govern the exercise of discretion and to which decision makers can be held accountable. In particular, standards governing discretionary decisions should expressly require consideration of the decision's broad ecological and social context and its long-term impacts, not just its relationships to immediately surrounding space or its short-term impacts. Decision makers should be required to consider cumulative and synergistic effects on ecosystem functions and services. Nonetheless, recognizing uncertainties and the potential for remote impacts, the standards should focus the decision makers on maintaining the adaptive capacity of ecosystems and social systems, including diverse communities.

Adaptive law would expressly authorize natural resource managers to use adaptive management methods. Grumbine (1994) summarizes adaptive management: "Adaptive management assumes that scientific knowledge is provisional and focuses on management as a learning process or continuous experiment where incorporating the results of previous actions allows managers to remain flexible and adapt to uncertainty." The central feature of adaptive management is "learning while doing" (Doremus 2007). Adaptive management is the preferred method of managing natural resources and ecosystems, given uncertainties about the effects of management actions and policies due to systemic complexities and dynamics and limits to the cognitive and management capacities of humans and human organizations (Holling 1978; Lee and Lawrence 1986). Adaptive management has foundations not only in the science of ecology and natural resources management but also in public administration theory about cognitive and organization constraints (Lindblom 1959) and in the public philosophy of Deweyan pragmatism (Karkkainen 2003). The concept of adaptive management applies not only to government agencies but also to private resource managers and could apply not only to management of the natural environment but also to management of human affairs, such as disaster response, financial systems, housing policies, and the like.

Environmental law scholars not only recognize the need for adaptive management but also the need for express legal authority for administrative agencies and natural resource managers to use adaptive management methods, particularly when they may conflict with mandates to plan, assess all environmental impacts upfront, or adhere to strict management directives (Craig 2010b; Ruhl 2010). Nonetheless, even when adaptive management is possible, managers may not engage in it fully or effectively, often failing to build in the essential feedback loops and adaptive adjustments to actions and policies according to the lessons learned from experimental management (Camacho 2007, 2009). Effective adaptive management design and accountability are critical.

Moreover, while adaptive management rejects conventional planning that is long-term, static, and certain in goals and methods, and linear in implementation (Camacho 2009), it is a mistake to throw out the role of planning altogether. Without any planning, adaptive management has the potential to become untethered from central goals and standards and become merely ad hoc reaction: a shift from "muddling through" to merely "drifting along" (Arnold 2010). Instead, adaptive legal regimes would embrace—both authorize and require—the use of adaptive planning methods. Adaptive planning, in contrast to conventional planning, "is an iterative and evolving process of identifying goals and making decisions for future action that are flexible, contemplate uncertainty and multiple possible scenarios, include feedback loops for frequent modification to plans and their implementation, and build planning and management capacity to adapt to change. It is planning that seeks to adapt to the complexity of systems and actors, conditions of uncertainty and unpredictability, and the dynamism of environments characterized by instability and rapid nonlinear changes." (Arnold 2010, 440). The features of adaptive planning are

- flexibility and expectations of difficult-to-predict change (i.e., tolerance for ambiguity and uncertainty);
- the use of multiple scenarios, multiple hypotheses, data replication and pseudoreplication, data analogues, testable and revisable thresholds, and similar scientific methods for reducing or accounting for uncertainty;
- continuous learning, monitoring, and feedback loops that affect plans (including maintaining a commitment to these processes);
- ongoing and iterative changes to plans (including some plan development during implementation);
- holistic and flexible design, including the embedding of options within the plan (e.g., in the forms of menus, branches, or sequels);
- integrated and interdisciplinary or transdisciplinary planning that addresses a range of interrelated scales, problems, and disciplinary insights;
- management or coordination of interdependent conditions;
- consideration of social, political, economic, cultural, institutional, and organizational complexities, as well as scientific, natural, and technical complexities when developing plans and management actions;
- participatory social interaction among multiple participants at various levels of organizational structure and through multi-organization networks (including scaling up and down and using dynamic decision making processes);
- planning of process (planning of planning) as well as planning of management activities. (Arnold 2010, 440–441)

For an extensive literature on adaptive planning theory and methods, spanning more than thirty years, see sources in Arnold (2010, notes 113–130).

A growing number of watershed planning efforts throughout the United States and Canada are utilizing adaptive planning methods for adaptation of watershed governance, plans, and/or management to climate change uncertainties, although much work remains to be done if adaptive planning is to be used fully and well (Arnold 2010). The question is whether coastal communities and coastal resource management agencies can or will engage in adaptive planning.

Improved yet flexible discretionary decision making and more and better use of adaptive management and planning methods will have only limited effect, unless legal institutions allow property law to evolve to changing conditions. Property law and its related constitutional jurisprudence on private property rights (e.g., takings, due process), play a critical role in how adaptive the legal system can be with respect to changing coastal conditions and coastal land use, for example. Currently, the fields of property, land use, and takings law favor: (1) human development and use of coastal lands and resources, even if coastal ecosystems are adversely affected (including redevelopment after destruction); (2) attempts to lock in existing property arrangements and resist change; and (3) government-funded infrastructure to facilitate development and government-funded compensation to property owners for losses. Among the problematic aspects of property law are: regulatory takings doctrines; judicial takings doctrines; disconnections among land use planning and regulation, water resources law, and environmental protection law; weak land use planning and regulatory regimes; antiquated distinctions between sudden avulsion and gradual accretion or erosion despite synergistic causes of change; the National Flood Insurance Program; and incomplete development of nuisance, trespass, and public trust doctrines (Arnold 2011a). They can result in decisions that are disconnected from the ecological and sociocultural conditions of coastal environments and communities, impeding adaptation to changes such as beach erosion, changes in hurricane and storm frequency and intensity, sea level rise, fisheries health, and water quality (Arnold 2011a).

Thus, property law—or at least certain aspects of it—are maladaptive for two reasons: (1) an anti-evolutionary bias, exemplified by U.S. Supreme Court Justice Antonin Scalia's assertion that the Takings Clause of the U.S. Constitution protects historic property arrangements (although the historical accuracy of these arrangements as Justice Scalia describes them is doubtful at best [Mulvaney 2011]); and (2) an elevation of artificial legal constructs and abstractions that are ill-matched to the ecological and social realities that law is intended to address. These two maladaptive features of property law are at odds with competing adaptive features of property law that have richly pervaded its development in the Anglo-American legal system throughout its history: (1) the evolution of property law principles and rules to adapt to changing conditions and needs; and (2) consideration of social and ecological context in developing and applying property law principles and rules (Freyfogle 1989; Arnold 2011a). Moreover, the complexities of interconnected and highly dynamic societal and ecological problems require evolution in property rights in order to build the adaptive capacity of society to meet these challenges (Doremus 2012). The context-regarding concept of property as a "web of interests"—a powerful and increasingly embraced alternative to the "bundle of rights" metaphor of property—calls for attention to the interconnected social and ecological conditions in which property issues arise and the impacts of various property alternatives on communities, social systems, and ecosystems (Arnold 2011a). Adaptation of property law to new conditions may be iterative and include both reform and resistance, but inflexible legal obstacles to change in rapidly evolving social and ecological environments, like coastal areas, make it more likely that all systems—ecological, economic, political, and sociocultural—will not be able to adapt to change and will suffer decline, perhaps over time or perhaps suddenly at a tipping point.

We offer two caveats about evolutionary, contextualized property standards. The first caveat is that adaptive law's goals of promoting resilience should not be used as a mere front for interest-based redistributive agendas that seek to alter property rules or arrangements. Legal institutions will have to pay particular attention to the potential for the misuse of adaptive law concepts in this manner. In their dissents in *Kelo v. City of New London* (2005, 499–504 and 506–521), U.S. Supreme Court Justices Sandra Day O'Connor and Clarence Thomas criticized flexible eminent-domain doctrines and legal changes in property allocations as: (1) resulting in unintended and perverse property outcomes; and (2) being tools of powerful interests to take property from less-powerful elements of society, particularly modest-income people and racial or ethnic minorities. The evolution of law to adapt to changing conditions can itself be turned into a legal abstraction that prevents consideration of the full context or potential ramifications of any change.

The second caveat is about the degree and pace of change. Some have argued that the degree and pace of change in environmental conditions and society, such as the many, substantial, and fast-paced impacts of climate change, require substantial and rapid adaptations in society and human communities, which in turn require substantial and rapid changes to property laws (Ruhl 2010; Doremus 2012). However, substantial, rapid changes in property rules, rights, and allocations are likely to have many reverberating effects throughout social systems and institutions, as well as unanticipated impacts on ecosystems. For example, sudden conversion of private beachfront property to public-trust beaches can lead to several possible results, including adverse impacts of resident out-migration on distant "destination" communities and ecosystems (e.g., What happens to Gainesville's water supply and growth patterns when landowners lose their property on St. Augustine Beach and move inland?), political and legal backlash (e.g., *Severance v. Patterson* 2010, 2012), the psychological and social association of already charged terms like "climate change" with negatively perceived terms like "retreat" or "abandonment," the replacement of limited private uses of beaches with nearly unlimited intensive public uses of beaches, or public pressure on government to continually fund and implement ecologically questionable beach renourishment. The last two were underlying issues in the *Stop the Beach Renourishment* case (*Stop the Beach Renourishment, Inc. v. Florida Department of Environmental Protection* 2010). The capacities of the legal system, government agencies, local communities, and private property owners/users to monitor and assess the impacts of property law changes and to make necessary adjustments in response to actual effects are greatly diminished if the changes are sweeping and swift. As Doremus (2010) has observed, human psychology and U.S. culture lead people and institutions to use property law to resist change, which they fear, and preserve the status quo, which they embrace, thus making legal transitions quite difficult. An adaptive law regime makes substantial, rapid changes to property rules when absolutely necessary for the resilience of ecosystems and social systems. However, an adaptive law regime favors incremental and gradual changes that transition experimentally to new property standards or arrangements, while monitoring, assessing, and adjusting these changes and their effects. Forms of compensation, risk sharing, or collaborative and innovative problem solving may be needed to ease transitions in property arrangements.

## Adaptive Law Processes

Adaptive law recognizes and embraces iterative processes with feedback loops among multiple participants, limits to human and organizational rationality, the effects of social and ecological forces on the ordering and management of human affairs, and accountability mechanisms for the conservation of capital. More traditional and less adaptive conceptions of law envision linear processes. For example, government agency actions that could have a substantial impact on the environment are expected to proceed from issue identification, to planning, to environmental impact assessment, to decision, to implementation, according to many statutory regimes (e.g., the national forest management statutes, Klein et al. 2009). Likewise, the common understanding of judicial processes proceed from the parties' actions or inactions, to conflict, to litigation, to initial judicial decision, to appeals and final judicial decision, to enforcement of the judicial decision and resolution of the conflict.

Understandings of legal processes as linear often conflict with the more complex reality of law–society–nature interrelationships. For example, government agencies may not implement decisions as planned and adopted, or their implementation may have unintended and/or unpredictable consequences in nature and/or society (Gunderson et al. 2006; Doremus 2007). The rationality of decision making and action presumed in legal requirements is often mismatched to constraints on human cognitive capacity, knowledge, and organizational behavior (Gunderson and Holling 2002). Judicial decisions often do not resolve the underlying conflict. The "aftermath" of major court cases typically consists of the dynamic impacts of forces in society (e.g., political, economic, social, cultural, psychological forces), natural forces (e.g., biological, physical, chemical), and the postlitigation actions of parties and non-parties that do not conform neatly to the resolution of the case that the courts intended (Arnold 2004). Some conflicts go through multiple iterations of litigation or other forms of decision making, such as legislative action, regulatory action, negotiation, mediation, arbitration, or nonlegal action in the private sector or civil society (Arnold 2004).

The iterations of legal process, though, do not necessarily strengthen the resilience of ecosystems and social systems. Some iterations may enhance the adaptive capacity of social institutions, including the law itself, through opportunities for additional or improved decisions and actions, accountability and feedback for initially maladaptive decisions and actions, or disturbances that stimulate the system to respond adaptively and strengthen its resilience in the process, much like grassland ecosystems depend on disturbance regimes of fire, drought, grazing, trampling, and wallowing in order to thrive (Reichman 1987; Knapp et al. 1998). Some iterations, though, may be disturbances that weaken certain systems, drains on limited resources by competitors or deceptively attractive "easy" pathways that lead decision makers away from adaptive but difficult solutions.

The problems in understanding whether process is adaptive or maladaptive can be seen in the twists and turns—iterations—of decisions about federal jurisdictional authority to regulate wetlands under Section 404 of the CWA (§ 1344). In order to broadly protect wetlands and their immense ecological functions in nature and society, the U.S. Army Corps of Engineers and the U.S. Environmental Protection Agency (EPA) interpreted the CWA to extend federal regulatory jurisdiction over a wide variety of wetlands in the United States, an approach that has been challenged in court by regulated property owners. Although the U.S. Supreme Court upheld federal jurisdiction over wetlands adjacent to navigable waters in the *United States v. Riverside Bayview Homes* case (1985), it struck down the agencies' interpretation of the CWA as extending federal jurisdiction to isolated wetlands in the *Solid Waste Agency of Northern Cook County v. U.S. Army Corps of Engineers* (*SWANCC*) case (2001) and to wetlands that are not immediately adjacent to navigable waters and have only an indirect connection to navigable waters in the *Rapanos v. United States* case (2006).

The *Rapanos* case produced an unclear collection of opinions from the court that has left both regulators and property owners confused about the scope of federal jurisdiction. In a plurality opinion, four justices stated that the CWA covers only those wetlands that are connected to traditional interstate navigable waters through an adjacent channel containing a relatively permanent body of water (i.e., not channels with merely intermittent or ephemeral flow) and that have a continuous surface connection with those waters making it difficult to determine where the water ends and the wetland begins (*Rapanos v. United States*

2006, 739–742). Justice Kennedy's concurring opinion, the necessary fifth vote on the U.S. Supreme Court to reach a decision, defined federal jurisdiction as extending to wetlands with a significant nexus to traditional navigable waters, with "navigable waters" defined as relatively permanent, standing or flowing bodies of water, and "significant nexus" defined as the ability to affect the chemical, physical, and biological integrity of traditional navigable water, as determined on a case-by-case basis, regardless of whether the wetlands have a direct surface connection to navigable waters or whether they are adjacent to tributaries (*Rapanos v. United States* 2006, 759). The remaining four justices in the dissent would have deferred to the agencies' expertise in interpreting the CWA.

The response to *Rapanos* has been muddled at best (Thomas 2008). Environmentalists decried the court's narrowing of federal regulatory jurisdiction inevitably leading to a loss of ecologically valuable wetlands in the United States and an elevation of artificial legal classifications over actual hydrological processes and structures. Development-seeking property owners were relieved to see federal jurisdiction limited but felt anxious, even paralyzed, by the lack of clarity as to the scope of federal jurisdiction. Federal regulators and courts required to interpret *Rapanos* in pending litigation struggled to understand whether to apply the plurality's test or Justice Kennedy's test. Eventually, the Army Corps of Engineers and the EPA jointly issued a draft guidance proposing to interpret the CWA to provide federal jurisdiction to wetlands meeting either of the two tests (U.S. EPA and Army Corps of Engineers 2011), but finalization of the guidance has been deferred in response to congressional legislation that would attempt to prevent its adoption on the theory that it interprets federal authority much more broadly than *Rapanos* allows. In addition, initial congressional consideration of the Clean Water Restoration Act (CWRA), which would statutorily define federal jurisdiction as broadly as or even more broadly than the pre-*Rapanos* agency interpretation, failed to advance far in the legislative process when the Democrats controlled both houses of Congress and is deemed essentially "dead" (Holm-Hansen 2012). Furthermore, the CWRA, even if adopted, might not survive a challenge to its constitutionality. In both *SWANCC* and *Rapanos*, justices in the majority or plurality hinted that if Congress truly had intended the CWA to apply as broadly as the agencies were interpreting it, Congress would have exceeded its constitutional authority under the Commerce Clause (e.g., *SWANCC v. U.S. Army Corps of Engineers* 2001, 172).

Compounding the issues of jurisdiction, federal courts have reined in federal wetlands regulatory authority in at least two other ways. In a 2012 case involving wetlands, the U.S. Supreme Court unanimously ruled against the EPA's arguments that its administrative compliance orders (ACOs) could not be reviewed by courts until they were enforced against violators (*Sackett v. EPA* 2012). The EPA uses ACOs against landowners with wetlands to define the scope of jurisdictional wetlands and prohibitions on landowners' actions with respect to the wetlands and to establish in the ACO the risk of penalties for failure to comply with the ACO. Nonetheless, the EPA argued that the courts did not have jurisdiction to review the ACOs until the EPA actually commenced enforcement actions against violators of their ACOs, which would mean that the only way for a landowner to challenge the agency's jurisdiction or prohibitions would be to violate the ACO and run the risk of substantial fines. The court held that the CWA allows judicial review of ACOs. It did not address the underlying concerns that administrative orders that cannot be reviewed unless violated deprive the regulated parties of basic due process rights under the U.S. Constitution, but the due process concerns were significant in the litigation itself (Adler 2012). Furthermore, some federal courts have found that particular prohibitions on the filling and development of wetlands constitute compensable takings under the Fifth Amendment Takings Clause, which arguably incentivizes the Army Corps of Engineers and the EPA to move more quickly and be more permissive with respect to private landowners' permit applications than careful study and minimization of ecosystem impacts might require (*Loveladies Harbor, Inc. v. United States* 1994).

Arguably, the iterations of legal process concerning contested federal regulatory authority over wetlands threaten the long-term viability and resilience of federal laws and programs to protect wetlands from loss to development and fill. The legal constraints on agency jurisdiction, enforcement tools, and fiscally feasible regulatory actions could very likely lead to the loss of wetlands that serve important ecosystem functions and values, producing significant ecological stressors at local scales or in the aggregate.

On the other hand, courts may be acting to protect the resilience of political and legal systems and their systemic features of federalism, due process, and private property rights; judicially defined limits to federal regulation may be akin to pressure-relief valves that reduce the kind of political backlash that has limited other federal regulatory regimes, such as ESA. In addition, legal boundaries to federal regulatory authority may help to stimulate nonfederal or nonregulatory methods of protecting wetlands from fill and development, such as strengthening state and local wetlands laws and programs; incentivizing private investment in wetland conservation; encouraging collaborative partnerships between government agencies and private landowners; and forcing legal, economic, and political systems to move faster to develop metrics and means for valuing the ecosystem services of wetlands. The current Section 404 system focuses primarily on a goal of no net loss in acreage of wetlands by allowing landowners to fill and develop wetlands in exchange for mitigation through wetlands that are created or restored elsewhere (NRC 1999). This system fails to give sufficient attention to whether the offsetting "mitigation" wetlands serve the same ecological functions, values, and scales as the developed wetlands or to the long-term resilience of protected, restored, and created wetlands under changing conditions (Gunderson et al. 2006). Some might argue that legal shocks to the existing regulatory system might stimulate federal wetlands policies to move toward more poly-resilient approaches.

Feedback loops are essential to aiding decision makers in assessing whether any particular decision or action is adaptive or maladaptive: to monitor, assess, learn from, and adapt to the impacts of the decision or action (Holling 1978; Walters 1986; Gunderson et al. 2006). All four elements are critical: (1) continuous monitoring of multiple indicators of system functions and resilience; (2) assessment of data from monitoring; (3) scientific and social learning from the lessons that the monitoring and assessment provide about the effects of particular decisions or actions; and (4) adaptation of plans, policies, programs, management, governance, and laws based on the lessons learned from monitoring and assessment.

Adaptive law authorizes or requires meaningful feedback loops in laws creating planning, management, regulatory, or governance authority and then holds planners, managers, regulators, and officials accountable for making use of these feedback loops. A standard adaptive management critique of many environmental and natural resource laws is that they lack meaningful feedback-loop processes or that the relevant actors are not employing the feedback-loop processes (Lee and Lawrence 1986; Gunderson et al. 2006). Incidentally, adaptive law can and should increasingly expect private sector managers of resources, including landowners, to engage in monitoring, assessment, learning, and adaptation processes, although it may have to help to build the capacity of some private sector managers, such as individual landowners, to do so, especially given the resources and knowledge that feedback-loop processes may demand. Regulatory permits should contain conditions, time limits (but allow for renewal), requirements for self-monitoring and regulator monitoring, and enforcement mechanisms, but regulators should use these tools to help permit holders to adapt effectively to changing conditions.

Moreover, the legal system itself should develop and improve its own feedback loops to evaluate and adapt to the impacts of legal decisions and actions. Both courts and legislatures, in particular, make many decisions based on anecdotal evidence, the normative power of abstract principles and rhetoric, poorly performing predictions of future impacts of the proposed course of action, and inattentiveness to how other forces in society and nature will interact with the proposed course of action to produce unexpected consequences. An adaptive legal system not only relies on multiple sources of feedback about the impact of legal decisions and adapts accordingly but also encourages systematic, multivariate, longitudinal study of the impacts of legal decisions, actions, and processes. Even experts often miss the real impact of "legal moments." For example, when major U.S. Supreme Court takings opinions, such as *Lucas v. South Carolina Coastal Council* (1992), *Nollan v. California Coastal Commission* (1987), and *Dolan v. City of Tigard* (1994), came down, environmental and land use experts critically predicted that the court's new takings jurisprudence would greatly limit environmental and land use regulation (Sax 1993; Starritt and McClanahan 1995). Studies have shown, though, that increased clarity about the lines between compensable regulation and noncompensable regulation actually improved regulators' capacity to design noncompensable regulations and led to increased, not decreased, regulation in some circumstances (Blumm and Ritchie 2005).

These observations highlight the need to evaluate impacts over time. Many impacts emerge over time, and many systems and institutions evolve over time (Gunderson et al. 1995). "Snapshot" assessments of the performance of particular laws or principles can lead to faulty lessons and ill-advised adjustments. Choice of temporal scale for assessing failure and making changes in law is a tricky matter, though. At what point is a rule of law, such as the rule of capture for groundwater, deemed not to be working and a change made? At what point does a government agency declare failure in a housing or financial market and intervene? At what point is a collaborative ecosystem management process, such as the California Bay-Delta Accord, the Chesapeake Bay Program, or the Comprehensive Everglades Restoration Plan, deemed unsuccessful and replaced with more conventional regulatory or litigation approaches? Adaptation is neither about patiently and futilely waiting for a failed approach to eventually work nor about constantly tinkering with an approach that has not been given time to work or is undergoing adaptive transformation. The complexity of this point is illustrated by the fact that many watershed-based problem-solving processes go through multiple iterations of cooperative and adversarial processes, informal and formal processes, and active and dormant processes, often morphing over time. "Dead" approaches can be "resurrected" when the conditions are right, and alternative approaches can stimulate a seemingly failed approach to respond adaptively and become successful.

One of the most challenging aspects of developing a legal system that both requires adaptation to changed conditions and allows for systems to evolve over time is that a primary function of the legal system is accountability. The legal system serves to hold people and entities accountable for their actions and accountable to boundaries on their actions. Defining the standards and boundaries to which people and entities will be held accountable can have significant impact on the resilience of ecosystems, social systems, and legal systems.

We believe that a general principle of adaptive law is to create mechanisms of accountability for the conservation of capital: natural capital, financial capital, human capital, social capital, and political capital. The basic premise is that present consumption of nonrenewable principal undermines the long-term resilience of the resource and its capacity to generate future benefits and functions, leading to systemic decline and/or collapse (Walker and Salt 2006).

One mechanism for capital-conservation accountability is the public trust doctrine. According to the basic structure of the public trust doctrine in the United States, the state is a trustee—a fiduciary—holding navigable waters, tidal waters, and the submerged lands of these waters (and in some cases interconnected waters such as feeder streams or groundwater) in trust for the public, who are the beneficiaries of the trust. Thus, the state cannot simply transfer or eliminate the multiple benefits of public trust waters and lands, because it must remain accountable to the public. Wood (2009, 2013) has argued articulately and forcefully for recognition of a public trust in the atmosphere, and a nationwide litigation strategy has been undertaken to compel federal and state governments to address climate change in order to protect the public's interest in the atmosphere (Wood 2009). Others have argued that the public trust should be applied to biodiversity (Johnson and Galloway 1994) and soils (Hannam and Boer 2002), both of which are commons on which all human life depends for survival.

Legal principles, such as the public trust doctrine, will not be sufficient by themselves to ensure the resilience of ecosystems and social systems. Craig (2007, 2010a) has shown that there is great variation in the scope and nature of the public trust doctrine from state to state. Arnold (2004) studied the aftermath of the landmark Mono Lake case, *National Audubon Society v. Superior Court* (1983), in which the California Supreme Court famously applied the public trust doctrine to Mono Lake's feeder streams, which were being diverted to Los Angeles for public water supplies pursuant to Los Angeles 's prior appropriation water rights, yet to the severe detriment of Mono Lake and its ecosystem. Despite the ruling that required the state's water board to balance Los Angeles's water rights with the public's trust-protected interests in Mono Lake and its ecosystem, the controversy took ten years to resolve. During this time, many postruling forces were needed to bring about conservation of the Mono Lake ecosystem: additional statutory-based litigation, scientific study, place-based conservation advocacy, public education and engagement, political leadership and change, federal and state financing for conservation and reclamation projects, and collaborative problem solving in which the Mono Lake Committee helped to identify conservation and reclamation solutions for Los Angeles that would meet Los Angeles 's water needs without requiring large diversions from Mono Lake or any substitute water body. Adaptive law principles require linkages between legal processes and nonlegal processes and forces, because law is not sufficient by itself to achieve environmental conservation. Interestingly, Owen (2012) has found that the Mono Lake case has also had limited impact on the California Water Resources Board's decision-making processes in subsequent water permit cases. Arnold (2009) has also argued that the public trust doctrine, at least with respect to water resources, needs to be amplified or modified to recognize public stewardship responsibilities in the public as a whole, not just fiduciary duties in the state. Legal reforms, including reforms aimed at making law and legal institutions more adaptive, must consider not only the many effects that they will have on nature and society but also the many effects that nature and society, with their complex, interdependent, dynamic interconnections (i.e., panarchy), have on law and legal reform.

## The Future: Studying and Developing Adaptive Legal Systems

The development of principles and methods of adaptive law is just beginning. We have aimed to set forth some general characteristics of adaptive law based in part on the scientific concepts of panarchy and resilience and in part on the sociolegal analyses of maladaptive characteristics of the current legal system. Much further study is needed. Four particular lines of inquiry should be developed.

First, we need more and better principle-based specifics of adaptive law. There is a very real risk that terms like "resilience" and "adaptive law" will become vague meaningless slogans that (1) lack specificity and measurement for implementation and evaluation and (2) can be used indiscriminately by many competing interests to advance widely divergent goals, which may be contrary to ecological values or justice (Wiersema 2008). These criticisms have been leveled against concepts such as sustainability (Freyfogle 2006), biodiversity (Goble 2007), ecosystem services (Goble 2007), and adaptive management (Pardy 2003).

The problem is finding the right mix of adaptive flexibility and principled accountability, an inquiry that will likely be highly contested. On one hand, neither law nor society nor nature is like a train on a track, headed linearly in a single direction. On the other hand, adaptive law should not be about countless groups splintering off in different directions and wandering around the wilderness without GPS. The task of developing principles, specifics, and metrics is yet incomplete.

Second, we need to study how particular features of or changes in the legal system affect the resilience of various ecosystems and social systems, including components such as species populations and human communities. This kind of research will be quite daunting, requiring the detailed field assessment of many variables and the longitudinal impacts of many systems that are dynamic and continue to evolve or transform even as the legal system is evolving and transforming. It will not be easy to design studies that are both accurate and useful.

For example, what effect does a decision like *Rapanos* have on wetlands, other ecosystems, the economy, environmental and natural resource politics, local communities, and the like? Scholars may be tempted to look at the effects on only one system, because of the conceptual challenges of evaluating a broad range of interconnected systems simultaneously. Narrowing the scope of the research question can make the gathering and analysis of data more manageable but less useful to understanding the legal system as a whole. Scholars may be tempted to engage in only a "snapshot" assessment of the effects of *Rapanos* as of the time of the study, instead of engaging in long-term longitudinal research to see how a legal decision has evolving effects over time on systems that are not static (i.e., with their own dynamic processes and changes that are not dependent on legal change).

Scholars may be tempted to design studies with preconceived theories or policy/legal preferences. Thus, a scholar who believes that *Rapanos* favors landowner destruction of wetlands is likely to study either aggregate data or examples of wetlands that were developed or filled post-*Rapanos* and that would have been protected by the pre-*Rapanos* assertion of federal regulatory jurisdiction over wetlands, perhaps surmising or demonstrating a variety of ecosystem harms from the loss of these wetlands. A scholar who believes that *Rapanos* was wrongly decided on federalism grounds will look at impacts on federal–state relations. A scholar who is critical of the uncertainty created by Supreme Court fifth-vote concurring opinions on different grounds than the plurality opinion will likely research the angst and paralysis of wetlands regulators, landowners, and even lower courts over not knowing which *Rapanos* jurisdictional "rule" to apply. A scholar who favors stronger private property rights and state power in comparison with federal regulatory power will likely look for indicia of positive effects of *Rapanos* on landowner wetlands stewardship, the land-development economy, local and state environmental policies, or the trajectory of federalism and/or property rights jurisprudence. How many legal scholars begin a research project without a preconceived idea of what they will find? How many legal scholars articulate surprise at the results of their studies, which cause them to rethink their ideas? The normative components of law and legal scholarship make it very difficult for legal scholars to be true scientists in their study of legal institutions and systems. Scientists—both in the ecological sciences and in the social sciences—can aid legal scholarship to become more scientifically rigorous, without sacrificing breadth of inquiry for narrowly defined research questions.

However, we acknowledge that scientists who study panarchy and resilience in complex adaptive systems may be tempted to impose their constructs and theories on our understanding of law or our reform of law, even though concepts of ecosystem dynamics might not map fully onto concepts of legal system dynamics. Thus, our third line of further study is about the extent to which the legal system manifests the dynamics discussed in the panarchy literature. Panarchy is characterized by nested sets of adaptive cycles with four stages: (1) an exploitation stage (r phase), with rapid garnering and exploitation of resources by system components; (2) a conservation stage (k phase), with a relatively longer period of accumulation of capital and other system elements, increasing connectivity, and increasing rigidity, which lead to decreased resilience and eventual collapse; (3) a release stage (omega phase), with a rapid unleashing of the energy accumulated and stored and collapse of the system; and (4) a reorganization stage (alpha phase), with a reorganization of the system into a new system or a reconstituted version of the prior system by rapidly assembling or reassembling components (Gunderson and Holling 2002). Ruhl (2012) has recently suggested that legal scholars and lawyers will need to join panarchy scientists to discover and design the features of the legal system that put panarchy theory into panarchy practice.

Finally, we believe that the starting point of a more robust analysis of adaptive law and its features is to improve the systematization and integration of in-depth case studies that evaluate the dynamic interplay of multiple systems, actions, and forces over time with respect to a particular legal phenomenon. Both of this chapter's authors have engaged in these kind of studies: for example Arnold's study of the Mono Lake conservation effort (2004) and Zellmer and Gunderson's study of the Everglades and Glen Canyon Dam ecosystem restoration projects (2009). Many other scholars have engaged in a similar type of research, among them Angelo's study of Lake Apopka's degradation and restoration (2009), Camacho's studies of HCPs and ecosystem management programs (2007, 2009), and studies of the Columbia River Basin by Cosens (Cosens 2010, 2012). Case studies may draw on both qualitative data and analyses and quantitative data and analyses, but they are essentially "field research" at intersections of law, society, and nature. Eventually they can and should lead both to increased synthesis of their results into a broader theoretical construct or model of adaptive law and resilience and also to more scientifically rigorous testing of hypotheses with quantitative methods. The theory and practice of adaptive law are still emerging in response to interconnected and complex transformations in both nature and society. We have much work ahead of us to refine our understanding and development of the features of adaptive law.

## References

Adelman, D. E., and K. H. Engel. "Adaptive Federalism: The Case Against Reallocating Environmental Regulatory Authority." *Minnesota Law Review* 92 (2006): 1796–1849.

Adler, J. "Wetlands, Property Rights, and the Due Process Deficit in Environmental Law." *CATO Supreme Court Review* 2011–2012: 139–165, 2012.

Adler, R. W. "Resilience, Restoration, and Sustainability: Revisiting the Fundamental Principles of the Clean Water Act." *Washington University Journal of Law and Policy* 32 (2010): 139–173.

Alliance for Community Trees. *Benefits of Trees and Urban Forests: A Research List* [resource list with 122 citations to scientific studies of urban tree benefits]. College Park, Md.: Alliance for Community Trees, 2011.

Angelo, M. J. "Stumbling Toward Success: A Story of Adaptive Law and Ecological Resilience." *Nebraska Law Review* 87 (2009): 950–1007.

Arnold, C. A. "Working Out an Environmental Ethic: Anniversary Lessons from Mono Lake." *Wyoming Law Review* 4 (2004): 1–55.

Arnold, C. A. "Clean-Water Land Use: Connecting Scale and Function." *Pace Environmental Law Review* 23 (2006): 291–350.

Arnold, C. A. *Fair and Healthy Land Use: Environmental Justice and Planning.* Chicago: American Planning Association, 2007a.

Arnold, C. A. "Structure of the Land Use Regulatory System in the United States." *Journal of Land Use and Environmental Law* 22 (2007b): 441–523.

Arnold, C. A. "Water Privatization Trends in the United States: Human Rights, National Security, and Public Stewardship." *William & Mary Environmental Law and Policy Review* 33 (2009): 785–849.

Arnold, C. A. "Adaptive Watershed Planning and Climate Change." *Environmental & Energy Law & Policy Journal* 5 (2010): 417–487.

Arnold, C. A. "Fourth-Generation Environmental Law: Integrationist and Multimodal." *William & Mary Environmental Law and Policy Review* 35 (2011a): 771–884.

Arnold, C. A. "Legal Castles in the Sand: The Evolution of Property Law, Culture, and Ecology in Coastal Lands." *Syracuse Law Review* 61 (2011b): 213–260.

Associated Press. "Federal Government Ending Protections for Wolves in Wyoming; Environmental Groups Vow to Fight." *Washington Post* August 31, 2012.

Babbitt, B. *Cities in the Wilderness: A New Vision of Land Use in America.* Washington, D.C.: Island Press, 2005.

Beatley, T., D. J. Brower, and A. K. Schwab. *An Introduction to Coastal Zone Management,* 2d ed. Washington, D.C.: Island Press, 2002.

Bell, T. W. "Polycentric Law." *Humane Studies Review* 7 (1991–1992): 1–10.

Benson, M. H., and A. S. Garmestani. "Embracing Panarchy, Building Resilience and Integrating Adaptive Management Through a Rebirth of the National Environmental and Policy Act." *Journal of Environmental Management* 92 (2011): 1420–1427.

Berkes, F., and C. Folke. *Linking Social and Ecological Systems: Management Practices and Social Mechanisms for Building Resilience.* Cambridge: Cambridge University Press, 1998.

Bird, E. *Coastal Geomorphology: An Introduction,* 2d ed. Chichester, UK: John Wiley, 2008.

Blumm, M. C., and L. Ritchie. "Lucas's Unlikely Legacy: The Rise of Background Principles as Categorical Takings Defenses." *Harvard Environmental Law Review* 29 (2005): 321–368.

Burke, M. A. "Klamath Farmers and Cappuccino Cowboys: The Rhetoric of the Endangered Species Act and Why It (Still) Matters." *Duke Environmental Law and Policy Forum* 14 (2004): 441–521.

Camacho, A. E. "Can Regulation Evolve?: Lessons from a Study in Maladaptive Management." *UCLA Law Review* 55 (2007): 293–358.

Camacho, A. E. "Adapting Governance to Climate Change: Managing Uncertainty Through a Learning Structure." *Emory Law Journal* 59 (2009): 1–77.

Camacho, A. E. "Assisted Migration: Redefining Nature and Natural Resource Law under Climate Change." *Yale Journal on Regulation* 27 (2010): 171–255.

Cole, L. W., and S. R. Foster. *From the Ground Up: Environmental Racism and the Rise of the Environmental Justice Movement.* New York: New York University Press, 2001.

Cosens, B. "Transboundary River Governance in the Face of Uncertainty: Resilience Theory and the Columbia River Treaty." *University of Utah Journal of Land Resources and Environmental Law* 30 (2010): 229–265.

Cosens, B. "Resilience and Law as a Theoretical Backdrop for Natural Resource Management: Flood Management in the Columbia River Basin." *Environmental Law* 42 (2012): 241–262.

Craig, R. K. "A Comparative Guide to the Eastern Public Trust Doctrine: Classifications of States, Property Rights, and State Summaries." *Penn State Environmental Law Review* 16 (2007): 1–113.

Craig, R. K. "A Comparative Guide to the Western States' Public Trust Doctrines: Public Values, Private Rights, and the Evolution Toward an Ecological Public Trust." *Ecology Law Quarterly* 37 (2010a): 53–197.

Craig, R. K. "'Stationarity Is Dead'—Long Live Transformation: Five Principles For Climate Change Adaptation Law." *Harvard Environmental Law Review* 34 (2010b): 9–73.

Daily, G. C., ed. *Nature's Services: Societal Dependence on Natural Ecosystems.* Washington, D.C.: Island Press, 1997.

Davenport, J., and J. L. Davenport. "The Impact of Tourism and Personal Leisure Transport on Coastal Environments: A Review." *Estuarine, Coastal and Shelf Science* 67 (2006): 280–292.

Delgado, R. "Our Better Natures: A Revisionist View of Joseph Sax's Public Trust Theory of Environmental Protection, and Some Dark Thoughts on the Possibility of Law Reform." *Vanderbilt Law Review* 44 (1991): 1209–1227.

Dolan v. City of Tigard, 512 U.S. 374 (1994).

Doremus, H. "Precaution, Science, and Learning while Doing in Natural Resource Management." *Washington Law Review* 82 (2007): 547–579.

Doremus, H. "CALFED and the Quest for Optimal Institutional Fragmentation." *Environmental Science and Policy* 12 (2009): 729–732.

Doremus, H. "The Endangered Species Act: Static Law Meets Dynamic World." *Washington University Journal of Law and Policy* 32 (2010): 175–235.

Doremus, H. "Climate Change and the Evolution of Property Rights." *UC Irvine Law Review* 1 (2012): 101–133.

Doremus, H., and A. D. Tarlock. *Water War in the Klamath Basin: Macho Law, Combat Biology, and Dirty Politics.* Washington, D.C.: Island Press, 2008.

Downey, L., E. Bonds, and K. Clark. "Natural Resource Extraction, Armed Violence, and Environmental Degradation." *Organization and Environment* 23 (2010): 417–445.

Driesen, D. M. *The Economic Dynamics of Environmental Law*. Cambridge, Mass.: MIT Press, 2003.

Dutt, K. "Governance, Institutions, and the Environment-Income Relationship: A Cross-Country Study." *Environment, Development and Sustainability* 11 (2009): 705–723.

Eagle, S. J. "Private Property, Development, and Freedom: On Taking Our Own Advice." *SMU Law Review* 59 (2006): 345–383.

Elmendorf, C. S. "Ideas, Incentive, Gifts, and Governance: Toward Conservation Stewardship of Private Land, in Cultural and Psychological Perspective." *University of Illinois Law Review* (2003): 423–505.

Endangered Species Act, 16 U.S.C. §§ 1531–1544.

Energy and Water Development Appropriation Act of 1980, Pub. L. No. 96–69, 93 Stat. 437 (1979).

Energy and Water Development Appropriations Act of 2004, Pub. L. No. 108–137, 117 Stat. 1827 (2003).

Farber, D. A. "Adaptation Planning and Climate Impact Assessment: Learning from NEPA's Flaws." *Environmental Law Reporter* 39 (2009a): 10605–10614.

Farber, D. A. "Climate Adaptation and Federalism: Mapping the Issues." *San Diego Journal of Climate and Energy Law* 1 (2009b): 259–286.

Federal Water Pollution Control Act of 1972 [commonly known as the Clean Water Act], 33 U.S.C. § 1251 et seq.

Feldman, D., and H. Ingram. "Making Science Useful to Decision Makers: Climate Forecasts, Water Management, and Knowledge Networks." *Weather, Climate and Society* 1 (2009): 9–21.

Freeman, J., and D. A. Farber. "Modular Environmental Regulation." *Duke Law Journal* 54 (2005): 795–912.

Freyfogle, E. T. "Context and Accommodation in Modern Property Law." *Stanford Law Review* 41 (1989): 1529–1556.

Freyfogle, E. T. *Why Conservation Is Failing and How It Can Regain Ground*. New Haven: Yale University Press, 2006.

Folke, C., S. Carpenter, B. Walker, M. Scheffer, T. Elmqvist, L. Gunderson, and C. S. Holling. "Regime Shifts, Resilience, and Biodiversity in Ecosystem Management." *Annual Review of Ecology Evolution and Systematics* 35 (2004): 557–581.

Galle, B. D., and J. K. Leahy. "Laboratories of Democracy? Policy Innovation in Decentralized Governments." *Emory Law Journal* 58 (2009): 1333–1400.

Garmestani, A. S., C. R. Allen, and H. Cabezas. "Panarchy, Adaptive Management and Governance: Policy Options for Building Resilience." *Nebraska Law Review* 87 (2009): 1036–1054.

Glicksman, R. L. "Ecosystem Resilience to Disruptions Linked to Global Climate Change: An Adaptive Approach to Federal Land Management." *Nebraska Law Review* 87 (2009): 833–892.

Glisson v. United State Forest Service, 51 F.3d 275 (7th Cir. 1995).

Goad, B., and D. R. Santschi. "Baca Bill Aims to Swat Bothersome Fly." *Press-Enterprise* March 21, 2011.

Goble, D. D. "What Are Slugs Good For? Ecosystem Services and the Conservation of Biodiversity." *Journal of Land Use and Environmental Law* 22 (2007): 411–440.

Goble, D. D., J. M. Scott, and F. Davis, eds. *The Endangered Species Act at Thirty,* vol. 1, *Renewing the Conservation Promise.* Washington, D.C.: Island Press, 2006.

Gold, S. "Rare Bugs Arrest a Development." *Los Angeles Times* March 28, 2005.

Goldstein, B. E. *Collaborative Resilience: Moving Through Crisis to Opportunity.* Cambridge, Mass.: MIT Press, 2012.

Gordon, O. "Saving the Salamanders: Conservation vs. Development." *State Impact* [Texas] July 27, 2012.

Grumbine, R. E. "What Is Ecosystem Management?" *Conservation Biology* 8 (1994): 27–38.

Guercio, L. D., and T. P. Duane. "Grizzly Bears, Gray Wolves, and Federalism, Oh My! The Role of the Endangered Species Act in de Facto Ecosystem-Based Management of the Greater Glacier Region of Northwest Montana." *Journal of Environmental Law and Litigation* 24 (2009): 285–366.

Gunderson, L. H., and C. S. Holling. *Panarchy: Understanding Transformations in Human and Natural Systems.* Washington, D.C.: Island Press, 2002.

Gunderson, L. H., C. S. Holling, and S. S. Light. *Barriers and Bridges to the Renewal of Ecosystems and Institutions.* New York: Columbia University Press, 1995.

Gunderson, L. H., S. R. Carpenter, C. Folke, P. Olsson, and G. Peterson. "Water RATs (Resilience, Adaptability, and Transformability) in Lake and Wetland Social-Ecological Systems." *Ecology and Society* 11, no. 1 (2006): 16. [online] http://www.ecologyandsociety.org/vol11/iss1/art16/.

Habitat Conservation Plan Assurances "No Surprises" Rule, 63 Fed. Reg. 8859 (Feb. 23).

Hannam, I., and B. Boer. *Legal and Institutional Frameworks for Sustainable Soils: A Preliminary Report.* IUCN Environmental Policy and Law Paper No. 45. Gland, Switzerland/Cambridge, U.K.: International Union for the Conservation of Nature and Natural Resources, 2002.

Hirokawa, K. H. "Driving Local Governments to Watershed Governance." *Environmental Law* 42 (2012): 157–200.

Holling, C. S. *Adaptive Environmental Assessment and Management.* London: Wiley, 1978.

Holm-Hansen, K. L. "A Stream Would Rise from the Earth, and Water the Whole Face of the Ground: The Ethical Necessity for Wetlands Protection post-*Rapanos.*" *Notre Dame Journal of Law, Ethics & Public Policy* 26 (2012): 621–644.

Hueiwang, A. C. J., A. J. Englandeb, R. M. Bakeerc, and H. B. Bradford. "Impact of Urban Stormwater Runoff on Estuarine Environmental Quality." *Estuarine, Coastal and Shelf Science* 63 (2005): 513–526.

Johnson, R. W., and W. C. Galloway. "Protection of Biodiversity under the Public Trust Doctrine." *Tulane Environmental Law Journal* 8 (1994): 21–32.

Karkkainen, B. C. "Toward a Smarter NEPA: Monitoring and Managing Government's Environmental Performance." *Columbia Law Review* 102 (2002): 903–972.

Karkkainen, B. C. "Adaptive Ecosystem Management and Regulatory Penalty Defaults: Toward a Bounded Pragmatism." *Minnesota Law Review* 87 (2003): 943–998.

Karleskint, G., R. Turner, and J. Small. *Introduction to Marine Biology*, 3d ed. Belmont, Calif.: Cengage Learning, 2009.

Karvonen, A. *Politics of Urban Runoff: Nature, Technology, and the Sustainable City*. Cambridge, Mass.: MIT Press, 2011.

Kelo v. City of New London, 545 U.S. 469 (2005).

Klein, C. A., F. Cheever, and B.C. Birdsong. *Natural Resources Law: A Place-Based Book of Problems and Cases*, 2d ed. Frederick, Md.: Wolters Kluwer, 2009.

Knapp, A. K., J. M. Briggs, D. C. Hartnett, and S. L. Collins. *Grassland Dynamics: Long-Term Ecological Research in Tallgrass Prairie*. New York: Oxford University Press, 1998.

Lee, K. N., and J. Lawrence "Adaptive Management: Learning from the Columbia River Basin Fish and Wildlife Program." *Environmental Law* 16 (1986): 431–460.

Lindblom, C. E. "The Science of 'Muddling Through.'" *Public Administration Review* 19 (1959): 79–88.

Long, A. "Tropical Forest Mitigation Projects and Sustainable Development: Designing U.S. Law for a Supporting Role." *William Mitchell Law Review* 36 (2010): 968–991.

Loveladies Harbor, Inc. v. United States, 28 F.3d 1171 (Fed. Cir. 1994).

Lucas v. South Carolina Coastal Council, 55 U.S. 1003 (1992).

Mann, K. H. *Ecology of Coastal Waters with Implications for Management*, 2d ed. Malden, Mass.: Blackwell Science, 2000.

Mooney, C. *The Republican War on Science*. New York: Basic, 2005.

Moore, W. S. "The Subterranean Estuary: A Reaction Zone of Ground Water and Seawater." *Marine Chemistry* 65 (1999): 111–125.

Mulvaney, T. M. "The New Judicial Takings Construct." *Yale Law Journal Online* 120 (2011): 247–266. http://yalelawjournal.org/2011/2/18/mulvaney.html.

Mutz, K. M., G. C. Bryner, and D. S. Kenney. *Justice and Natural Resources: Concepts, Strategies, and Applications*. Washington, D.C.: Island Press, 2002.

National Audubon Society v. Superior Court, 658 P. 2d 709 (Cal. 1983).

National Research Council. *New Strategies for American's Watersheds*. Washington, D.C.: National Academies Press, 1999.

Nelson, R. H. "Shoot, Shovel, and Shut Up." *Forbes* (Dec. 4, 1995): 82.

New State Ice Co. v. Liebmann, 285 U.S. 262 (1932).

Nollan v. California Coastal Commission, 43 U.S. 825 (1987).

Nordstrom, K. F. *Beaches and Dunes of Developed Coasts.* Cambridge: Cambridge University Press, 2000.

Osofsky, H. M. "The Future of Environmental Law and Complexities of Scale: Federalism Experiments with Climate Change under the Clean Air Act." *Washington University Journal of Law and Policy* 32 (2010): 79–97.

Osofsky, H. M., and H. J. Wiseman. *Hybrid Energy Governance.* FSU College of Law, Public Law Research Paper No. 608, Tallahassee, Fla.; Minnesota Legal Studies Research Paper No. 12–49, Minneapolis, Minn., 2012.

Ostrom, E. *A Polycentric Approach for Coping with Climate Change.* World Bank Research Working Paper 5095, Washington, D.C.: World Bank, 2009.

Ostrom, E., M. A. Janssen, and J. M. Anderies. "Going Beyond Panaceas." *Proceedings of the National Academy of Sciences USA* 104 (2007): 15176–15178.

Ostrom, V., C. M. Tiebout, and R. Warren. "The Organization of Government in Metropolitan Areas: A Theoretical Inquiry." *American Political Science Review* 55 (1961): 831–842.

Otter v. Salazar, Memorandum Decision and Order, Case No. 1:11-cv-00358-CWD (U.S. District of Idaho, Aug. 8, 2012).

Owen, D. "The Mono Lake Decision, the Public Trust Doctrine, and the Administrative State." *UC Davis Law Review* 45 (2012): 1099–1153.

Pardy, B. "Changing Nature: The Myth of the Inevitability of Ecosystem Management." *Pace Environmental Law Review* 20 (2003): 675–692.

Pielke, R. A., Jr. *The Honest Broker. Making Sense of Science in Policy and Politics.* Cambridge: Cambridge University Press, 2007.

Rapanos v. United States, 547 U.S. 715 (2006).

Reichman, O. J. *Konza Prairie: A Tallgrass Natural History.* Lawrence, Kans.: University Press of Kansas, 1987.

Reid, W. V., F. Berkes, T. J. Wilbanks, and D. Capistrano, eds. *Bridging Scales and Knowledge Systems: Concepts and Applications in Ecosystem Assessment.* Washington, D.C.: Island Press, 2006.

Roberts, T. M. "Innovations in Governance: A Functional Typology of Private Governance Institutions." *Duke Environmental Law and Policy Forum* 22 (2011): 67–144.

Ruhl, J. B. "Climate Change and the Endangered Species Act: Building Bridges to the No-Analog Future." *Boston University Law Review* 88 (2008): 1–62.

Ruhl, J. B. "Climate Change Adaptation and the Structural Transformation of Environmental Law." *Environmental Law* 40 (2010): 363–431.

Ruhl, J. B. "General Design Principles for Resilience and Adaptive Capacity in Legal Systems—With Applications to Climate Change Adaptation." *North Carolina Law Review* 89 (2011): 1373–1401.

Ruhl, J. B. "Panarchy and the Law." *Ecology and Society* 17, no. 3 (2012): 31. [online] http://www.ecologyandsociety.org/vol17/iss3/art31/.

Ruhl, J. B., S. E. Kraft, and C. L. Lant. *The Law and Policy of Ecosystem Services.* Washington, D.C.: Island Press, 2007.

Ruhl, J. B. and J. Salzman. "Climate Change, Dead Zones, and Massive Problems in the Administrative State: Guidelines for Whittling Away." *California Law Review* 98 (2010): 59–120.

Ryan, E. "Federalism and the Tug of War Within: Seeking Checks and Balance in the Interjurisdictional Gray Area." *Maryland Law Review* 66 (2007): 503–667.

Sackett v. EPA, 132 S. Ct. 1367 (2012).

Sax, J. L. "Some Thoughts on the Decline of Private Property." *Washington Law Review* 58 (1983): 481–496.

Sax, J. L. "Property Rights and the Economy of Nature: Understanding Lucas v. South Carolina Coastal Council." *Stanford Law Review* 45 (1993): 1433–1455.

Scharpf, C. "Politics, Science, and the Fate of the Alabama Sturgeon." *American Currents* (Summer [Aug.] 2000): 6–14.

Severance v. Patterson, 345 S.W.3d 18 (Tex. 2010) and 682 F.3d 360 (5th Cir. 2012).

Solid Waste Agency of Northern Cook County v. U.S. Army Corps of Engineers, 531 U.S. 159 (2001).

Southview Associates v. Bongartz, 980 F.2d 84 (2nd Cir. 1992).

Sprankling, J. G. "The Antiwilderness Bias in American Property Law." *University of Chicago Law Review* 63 (1996): 519–590.

Starritt, S. D., and J. H. McClanahan. "Land-Use Planning and Takings: The Viability of Conditional Exactions to Conserve Open Space in the Rocky Mountain West after Dolan v. City of Tigard, 114 S. Ct. 2309 (1994)." *Land and Water Law Review* 30 (1995): 415–464.

Stop the Beach Renourishment, Inc. v. Florida Department of Environmental Protection, 130 S. Ct. 2592 (2010).

Tennessee Valley Authority v. Hill, 437 U.S. 153 (1978).

Thomas, J. C. "Clearing the Muddy Waters? Rapanos and the Post-Rapanos Clean Water Act Jurisdictional Guidance." *Houston Law Review* 44 (2008): 1491–1534.

Tulare Lake Basin Water Storage District v. United States, 49 Fed. Cl. 313 (2001).

United States v. Riverside Bayview Homes, Inc., 474 U.S. 121 (1985).

U.S. Environmental Protection Agency and Army Corps of Engineers. 2011. Draft Guidance Regarding Identification of Waters Protected by the Clean Water Act. 76 Fed. Reg. 24,479.

Walker, B., and D. Salt. *Resilience Thinking: Sustaining Ecosystems and People in a Changing World.* Washington, D.C.: Island Press, 2006.

Walters, C. *Adaptive Management of Renewable Resources.* New York: Macmillan, 1986.

Wiersema, A. "A Train Without Tracks: Rethinking the Place of Law and Goals in Environmental and Natural Resources Law." *Environmental Law* 38 (2008): 1239–1300.

Wildavsky, A. *Speaking Truth to Power: The Art and Craft of Policy Analysis.* New Brunswick, N.J.: Transaction, 1987.

Wood, M. C. "Atmospheric Trust Litigation." In *Adjudicating Climate Change: Sub-National, National, and Supra-National Approaches*, eds. W. C. G. Burns and H. M. Osofsky, 99–124. Cambridge: Cambridge University Press, 2009.

Wood, M. C. *Nature's Trust: Environmental Law for a New Ecological Age.* Cambridge: Cambridge University Press, 2013.

Zellmer, S., and L. Gunderson. "Why Resilience May Not Always Be a Good Thing: Lessons in Ecosystem Restoration from Glen Canyon and the Everglades." *Nebraska Law Review* 87 (2009): 893–949.

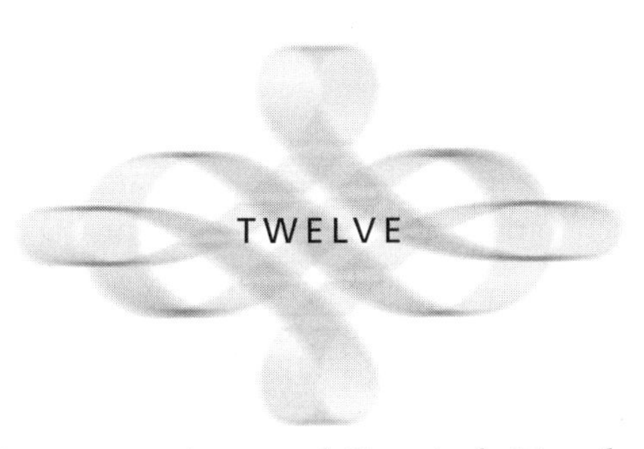


# The Integration of Social-Ecological Resilience and Law

AHJOND S. GARMESTANI, CRAIG R. ALLEN,
J. B. RUHL, AND C. S. HOLLING

A dramatic paradigm shift in American law occurred in 1970, as Congress began to target hazardous waste, water pollution, and protection of endangered species with sweeping new legislation (Lazarus 2004). Preceding this new era of environmental protection, federal policies had already begun to shift resource use from private interests for economic development to conservation and preservation by and for the public. This shift in U.S. policy was preceded by subtle shifts in the way that scientists, policy makers, and the public viewed and conceptualized the natural environment. In particular, changing conceptions of the naturalness of ecosystems, humans' ability to affect ecological processes, and the manner in which ecological systems function preceded the emergence of new environmental laws (Lazarus 2004). Early environmental laws in the United States were dominated by optimism and the belief that technology would be able to "fix" environmental problems, and many of these laws were successful at addressing numerous environmental issues (Lazarus 2004). Growing recognition of the inherent uncertainty associated with the dynamics of ecological systems and their often nonlinear and surprising behavior, however, presents a set of problems outside the scope of classic environmental law and has led to a fundamental understanding about the interaction of environmental law and ecological systems: rigid legal standards are largely incompatible with our current understanding of the dynamics of social-ecological systems. Nature is not static, and thus environmental law should be adapted to reflect our current understanding of nature. The chapters in this volume have deeply considered this nexus of law and social-ecological resilience. The authors have considered how legal systems can be made more resilient, and how laws might be configured to foster resilience in social, ecological and social-ecological systems.

Zellmer and Anderies (chapter 1) opened the volume with an assessment of the history and current state of wilderness preserves in the United States. They characterized wilderness laws and policies within the context of resilience theory. Adaptation, a key attribute of resilience, will sometimes require active intervention in "untouched" wilderness preserves. Zellmer and Anderies argued that intervention in wilderness areas should only proceed if there is sufficient understanding of the system; if the intervention will likely improve system function; and if the human intervention is restricted to a limited period of time. Allowing active intervention only after meeting these three conditions will enable wilderness preserves to maintain their significance and also serve as part of a broader strategy to manage for resilience.

Benson and Hopton (chapter 2) analyzed the legal frameworks and institutions that can accommodate a resilience-based approach to biodiversity protection. They accomplished this by conducting an overview of the history and the current state of wildlife management and biodiversity protection in the United States. Benson and Hopton asserted that the Endangered Species Act of 1973 is one of the most controversial environmental laws because of hard deadlines built into its process. Each step in the process (e.g., critical habitat designation) is subject to litigation if the government does not meet hard deadlines. Such a system of hard deadlines could be a starting point for setting thresholds for systems-based regulations that foster resilience, as opposed to species-specific rules. Importantly, legal thresholds must be linked to monitoring thresholds that explicitly incorporate leading indicators, modeling, and scenario planning. Benson and Hopton concluded that the current legal framework in the United States is in need of reform, and they provided recommendations for legal and institutional reform that allows for resilience-based management of biodiversity.

Glicksman and Cumming (chapter 3) provided an analysis of the management of parks, refuges, and preserves for resilience. They argued that there is now broad consensus that resource management laws need to shift to fostering and managing for resilience. Their analysis found that the emphasis in the law on preservation acts as a barrier to managing for resilience. The obstacles to cooperative environmental management between U.S. federal agencies that may have different statutory mandates further complicates the capacity of the current legal milieu in the U.S. for resilience-based governance. Glicksman and Cumming concluded that while there is some flexibility in current law that is being utilized, the current legal framework was developed under the assumption of a "balance of nature" that is not in concert with ecological reality. Thus, statutory and administrative reforms are necessary for resilience-based management of parks, refuges, and preserves. For example, they suggested that thresholds linked explicitly to requirements for management interventions are necessary and made reference to the state of Oregon's storm sewer program. In the Oregon program, storm sewer permit holders must meet threshold water quality standards, and if they do not, an adaptive management process is triggered that requires active management action to meet storm water quality goals (Dunn and Burchmore 2007). Another possible reform is for the federal government to make payments to private landowners to guarantee that landowners will not engage in activities on their land that have the capacity to erode resilience. With any legal reform, the reform will need to be broad-scale in nature in order to cope with cross-scale environmental problems and manifest resilience-based governance.

Craig and Hughes (chapter 4) explored the governance of the oceans as it relates to marine protected areas, spatial planning, and resilience. The oceans, and the larger marine environment, have largely been neglected by researchers and policy makers, perhaps because they are so expansive and are seen as incapable of being driven into degraded regimes by the variety of threats they now experience. Marine systems are also largely "out of sight, out of mind," and that may contribute to the lack of attention on the mounting problems faced by these systems. Craig and Hughes identified the fragmentation of regulatory authority as the critical governance challenge for these systems. They concluded that coastal nations could foster resilience in marine ecosystems by incorporating place-based marine management (e.g., marine protected areas) into their governance regimes.

Cosens and Stow (chapter 5) asserted that there are two areas of law that must be reformed if the goals of the Clean Water Act of 1972 (Federal Water Pollution Control Act 1972) are to be met: fragmentation of policy and addressing uncertainty in the data that policy is based upon. Rather than proposing new forms of governance, Cosens and Stow proposed process changes and legal reform as a means of integrating water governance. One suggested reform is to allow experimentation at small scales, which does not have the same degree of risk as implementing new policy at large scales. Second, while administrative law and institutional structure cannot require the types of individuals involved in a process, they can be reformed to maximize diversity, which provides a greater chance of including leaders and connectors in the process. Finally, Cosens and Stow argued that coordination and communication should be a legal requirement assigned to a specific position within an organization's part of the social-ecological system and that funding must be provided for monitoring of these proposed changes to watershed governance.

Green and Perrings (chapter 6) summarized the principles governing international water allocation, analyzed several water allocation frameworks, discussed other mechanisms for treaty design, and offered the concept that treaties can integrate adaptive management to foster resilience. Their analysis of social-ecological resilience concluded that treaties must create the institutional capacity to manage conflicts and integrate iterative governance mechanisms. Green and Perrings suggested that this reform is likely best served by the establishment of joint water commissions bound to implementing adaptive management.

Ruhl and Chapin (chapter 7) emphasized that the resilience of ecosystems and the resilience of policy are two different animals, and the resilience of one does not guarantee the resilience of the other. They examined whether the emerging theory of ecosystem services provides a useful approach for building resilience into ecosystems and policy. Ecosystem services theory expands the economic understanding of ecosystems and ecosystem processes, and as this becomes more important, social-ecological resilience will likely be fostered. Ruhl and Chapin concluded by offering suggestions regarding areas in which ecosystem services theory can support the resilience of ecosystems and the resilience of ecosystem policy, but they also noted areas in which it may be counterproductive.

Camacho and Beard (chapter 8) contended that existing government institutions lack the adaptive capacity to effectively manage a problem as complex as climate change. Regulators and managers lack information on effects of and management strategies for climate change and the institutional infrastructure to obtain this critical information. Camacho and Beard suggest that a federal, publicly accessible, system-wide information portal and clearinghouse will improve the institutional capacity to adapt to climate change. This reform, combined with incentives to encourage the adoption of adaptive management and institutional monitoring (i.e., monitoring of administrative, legal, and management institutions), should help to improve adaptive capacity and therefore foster social-ecological resilience.

Ebbesson and Folke (chapter 9) made the case that the legal jurisdictions currently drawn do not allow for the adequate management of transboundary international matters, many of which are global in nature. They contended that in order to account for cross-scale interactions, which have the capacity to affect the entire biosphere, it may be necessary to relax the distinction between private and public interests for access to judicial review. Ebbesson and Folke also touted international treaties and institutions, despite their shortcomings, as demonstrating that adaptation in response to the scale and scope of environmental matters is possible. The establishment of effective international regimes is critical for fostering social-ecological resilience across the borders of nation-states. Ebbesson and Folke contended that polycentric governance, while not perfect, is also a key aspect to legal resilience building, as it allows for a diversity of actors and stakeholders (e.g., civil society organizations) to be part of the process.

Eason, Flournoy, Cabezas, and Gonzalez (chapter 10) proposed two new laws to incorporate resilience and innovation into American law and policy: the National Environmental Legacy Act (NELA) and the Environmental Competition Statute (ECS). They provided an analysis of both model laws based upon recent sustainability research in the United States. These authors looked at the problem of resilience and law as one that will require new law in order to manifest a transition to sustainability. They analyzed the two model statutes (NELA and ECS;

Flournoy and Driesen 2010) by testing the specifics of the acts against publicly owned natural resources (NELA) and decision making in the private sector (ECS). NELA is designed to create the conditions necessary to allow for sustainable use of public natural resources. Eason and colleagues built upon NELA by assessing how quantitative indicators and metrics of resilience and sustainability can aid in meeting the goals of NELA. The authors also characterized how ECS could promote sustainability in the chemical sector of the economy by encouraging competition and innovation to further a sustainability paradigm.

Arnold and Gunderson (chapter 11) undertook the task of proposing a system of adaptive law, necessary because legal decision-making detached from the dynamics of social-ecological systems is likely to result in adverse consequences. They contended that an adaptive system of law has multiple goals, is multimodal and integrated, has the capacity to adapt to context, and has iterative legal processes with accountability. They proposed that their adaptive law framework should be tested via case studies. Case studies draw upon qualitative and quantitative research and should be transdisciplinary in structure in order to generate the type of information necessary to affect resilience-based governance.

The contributors to this volume have offered a variety of proposals that seek to foster the resilience of social-ecological systems and the legal system. These proposals range from small-scale to large-scale reform of law to entirely new law for resilience-based governance. The common ground among the authors' views is the recognition of the importance of adaptability (i.e., adaptive management and adaptive governance), the significance of thresholds for resilience-based governance, and the need for multiscalar resilience strategies. In the following sections, we expand upon those themes and synthesize our perspective with the chapters in this volume, and then offer recommendations for the integration of social-ecological resilience and law.

## The Critical Role of Adaptive Management and Adaptive Governance

Adaptive management is a method of integrating learning into the management process by designing management to test theories of ecological causation and response. Adaptive management follows from resilience theory, because it provides a framework to learn about the systems being managed while they are being managed, in a way that is safe to fail, that is, in a manner that is unlikely to push the system being managed over a tipping point. This entails an iterative process of decision making designed to identify and reduce uncertainty and the inevitability of "surprise" (i.e., nonlinear change) into the management process (Benson and Garmestani 2011). Monitoring is an essential aspect of adaptive management, as information from the system (i.e., monitoring data) feeds back into the management process in an iterative manner that allows managers to adapt to changing circumstances associated with managing ecosystems. Thus, management interventions are hypotheses to be put "at risk" in an adaptive management framework, and information that allows for learning is generated to improve management decisions (Garmestani et al. 2009).

Adaptive management is not a front-end exercise but rather an interdisciplinary iterative process of identifying uncertainties in management actions and feeding that information back into the management process to improve environmental management (Karkkainen 2005). Within this context, local-scale innovations have the capacity to scale up and generate larger cross-scale change in environmental management (Karkkainen 2005). As Karkkainen (2005) so eloquently states, the question is not how much we allow agencies to depart from a management action, but when do we allow agencies to depart from a management action. The answer is that departures are allowed with improvements in our understanding of system dynamics. Thus, administrative law should be reformed in order to create synergy between science and the law (Karkkainen 2005). In particular, he offers that administrative law could allow for an "adaptive management track" for adaptive management projects, in which a new set of administrative law standards specific to adaptive management would hold precedence.

Adaptive governance theory builds on that theme by incorporating formal institutions, informal groups (e.g., networks), and individuals at multiples scales for integrated environmental management (Folke et al. 2005), and by requiring collaboration amongst diverse stakeholders; communication between stakeholders; flexible, nested institutions; and adaptation in response to social and ecological monitoring (Plummer and Armitage 2010). Adaptive governance links collaboration with an iterative process of learning and recalibrating the governance process (Plummer and Armitage 2010). Law, policy, bridging organizations, and polycentric institutions are also important aspects of adaptive governance, producing networks that can increase political and financial support critical for fostering adaptive management (Folke et al. 2005; Olsson et al. 2006).

The purpose of adaptive management and adaptive governance is to build adaptive capacity, which should be explicitly addressed in law and resilience. Adaptive capacity is dependent on the ability to tap local resources by the mechanisms and decisions of governments (Sydneysmith et al. 2010). For example, on the Ivalo River in Finland, response to flooding events has been improved due to closer ties between local officials in higher-level agencies (Sydneysmith et al. 2010). A shock to the system, in the form of a major flood in 2005, forced improved communication between local, regional, and higher-level agencies, whereas in the past, institutional barriers between the leaders and bureaucrats in Helsinki limited the coordination of emergency response (Sydneysmith et al. 2010). Other key aspects of adaptive capacity are social memory and social capital (Adger et al. 2005). Social memory is generated from local knowledge and the cross-scale capacity of organizations to adapt to change. Social capital is the actual and potential resources generated from the scale of interest, the networks of stakeholders associated with a project, and community engagement. Thus, cultivating local knowledge and stakeholder involvement is critical to managing for resilience.

Due to the complexity of social-ecological systems, environmental governance must account for cross-scale dynamics (Garmestani et al. 2009). Communication and the flow of information across scales are essential, and bridging organizations have the capacity to foment this critical aspect for sound environmental management (Garmestani et al. 2009). Adaptive governance revolves around decentralized governance, stakeholder input, and informal networks. Adaptive management and adaptive governance are both meant to enhance learning while enabling continued action in either resource management or governance. A key area of learning and uncertainty is where, and what, are the critical thresholds. Resilience theory differs from equilibrium-based understanding of complex systems in that it explicitly recognizes that alternative states of systems are separated by thresholds. When a system is in a desirable state, and the exact nature of alternative states is unknown, it is in humankind's best interest to avoid crossing the thresholds that separate one state from another. In the following section, we expand upon the connection between thresholds and resilience.

## Thresholds and Resilience

Another essential component for fostering resilience in social-ecological systems is the identification of critical slow variables (Chapin et al. 2009). Critical slow variables are those that have strong influence over systems but remain relatively constant over time due to system self-organization that generates resilience (Chapin et al. 2009). For example, landscape functions as a critical slow variable, as there is a threshold of necessary landscape function that, if crossed, can result in the system losing resilience no matter what mitigation measures are taken (Rietkerk et al. 2004; Smith et al. 2009). Other examples of slow variables include soil fertility, which can be negatively affected by accelerated crop rotation cycles, which in turn can result in significant negative consequences for crop production and human welfare (Cumming 2011), and soil organic matter that retains pulses of nutrients and releases these resources, which are then utilized by plants (Chapin et al. 2009). Critical slow variables are characterized by functional types and disturbance regimes and may "slave" fast variables at the same scale (Chapin et al. 2009). If critical slow variables can be identified, environmental management has a much better chance of designing management actions that result in good outcomes. For example, with respect to climate change, the slow variable of carbon dioxide levels should not be as great of a threat as it is now becoming, if the fast variable (i.e., the political process) were acting as quickly as it could to mitigate climate change (Cumming 2011). Note, however, that panarchy theory—and reality we believe—suggests that there are multiple scales of structure and process, and therefore there is no single critical slow variable, but rather a limited set of critical variables that are largely responsible for the resilience of complex systems (Gunderson and Holling 2002).

These critical variables are defined by thresholds bounding the upper and lower levels of a domain of scale. Potential thresholds of concern (Biggs and Rogers 2003) can be defined as management goals that represent the current understanding of the limits of the resilience of a system (Smith et al. 2009). When a threshold is approached, management action can be applied, and thresholds can be recalibrated in an adaptive manner if new information suggests the threshold is incorrect (Smith et al. 2009). Thresholds should be treated as hypotheses to be put at risk with monitoring data and represent the multidimensional regime in which variation is acceptable for system resilience (Smith et al. 2009). The identification of thresholds is difficult and fraught with uncertainty, but it is an essential component in resilience science (Walker et al. 2009). Thus, when working to establish thresholds for specific slow variables, we must categorize thresholds (i.e., known, strongly suspected, and possible) in order to establish the degree of confidence in the threshold estimates (Walker et al. 2009), and adaptive management or other learning approaches should be embraced to reduce the uncertainty associated with the knowledge of variables exhibiting threshold behavior and the location of those thresholds.

Linking thresholds to an adaptive management framework can be an effective tool for managing for resilience, but it must operate under the constraints of the law. Legal criticisms of adaptive management have centered around agencies' using the term "adaptive management" as a means to allow for informal management, to avoid making hard decisions, or to shirk management responsibility altogether. Linking the iterative aspects of adaptive management to specific thresholds at which changes would be warranted in the process would go a long way toward addressing those criticisms. For example, researchers in Africa identified thresholds in a national park system and have focused monitoring on critical slow variables (Cumming 2011). If monitoring data indicates a threshold is being approached, a management response is triggered. With respect to the national park, if elephant density reaches a level that cannot be sustained by the park due to landscape constraints (e.g., vegetation), actions to reduce the population (e.g., culling, relocation) must be taken (Cumming 2011).

## Multiscalar Resilience Strategies

In order to manage for resilience, reform of law is likely necessary. Reforms such as integration of an iterative governance process and requiring monitoring have the potential to allow our current legal framework to account for the dynamic nature of social-ecological systems (Benson and Garmestani 2011). Reforms have the capacity to provide agencies the flexibility they need to actualize a sustainability transition (Benson and Garmestani 2011). Purely science-based management is incomplete in that it may not account for hard to quantify aspects of human behavior, so it should not be the sole basis for policy (Brunner et al. 2005). Sound policy should utilize qualitative and quantitative research, as well as often overlooked sources of information about the system of interest (e.g., local knowledge; Brunner et al. 2005; Brunner and Steelman 2005). Panarchy provides a framework to link ecological and social systems in a manner that allows researchers and managers to better understand the cross- and within-scale linkages that are necessary for improved environmental management (Garmestani and Benson 2013). Janssen et al. (2007) show from case studies that top-down interventions that do not account for or integrate information and local knowledge from smaller scales lead to environmental management failures and often make the system more vulnerable to perturbations. For example, the Oregon Plan, which established watershed councils at small scales in order to release some of the tension in the state associated with salmon recovery efforts, demonstrated that smaller-scale initiatives can be successful when supported by larger-scale organizations that do not dictate the end results (Coe-Juell 2005).

One of the key lessons of the legal analysis of resilience is that scale matters, and thus, subnational governments must play a key role in actualizing sustainability (Dernbach and Mintz 2011). This is readily apparent when laws are not harmonized across scales (Dernbach and Mintz 2011). For example, the state of California implemented climate change legislation in 2006 to reduce greenhouse gas emissions, but the law did not address local land use laws (Medina and Tarlock 2010). Litigation challenged the local land use laws, which resulted in the California legislature passing a law in 2008 to encourage less automobile use. Thus, command-and-control hierarchical regulation needs to be supplemented by a suite of scale-specific policy instruments, including information disclosure and market instruments (Schoenbrod et al. 2010). Craig and Ruhl (2010) advocate for scale-specific management strategies that utilize a suite of legal and policy instruments, including collaborative governance, reflexive law (e.g., information reporting), and economic incentives. Scale-specific or place-based governance would require reform of the law to allow for novelty in environmental management (Craig and Ruhl 2010). These legal reforms are necessary, because economic, environmental, and political conditions; local preferences; and organizations and institutions vary with scale (Craig and Ruhl 2010).

Some legal scholars suggest that environmental laws can be made more effective (i.e., scale-specific) by integrating regulation mechanisms such as community-based social marketing, which is an approach that seeks to change behavior at small scales via direct contact with people at the local scale (Kennedy 2010). For example, in a survey of community-based social-marketing case studies, Kennedy (2010) found that compliance with regulation was increased via community-based social-marketing approaches. Kennedy (2010) suggests that affirmative motivations combined with deterrence and enforcement has the capacity to enhance the effectiveness of regulation. Community-based social marketing uses a structured program designed to remove barriers to change, change environmental behavior via behavior-changing tools, and engage citizens (Kennedy 2010). Kennedy (2010) points to an example from Portland, Oregon, as an example of a successful community-based social-marketing program. In this program, Portland failed to meet standards under the Clean Air Act, and implemented a non-regulatory incentive program to meet the legal standard. The program targeted vehicles and lawnmowers, as well as paint and other consumer products (Kennedy 2010). The program was primarily funded by U.S. Environmental Protection Agency, but relied upon additional private donations, as well as the critical component of developing community partnerships, and resulted in a significant decrease in the emission of volatile organic compounds (Kennedy 2010). Importantly, regulation provides community-based social-marketing mechanisms the necessary "teeth" to ensure compliance with the legal standard (Kennedy 2010). The tools to foster change include: getting commitments from stakeholders, developing and providing incentives for "good" environmental behavior, utilizing captivating communication strategies, and offering prompts as mental cues to encourage change (Kennedy 2010). A critical aspect in this process is to monitor the effectiveness of incentives for changing behavior.

The U.S. Constitution has been characterized as ill-suited for environmental law due to its preference for decentralized, fragmented, and incremental law making (Lazarus 2004). But decentralized law making may offer an avenue for fostering flexibility in law. Decentralization could act as a mechanism to allow for holistic environmental management, as the spatiotemporal dimensions of governing linked social-ecological systems demands creative solutions. Local governments are more reactive to public opinion at small scales, while federal (and state) agencies operate at larger scales (Garmestani et al. 2009). Cross-scale interaction is essential to managing for resilience, and a more diverse set of management options can manifest when scale is taken into account in governance (Garmestani et al. 2009). Communication and information flow across and between scales contributes to the adaptive capacity of organizations to effectively manage natural resources (Gunderson et al. 1995). Intermediaries (e.g., bridging organizations) have the capacity to act as a conduit between organizations operating at different scales (Garmestani et al. 2009), and bridging organizations, for instance, can create opportunities for collaboration between diverse stakeholders, which is key for sound environmental governance (Brown 1993).

Fostering redundancy of function in the roles of management entities may appear inefficient, but in light of the likelihood of increasing nonlinear change, this redundancy provides cross-scale resilience by affording more policy space for experimentation and could promote synergy between agencies working on similar problems at different scales (Ruhl 2011). As previously mentioned, a variety of policy instruments should be developed and implemented at the intended scale of the policy outcome (Garmestani et al. 2009). For example, Flournoy and Driesen (2010) have proposed the ECS, which allows for anyone who makes an environmental improvement to collect the cost of making that improvement (plus a premium set by the statute) from competitors who pollute more. This statute would create a "race to the top" for environmental innovators, as those who perform best collect money, and those who do poorly pay for their competitor's environmental innovations (Flournoy and Driesen 2010). This statute could harness the nimbleness and innovative capacity of the private sector to create the conditions for discontinuous leaps in environmental innovation. Of course, this is easier said than done, due to cross-scale dynamics, but treating policies as hypotheses to be put at risk could help to improve environmental governance (Garmestani et al. 2009). As we have yet to develop the capacity to directly measure system resilience (but see Allen et al. 2005), "surrogates" for systems should be developed when conducting resilience science (Carpenter et al. 2005). Resilience surrogates are based upon stakeholder assessments, models, historical profiling, and case studies (Carpenter et al. 2005). In particular, researchers should use models and scenario analysis to reveal processes that act to stabilize or destabilize a system; this practice could lead to identification of resilience surrogates (Carpenter et al. 2005). Examples of resilience surrogates include: distance of a system variable from a system threshold, the rate at which a system variable is moving toward or away from a threshold, and external perturbations (e.g., shocks, controls) that could alter the rate of change of a system variable (Bennett et al. 2005).

## Conclusion

Barring the development of entirely new law (e.g., National Adaptive Management Act; Ruhl 2005), legal reform is needed, as is identification of ecological thresholds that capture the resilience of an ecosystem. Ecological thresholds can then be linked to legal thresholds, which would allow for an iterative process of recalibrating thresholds in light of new information, what has been called a "rolling rule" (Karkkainen 2005) approach to ecosystem management. The lessons learned from this volume allow us to condense our understanding into the following key aspects for integrating social-ecological resilience and law:

1. Reform of law
    - Reduce front-end decision making in law making and replace it with an iterative back-end process; this will require reform of administrative law.

   - In addition, reform administrative law to maximize diversity, increasing the chances of including leaders and connectors in the process.
2. Adaptive management and adaptive governance
   - Utilize adaptive management and adaptive governance, which allow for a variety of mechanisms (e.g., stakeholder participation, intermediaries, market tools, information disclosure, land trusts [Garmestani et al. 2012]) to reach environmental goals.
   - Account for scale: Allow for creativity in local and regional environmental governance (place-based governance, context-dependent).
   - Require communication and information flow between management entities and nonmanagement entities in order to foster social-ecological resilience.
3. Monitoring
   - Require and fund monitoring of ecological systems and agency actions (organizational learning).
   - Employ scale-specific metrics for environmental management (e.g., leading indicators, sustainability metrics).
   - Link ecological thresholds to required management action (legal requirements) to implement adaptive management in the system of interest.

Humankind's rich heritage of experimentation in governance and management offers hope that some of the ideas that have been put forward here will be embraced. The critical lesson of resilience theory is that barring some significant actions on humankind's part, the further accrual of more and more rigid regulations and laws will inevitably result in adverse consequences for social-ecological systems. The approaches put forward here offer a different path and outcome.

## References

Adger, W. N., T. P. Hughes, C. Folke, S. R. Carpenter, and J. Rockström. "Social-Ecological Resilience to Coastal Disasters." *Science* 309 (2005): 1036–1039.

Allen, C. R., L. Gunderson, and A. R. Johnson. "The Use of Discontinuities and Functional Groups to Assess Relative Resilience in Complex Systems." *Ecosystems* 8 (2005): 958–966.

Bennett, E. M., G. S. Cumming, and G. D. Peterson. "A Systems Model Approach to Determining Resilience Surrogates for Case Studies." *Ecosystems* 8 (2005): 945–957.

Benson, M. H., and A. S. Garmestani. "Embracing Panarchy, Building Resilience and Integrating Adaptive Management Through a Rebirth of the National Environmental Policy Act." *Journal of Environmental Management* 92 (2011): 1420–1427.

Biggs, H. C., and K. H. Rogers. "An Adaptive System to Link Science, Monitoring, and Management in Practice." In *The Kruger Experience: Ecology and Management of Savanna Heterogeneity*, eds. J. T. du Toit, K. H. Rogers, and H. G. Biggs, 59–80. Washington, D.C.: Island Press, 2003.

Brown, L. D. "Development Bridging Organizations and Strategic Management for Social Change." In *Advances in Strategic Management*, eds. P. Shrivastava, A. Huff, and J. Dutton, 381–406. London: JAI, 1993.

Brunner, R. D., and T. D. Steelman. "Beyond Scientific Management." In *Adaptive Governance: Integrating Science, Policy, and Decision Making*, eds. R. D. Brunner, T. A. Steelman, L. Coe-Juell, C. M. Cromley, C. M. Edwards, and D. W. Tucker, 1–46. New York: Columbia University Press, 2005.

Brunner, R. D., T. A. Steelman, L. Coe-Juell, C. M. Cromley, C. M. Edwards, and D. W. Tucker. *Adaptive Governance: Integrating Science, Policy, and Decision Making*. New York: Columbia University Press, 2005.

Carpenter S. R., F. Westley, and M. G. Turner. "Surrogates for Resilience of Social-Ecological Systems." *Ecosystems* 8 (2005): 1–5.

Chapin, F. S., C. Folke, and G. P. Kofinas. "A Framework for Understanding Change." In *Principles of Ecosystem Stewardship: Resilience-based Natural Resource Management in a Changing World*, eds. F. S. Chapin, G. P. Kofinas, and C. Folke, 3–28. Heidelberg: Springer, 2009.

Clean Air Act, 42 U.S.C. §7401 et seq.

Coe-Juell, L. "The Oregon Plan: A New Way of Doing Business." In *Adaptive Governance: Integrating Science, Policy, and Decision Making*, eds. R. D. Brunner, T. A. Steelman, L. Coe-Juell, C. M. Cromley, C. M. Edwards, and D. W. Tucker, 181–220. New York: Columbia University Press, 2005.

Craig, R. K., and J. B. Ruhl. "Governing for Sustainable Coasts: Complexity, Climate Change, and Coastal Ecosystem Protection." *Sustainability* 2 (2010): 1361–1388.

Cumming, G. C. *Spatial Resilience in Social-Ecological Systems*. Dordrecht: Springer, 2011.

Dernbach, J. C., and J. A. Mintz. "Environmental Laws and Sustainability: An Introduction." *Sustainability* 3 (2011): 531–540.

Dunn, A.D., and D. W. Burchmore. "Regulating Municipal Separate Storm Sewer Systems." *Natural Resources and Environment* 21 (2007): 3–6.

Endangered Species Act of 1973, 16 U.S.C. §§ 1531–1544.

Federal Water Pollution Control Act [commonly known as the Clean Water Act of 1972] 33 U.S.C. §1251–1387.

Flournoy, A. C., and D. M. Driesen. *Beyond Environmental Law: Policy Proposals for a Better Environmental Future.* Cambridge: Cambridge University Press, 2010.

Folke, C., T. Hahn, P. Olsson, and J. Norberg. "Adaptive Governance of Social-Ecological Systems." *Annual Review of Environment and Resources* 30 (2005): 441–473.

Garmestani, A.S. and M.H. Benson. "Framework for Resilience-Based Governance of Social-Ecological Systems." *Ecology and Society* 18, no. 1 (2013): 9. http://www.ecologyandsociety.org/vol18/iss1/art9/.

Garmestani, A. S., C. R. Allen, and H. Cabezas. "Panarchy, Adaptive Management and Governance: Policy Options for Building Resilience." *Nebraska Law Review* 87 (2009): 1036–1054.

Garmestani, A. S., M. E. Hopton, and M. T. Heberling. "Actualizing Sustainability: Environmental Policy for Resilience in Ecological Systems." In *Sustainability: Multi-Disciplinary Perspectives,* eds. H. Cabezas and U. Diwekar, 65–87. London: Bentham, 2012.

Gunderson L. H., and C. S. Holling. *Panarchy: Understanding Transformations in Human and Natural Systems.* Washington, D.C.: Island Press, 2002.

Gunderson, L. H., C. S. Holling, and S. S. Light. *Barriers and Bridges to the Renewal of Ecosystems and Institutions.* New York: Columbia University Press, 1995.

Janssen, M. A., J. M. Anderies, and E. Ostrom. "Robustness of Social-Ecological Systems to Spatial and Temporal Variability." *Society and Natural Resources* 20 (2007): 1–16.

Karkkainen, B.C. "Panarchy and Adaptive Change: Around the Loop and Back Again." *Minnesota Journal of Law, Science & Technology* 7, no. 1 (2005): 59–77.

Kennedy, A. L. "Using Community-Based Social Marketing Techniques to Enhance Environmental Regulation." *Sustainability* 2 (2010): 1138–1160.

Lazarus, R. J. *The Making of Environmental Law.* Chicago: University of Chicago Press, 2004.

Medina, R., and A. D. Tarlock. "Addressing Climate Change at the State and Local Level: Using Land Use Controls to Reduce Automobile Emissions." *Sustainability* 2 (2010): 1742–1764.

Olsson, P., L. H. Gunderson, S. R. Carpenter, P. Ryan, L. Lebel, C. Folke, and C. S. Holling. "Shooting the Rapids: Navigating Transitions to Adaptive Governance of Social-Ecological Systems." *Ecology and Society* 11, no. 1 (2006): 18. http://www.ecologyandsociety.org/vol11/iss1/art18.

Plummer, R., and D. Armitage. "Sociobiology and Adaptive Capacity: Evolving Adaptive Strategies to Build Environmental Governance." In *Adaptive Capacity and* Environmental Governance, eds. D. Armitage and R. Plummer, 243–261. Heidelberg: Springer, 2010.

Rietkerk, M. G., S. C. Dekker, P. C. Ruiter, and J. van de Koppel. "Self-Organized Patchiness and Catastrophic Shifts in Ecosystems." *Science* 305 (2004): 1926–1929.

Ruhl, J. B. "Regulation by Adaptive Management—Is It Possible?" *Minnesota Journal of Law, Science & Technology* 7 (2005): 21–57.

Ruhl, J. B. "General Design Principle for Resilience and Adaptive Capacity in Legal Systems—With Applications to Climate Change Adaptation." *North Carolina Law Review* 89 (2011): 1373–1401.

Schoenbrod, D., R. B. Stewart, and K. M. Wyman. *Breaking the Logjam: Environmental Protection That Will Work.* New Haven: Yale University Press, 2010.

Smith, D. M. S., N. Abel, B. Walker, and F. S. Chapin. "Drylands: Coping with Uncertainty, Thresholds, and Changes in State." In *Principles of Ecosystem Stewardship: Resilience-based Natural Resource Management in a Changing World,* eds. F. S. Chapin, G. P. Kofinas, and C. Folke, 171–196. Heidelberg: Springer, 2009.

Sydneysmith, R., M. Andrachuk, B. Smit, and G. K. Hovelsrud. "Vulnerability and Adaptive Capacity in Arctic Communities." In *Adaptive Capacity and Environmental Governance,* eds. D. Armitage and R. Plummer, 133–156. Heidelberg: Springer, 2010.

Walker, B. H., N. Abel, J. M. Anderies, and P. Ryan. "Resilience, Adaptability, and Transformability in the Goulburn-Broken Catchment, Australia." *Ecology and Society* 14, no. 1 (2009): 12. http://www.ecologyandsociety.org/vol14/iss1/art12.

# Contributors

CRAIG R. ALLEN. U.S. Geological Survey–Nebraska Cooperative Fish & Wildlife Research Unit, School of Natural Resources, University of Nebraska–Lincoln, Lincoln, Nebraska.

JOHN M. ANDERIES. School of Human Evolution and Social Change, and School of Sustainability, Arizona State University, Tempe, Arizona.

CRAIG ANTHONY (TONY) ARNOLD. Louis D. Brandeis School of Law, University of Louisville, Louisville, Kentucky.

T. DOUGLAS BEARD. U.S. Geological Survey–National Climate Change & Wildlife Center, Reston, Virginia.

MELINDA HARM BENSON. Department of Geography and Environmental Studies, University of New Mexico, Albuquerque, New Mexico.

HERIBERTO CABEZAS. U.S. Environmental Protection Agency, National Risk Management Research Laboratory, Cincinnati, Ohio.

ALEJANDRO E. CAMACHO. School of Law, University of California–Irvine, Irvine, California.

F. STUART CHAPIN III. Department of Biology and Wildlife, University of Alaska–Fairbanks, Fairbanks, Alaska.

BARBARA A. COSENS. College of Law, University of Idaho, Moscow, Idaho.

ROBIN KUNDIS CRAIG. S. J. Quinney College of Law, University of Utah, Salt Lake City, Utah.

GRAEME S. CUMMING. Percy FitzPatrick Institute of African Ornithology, University of Cape Town, Cape Town, South Africa.

TARSHA EASON. U.S. Environmental Protection Agency, National Risk Management Research Laboratory, Cincinnati, Ohio.

JONAS EBBESSON. Department of Law, Stockholm University, Stockholm, Sweden.

ALYSON C. FLOURNOY. Levin College of Law, University of Florida, Gainesville, Florida.

CARL FOLKE. Beijer Institute of Ecological Economics and Stockholm Resilience Centre, Stockholm, Sweden.

AHJOND S. GARMESTANI. U.S. Environmental Protection Agency, National Risk Management Research Laboratory, Cincinnati, Ohio.

ROBERT L. GLICKSMAN. George Washington University Law School, Washington, D.C.

MICHAEL A. GONZALEZ. U.S. Environmental Protection Agency, National Risk Management Research Laboratory, Cincinnati, Ohio.

OLIVIA ODOM GREEN. U.S. Environmental Protection Agency, National Risk Management Research Laboratory, Cincinnati, Ohio.

LANCE H. GUNDERSON. Department of Environmental Studies, Emory University, Atlanta, Georgia.

C. S. HOLLING. Department of Biology, University of Florida, Gainesville, Florida.

MATTHEW E. HOPTON. U.S. Environmental Protection Agency, National Risk Management Research Laboratory, Cincinnati, Ohio.

TERRY P. HUGHES. ARC Centre of Excellence for Coral Reef Studies, James Cook University, Townsville, Queensland, Australia.

CHARLES PERRINGS. School of Life Sciences, Arizona State University, Tempe, Arizona.

J. B. RUHL. Vanderbilt University Law School, Nashville, Tennessee.

CRAIG A. STOW. NOAA Great Lakes Environmental Research Laboratory, Ann Arbor, Michigan.

SANDRA B. ZELLMER. University of Nebraska College of Law, Lincoln, Nebraska.

# Index